Quantum ESPRESSO Course for Solid-State Physics

Quantum ESPRESSO Course for Solid-State Physics

Nguyen Tuan Hung
Ahmad R. T. Nugraha
Riichiro Saito

JENNY STANFORD
PUBLISHING

Published by

Jenny Stanford Publishing Pte. Ltd.
101 Thomson Road
#06-01, United Square
Singapore 307591

Email: editorial@jennystanford.com
Web: www.jennystanford.com

British Library Cataloguing-in-Publication Data
A catalogue record for this book is available from the British Library.

Quantum ESPRESSO Course for Solid-State Physics

ISBN 978-981-4968-37-9 (Hardcover)
ISBN 978-981-4968-63-8 (Paperback)
ISBN 978-1-003-29096-4 (eBook)

Contents

Preface

This book is based on a first-principles workshop "An introduction and hands-on tutorials with Quantum ESPRESSO" held at Tohoku University, Japan, in 2016. Since many participants of the workshop asked us about any textbook on Quantum ESPRESSO, and since the publisher accepted to publish such a book, we started writing this book in 2019 and kept adding latest information on density-functional theory and solid-state physics. We expect the following two groups as the potential readers: Group 1 is experimental researchers who wish to run the first-principles calculation themselves to compare experimental results of new materials directly. Nowadays, it is not surprising to know that experimental researchers calculate Quantum ESPRESSO by themselves. Group 2 is theoretical researchers who wish to know details of the first-principles calculation for running Quantum ESPRESSO. It is not easy for a student to run Quantum ESPRESSO and learn density-functional theory at the same time.

Quantum ESPRESSO is one of the most used packages for first-principles calculations, and it has been developing continuously; thanks to its open-source and excellent community support. The abbreviation ESPRESSO stands for "**ESPRESSO** = op**E**n **S**ource **P**ackage for **R**esearch in **E**lectronic **S**tructure, **S**imulation, and **O**ptimization." First and foremost, we are neither developers nor experts. We are just Quantum ESPRESSO users. We started to learn Quantum ESPRESSO at a workshop held at the Tokyo Institute of Technology, Japan, in 2014. After that, in 2016, Tohoku University Program for Leading Graduate Schools supported us to organize the first-principles workshop by NTH and ARTN as mentioned above. It was a very successful workshop with 51 participants, including students and even professors. We were surprised that all participants were happy to run Quantum ESPRESSO on their notebook PC only after one day workshop. We hope the readers can have the same experience by reading this book. The contents of this book have also been used in first-principles workshops organized by RS at Zhejiang University (China) in March 2019 and by NTH at the Vietnam School of Physics (Vietnam) in July 2019.

This book is a step-by-step guide to practice first-principles calculations with Quantum ESPRESSO. It is organized into three parts. The first part gives instructions to install Quantum ESPRESSO on your computer in Chapter 2 and detailed hands-on tutorials of Quantum ESPRESSO in Chapter 3, which is the main part of the workshop. The second part provides the concepts of the density-functional theory that the readers want to know for understanding the keywords of Quantum ESPRESSO in Chapter 4. The third part consists of solid-state physics in Chapter 5 and productivity tools in Chapter 6, such as graphics and Linux command scripts, which are used in Chapter 3. The book's text is suitable for researchers not only in physics but also in chemistry or materials science.

Even though the first-principles calculations have been introduced by many reviews and books, this book will be useful for many new scientists and students that are interested in Quantum ESPRESSO. The book has a reasonable balance between practical examples and fundamental theories for learning Quantum ESPRESSO. Therefore, we suggest the readers to first go through the examples in Quantum ESPRESSO and then learn the details from this book. All examples, codes, and scripts in this book are available on the following web page: `https://github.com/nguyen-group/QE-SSP` and therefore, the readers will not require to type the examples.

The authors acknowledge the efforts of the students and collaborators who assisted in checking the examples and contents of this book. Finally, we wish to tell the readers, "Welcome to Quantum ESPRESSO."

Nguyen Tuan Hung
Ahmad R. T. Nugraha
Riichiro Saito
Sendai, 2022

Chapter 1

Introduction

In this chapter, we will give a brief overview of book. Then, we will explain the goal of the book and how to read the book for your purpose.

1.1 How to read and use the book?

First-principles calculation in solid-state physics is a computer program for calculating the energy band structure of a solid. Here, the word "first-principles" means that we can calculate the energy band with no adjustable parameters but only by giving coordinates of atoms in solid and their atomic numbers.[1] Once we know how to run the first-principles calculation programs, we can get the ground-state properties from which we discuss many properties of solids.

In the 20th century, physics and computer science of the first-principles calculations were developed by many researchers. Only a few people could run the program in the early times because: (1) it was generally not open for the public, (2) it required a

[1] It is noted that we have many principles in the first-principles. Thus, we do not say "first-principle calculation" but "first-principles calculation".

Quantum ESPRESSO Course for Solid-State Physics
Nguyen Tuan Hung, Ahmad R. T. Nugraha, and Riichiro Saito

ISBN 978-981-4968-37-9 (Hardcover), 978-981-4968-63-8 (Paperback), 978-1-003-29096-4 (eBook)
www.jennystanford.com

supercomputer, and (3) it used to take a lot of experience to operate the program. However, the situation has changed in the 21st century: (1) many computer programs are open to the public and free of charge on the internet, (2) a desktop computer has continuously increased its performance, and (3) many users can share the possible problems for using, improving, and developing the software. Thus even the experimentalists, who do not have any experience of running the software, can do the first-principles calculation. Conveniently, we can compare the calculated results with the experimentally observed ones, which we find in the database on the internet.

Quantum ESPRESSO [Giannozzi *et al.* (2009)] is a first-principles calculation package, which is now used by many people in the world. We currently teach how to run Quantum ESPRESSO in the class of solid-state physics in the university's graduate school. It is impressive that any student can download Quantum ESPRESSO and run the program on their note PC. It means that the first-principles calculation is no more a special program that only an expert of computational physics can use. Quantum ESPRESSO is a kind of tool that anybody can use for understanding materials and physics. However, it might not be easy for the reader to be accustomed to Quantum ESPRESSO by oneself even though many documentations are on the internet.

This book aims to give sufficient information on Quantum ESPRESSO for the readers to run the sample programs. Furthermore, since most readers are not always physics students or computer experts, we will add the minimum concept of solid-state physics and computers relevant to the first-principles calculation.

To use this book, please do as follows: (1) set up computers and download sample inputs (see Chapter 2), (2) run Quantum ESPRESSO by using the sample inputs and commands (see Chapter 3), (3) learn density-functional theory (see Chapter 4), related solid-state physics (see Chapter 5), and supporting Linux commands and Python language (see Chapter 6). After that, you can run Quantum ESPRESSO to obtain the properties of your materials.

1.2 What do we need to run a program?

What we need to run Quantum ESPRESSO is (1) a personal computer (PC) connected to the internet and (2) this book. We do not ask for any additional cost to run the program.

Downloading the operating system called Linux, the package of Quantum ESPRESSO, and the input files, you can quickly practice some calculations of energy bands for graphene or monolayer molybdenum disulfide (MoS_2) and their physical properties. For other known materials that you want to run the program, you can find the input files of any existing materials on the internet with Automatic FLOW for Materials Discovery (`http://www.aflowlib.org`) or Materials Project (`https://materialsproject.org`).

The most crucial points that you need are your curiosity and motivation to run Quantum ESPRESSO. Although the book contains minimum information, it would be nice to learn Quantum ESPRESSO with some friends. This way may reduce the height of possible barriers of computer and physics and keep your motivation.

1.3 What we get, and what we do not get?

You will be surprised if you know what we can get from Quantum ESPRESSO. The following concepts are what we get from Quantum ESPRESSO: (1) lattice structure of solid and X-ray diffraction spectra, (2) electronic band structure and density of states, (3) phonon dispersion, (4) optical absorption spectra, (5) Raman spectra, (6) electronic transport properties such as carrier mobility, and (7) thermal properties such as thermal expansion and thermal conductivity. These concepts are subjects of solid-state physics that you do not need to know now. For the readers who are not familiar with physics, we will give minimum information about solid-state physics by using some equations to understand the special words. Although solid-state physics contents are minimal, you can use this book as the first step of understanding solid-state physics. It is highly desired for you to read some solid-state physics textbooks, too.

It is important to note that running Quantum ESPRESSO does NOT mean that (1) we can explain the phenomena of solid-state physics, (2) we can write a scientific paper by showing only the calculated results of Quantum ESPRESSO, (3) all calculated results are correct without knowing the detail of the method, and (4) we can find new physics from the calculated results. They are all possible misunderstandings of usage of the first-principles calculation. The situation is the same: we cannot be an expert on the computer even though we buy an expensive computer. Nevertheless, if you can use Quantum ESPRESSO, you will have a powerful tool for applying the physical concepts to real materials. If you are an experimental researcher of physics, you will be a strong researcher that can run Quantum ESPRESSO as the third tool for understanding the observed properties.

1.4 Organization of the book

In Chapter 2, we explain how to install a software on your PC, including the operating system (Linux) and Quantum ESPRESSO in Linux. In Chapter 3, we explain how to run a program for each purpose of calculating materials' properties. In Chapter 4, we briefly explain the density-functional theory and some keywords to understand the description in the input files of Quantum ESPRESSO in Chapter 3. In Chapter 5, we explain the solid-state physics related to the calculated results of Quantum ESPRESSO. Finally, in Chapter 6, we briefly introduce Linux commands and Python language, which are used in Chapter 3.

Please enjoy running Quantum ESPRESSO!

Chapter 2

Software Installation

This chapter will guide the readers to install all necessary software for running Quantum ESPRESSO. We expect that the readers already have a computer with an operating system (OS) such as Linux, Windows, and macOS. Firstly, we will show you the minimum requirements for each OS to avoid start-up problems in installing Quantum ESPRESSO. Next, we will show the main method to install Quantum ESPRESSO and its supporting software. Finally, we will show you how to run the simplest running instance of Quantum ESPRESSO.

2.1 Preparing the operating systems

Installing Quantum ESPRESSO could be challenging for beginners, but we are confident that by following all the steps given in this chapter, the readers will have a smooth, working environment of Quantum ESPRESSO. The main reason for the challenge is that Quantum ESPRESSO is an open-source software that consists of various packages originally targeted for Linux (or Unix-like OSs), which might sound unfamiliar for most Windows users in the world. However, current technology allows us to install Linux (or any other

Quantum ESPRESSO Course for Solid-State Physics
Nguyen Tuan Hung, Ahmad R. T. Nugraha, and Riichiro Saito

ISBN 978-981-4968-37-9 (Hardcover), 978-981-4968-63-8 (Paperback), 978-1-003-29096-4 (eBook)
www.jennystanford.com

OS) as a "guest" OS on Windows using either Windows subsystem for Linux (WSL) or a virtual machine (such as VirtualBox). Therefore, it is currently possible for Windows users to "run Quantum ESPRESSO in Windows" by using the WSL or VirtualBox.

Since we realize that the readers of this book use various operating systems in their daily life, we will show you how to install Quantum ESPRESSO in Linux, Windows, and macOS. We chose these OSs simply due to their popularity although we will not explain how to install the OSs. We assume the readers have such a specific OS and **internet access** in their own computers.

It should be noted that the installation guide for Windows users is mostly arranged in a way to show how we can utilize the virtualization feature. It means that after installing the "virtual" Linux on Windows, the remaining Quantum ESPRESSO installation procedure should be the same as that on Linux. Before proceeding to the Quantum ESPRESSO installation guide, we recommend you ensure that your OS is updated to the latest version. We will show you how to prepare three specific OSs, i.e., Ubuntu Linux, Windows, and macOS Catalina.

2.1.1 Ubuntu Linux

In Ubuntu, please check that you can open a `Terminal` window from the `Activities` panel. `Terminal` is a place in the OS where we can interact with the "shell". Simply saying, the shell is a program that takes commands from the keyboard and returns them to the operating system to perform.

In the old days, the shell was the only user interface available on a Unix-like system. Nowadays, we have graphical user interfaces (GUIs) in addition to the shell, which is a kind of command-line interfaces (CLIs). To interact with the shell in the terminal, we can type a line of "command" or (lines of commands) corresponding to some computer operations.

The basic structure of Ubuntu `Terminal` is shown in Fig. 2.1. We can see there is a prompt showing the login username, machine hostname, and a current working directory ("~" sign, indicating the so-called "home" directory of the user). The prompt ends with a dollar ($) sign. The cursor next to the $ sign is the starting point

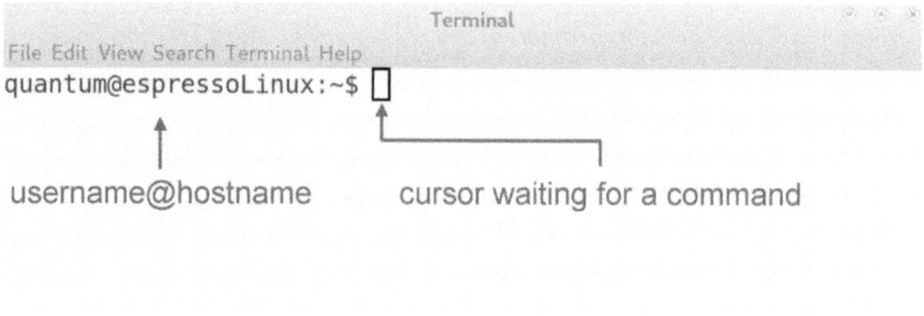

Figure 2.1 Terminal in Ubuntu.

where we can type the actual Linux commands. There are so many commands for various purposes, but we will only use a few of them to install software and manage our Quantum ESPRESSO jobs and files.

To update all software in Ubuntu, on the terminal, type the following commands (next to the $ sign) and press enter after each line:

```
$ sudo apt update
$ sudo apt upgrade
```

You will be asked for "your password" (not root password) when executing the above commands because the administrator (often called "root") privileges are required when installing or updating the software in Ubuntu. The root privileges are represented by the `sudo` command.

The `apt` command corresponds to the "Advanced Packaging Tool", a command-line tool that helps in handling packages in Ubuntu. Its main task is to retrieve the information and packages from the authenticated sources (mostly from the internet) for installation, upgrade, and removal of packages along with their dependencies. In the above example, the `update` and `upgrade` texts are a set of commands to ensure your Ubuntu software is all updated.

We also need some development tools and libraries, such as Git (version control system), GNU Wget (file retrieval software), GCC/G++(C/C++ compiler), GFortran (Fortran compiler), LAPACK (linear algebra package), FFTW (Fourier transform library), and Open MPI (parallel computing implementation). All of them are necessary to compile Quantum ESPRESSO. On the terminal, execute the following commands one by one:

```
$ sudo apt install git wget build-essential
$ sudo apt install g++ gfortran liblapack-dev
$ sudo apt install libfftw3-dev libopenmpi-dev
```

If you are prompted by a "Yes/No" question, just type "y" (or "Yes") and press enter to agree with the installation.

2.1.2 Windows

For Windows users, the OS should be updated to at least Windows 10 version 1903. If you want a native support for Linux GUI applications, we recommend to use or update Windows 11 with build number at least 22000. To perform the update, we should open `Windows Update` from the `Start` menu in Windows, and we will either see the option for `Feature update`, or we will have to click `Check for updates` (see Fig. 2.2).

If you still do not see the option to update Windows to the latest version, click `Check for updates`, reboot your computer, and try the process again. There is also a possibility that you already have this update without realizing it. You can find the OS information in Windows by typing → `About your PC` in the Windows `Start` menu. Under Windows specifications, check which edition and version of Windows your computer is running. If you see version 1903 or above (see Fig. 2.3), you can continue to the next step.

Even if Windows Update does not offer the update, we can download Microsoft's Update Assistant tool to install it manually. This tool will give you the update even if Microsoft is not confident it is ready for your computer yet. With a version of more than 1903, it is possible for us to install Windows subsystem for Linux (WSL) with the ability to access Linux files from File Explorer or other file manager applications. Unlike older ways of accessing virtual Linux files, this version of Windows offers full read-write access without the worry of breaking anything.

Next, we have to check whether or not our computer supports virtualization. In most computers, the virtualization can be enabled

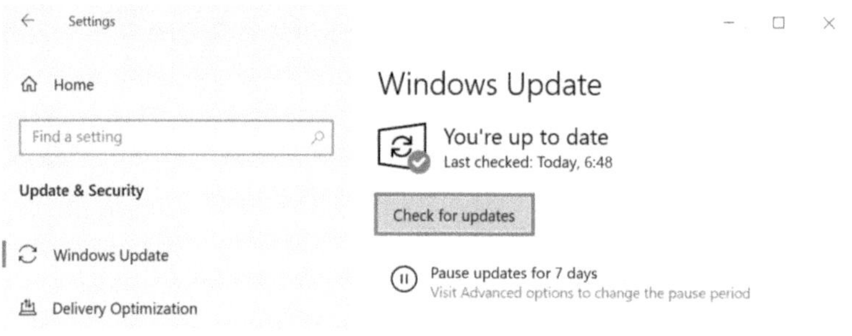

Figure 2.2 Windows Update. You can click `Check for updates` and `Download and install now` to guarantee your Windows is in the latest version.

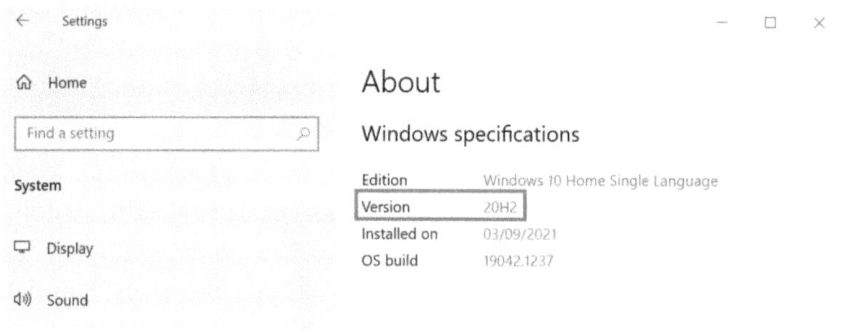

Figure 2.3 Windows specifications in `About your PC`. Make sure that you are running at least Windows 10 version 1903. Again, for a native support of Linux graphical applications in WSL, we recommend at least Windows 11 build 22000.

from BIOS, which can be accessed before booting your OS. The key you should press on the keyboard to access BIOS will depend on the manufacturer of the computer. You can check from the computer screen when you turn it on. Assuming the key is `F2` (or `DEL`), below are the steps to enable the virtualization from the BIOS settings:

- Turn on or restart the computer.
- Press `F2` (or `DEL`) key at startup BIOS Setup.
- Press the right arrow key to `System Configuration` tab, select `Virtualization Technology` or a similar option and then press the `Enter` key.
- Select `Enabled` and press the `Enter` key.

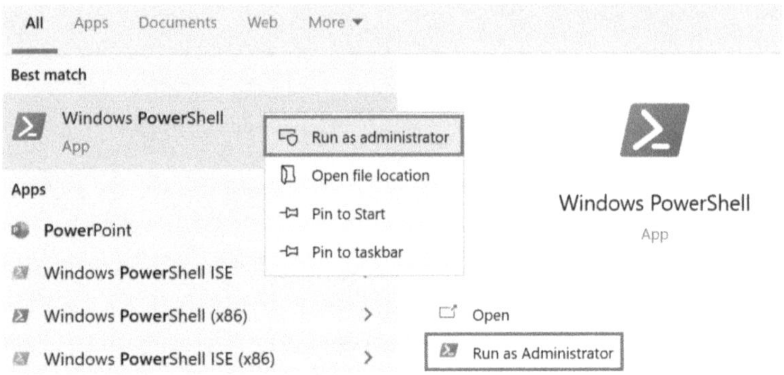

Figure 2.4 Opening PowerShell as the administrator.

- Press the F10 key and select `Yes` and press the `Enter` key to save changes and Reboot into Windows.

After performing the above steps, we are ready to install Windows Subsystem for Linux (WSL).

WSL is a Windows feature that enables us to run native Linux command-line tools directly on Windows. The merit of using WSL over other virtual machines is that WSL requires fewer resources (CPU, memory, and storage). WSL also allows us to run Linux command-line tools and applications (apps) alongside Windows command-line, desktop, and store apps and access Windows files from Linux. This technique allows us to use Windows apps and Linux command-line tools on the same set of files.

To install WSL and Ubuntu Linux automatically on Windows, first open PowerShell as an administrator by searching "PowerShell" in the Windows menu and right-click PowerShell to get access as an administrator (see Fig. 2.4). Inside PowerShell, type the following command and press enter:

```
wsl --install
```

Follow the instructions on the screen and restart the computer when prompted. This reboot is required in order to ensure that WSL can initiate a trusted execution environment.

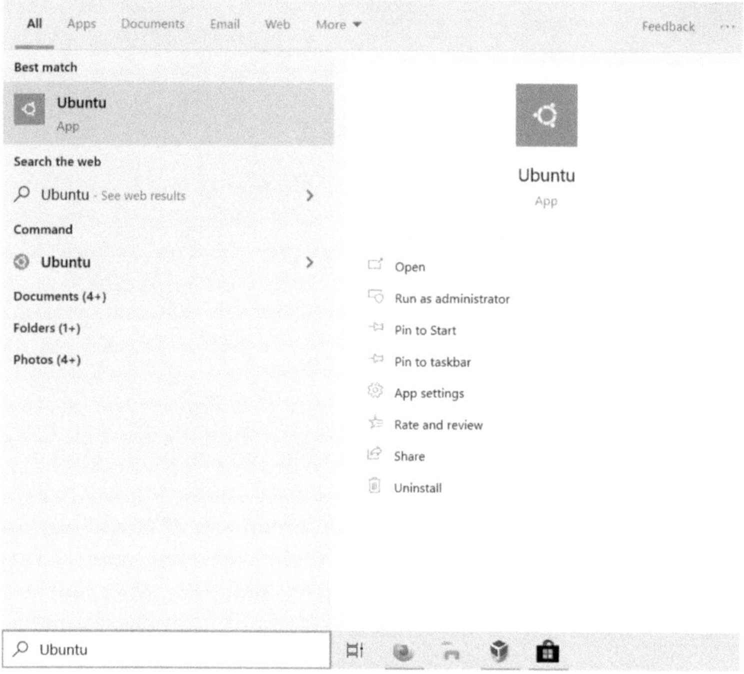

Figure 2.5 Launching Ubuntu from the Start menu.

After reboot, WSL should initialize the Ubuntu instance once before the system can be used properly. If everything goes normally, Ubuntu initialization should launch automatically when we re-login to Windows. *If not*, we should launch a new instance of Ubuntu by clicking the "launch" button in the Microsoft Store app for Ubuntu or launching Ubuntu from the Start menu (Fig. 2.5). Note that if you cannot perform this kind of WSL installation, please follow the detailed manual instructions in Microsoft Docs: `https://docs.microsoft.com/en-us/windows/wsl/install-manual`

The first time a new Ubuntu instance runs, a console window will open, and we will be asked to wait for several minutes for the installation to complete. During this final stage of installation, the Ubuntu files are extracted and stored on your PC. It may take a couple of minutes, depending on the performance of your computer's storage devices. This initial installation phase is only required when a Linux distribution in Windows is clean-installed. All future launches should take less than a second.

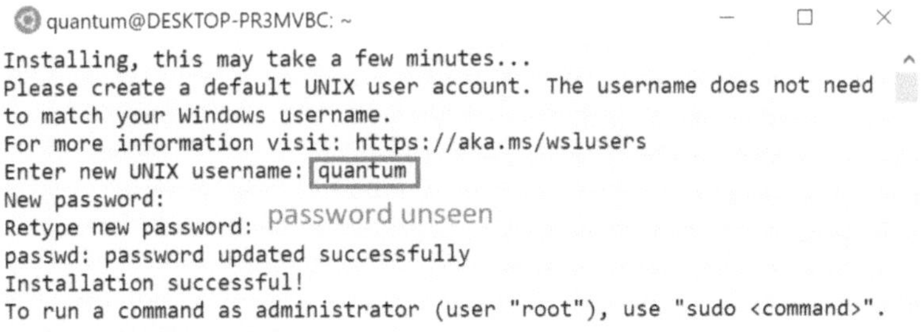

Figure 2.6 First Ubuntu instances through WSL and creation of a user account.

Once installation is complete, we will be prompted to create a new user account (and its password), as shown in Fig. 2.6. This user account is for the normal non-admin user that will log in as default when launching Ubuntu. We can choose any username and password. There is no necessity to match the Windows username.

In our example, we set "quantum" as the username. In Fig. 2.6, you should notice that the password is not displayed by the system. Therefore, be careful when typing it for the first time. Next time we open a new Ubuntu instance, we will not be prompted for the password. Only when we want to elevate a process using sudo to access root privilege, we need to enter the user password, so make sure to choose a password that is difficult to forget as well as difficult to break.

What you see in Fig. 2.6 is essentially the same shell as in Fig. 2.1 that shows the Ubuntu terminal. Therefore, hereafter, we loosely refer to the shell as the Ubuntu terminal. You can open a File Explorer window directly from the terminal in the current directory, which is the home (also denoted $HOME) directory of Linux. The corresponding command is:

```
$ explorer.exe .
```

You can work with files normally from here (see Fig. 2.7). Use drag and drop, copy and paste them, or even open them directly in Windows applications.

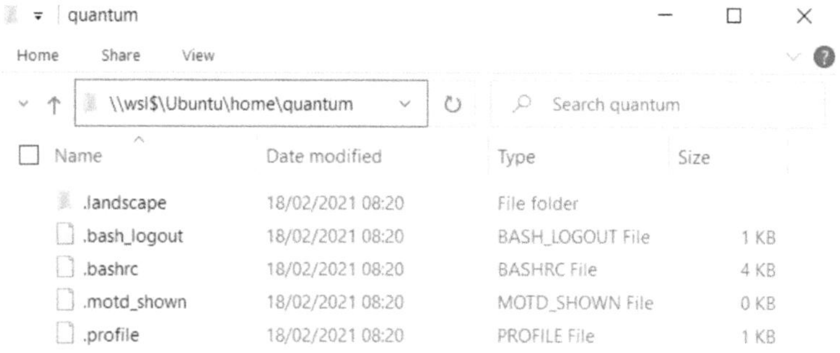

Figure 2.7 WSL `$HOME` directory opened in Windows.

Besides typing the above command, you can also open the `$HOME` directory from Windows Explorer by directing the explorer to the following address:

```
\\wsl$\Ubuntu\home\quantum
```

(Change "quantum" with your username of Ubuntu in WSL.)

As in Ubuntu Linux, we should update all the default software and install development tools with libraries necessary to compile Quantum ESPRESSO. We type the following commands one by one in the Ubuntu terminal on WSL:

```
$ sudo apt update
$ sudo apt upgrade
$ sudo apt install git wget build-essential
$ sudo apt install g++ gfortran liblapack-dev
$ sudo apt install libfftw3-dev libopenmpi-dev
```

Note again, the $ sign should not be typed/copied to the terminal. You will also be asked for your password when executing the above commands because administrator privileges are required. The detailed explanation of the meaning of each line above can be read in the previous section (Sec. 2.1.1).

By completing all the above steps, we consider that you are already "having a Linux OS" on Windows. Therefore, the remaining procedure to install Quantum ESPRESSO and its supporting software is the same as that in the "real" Linux OS, which will be explained in Sec. 2.2.

2.1.3 macOS Catalina

macOS is one of Unix-based OS, so that it already has the terminal tool. However, some additional packages are needed before we can install Quantum ESPRESSO. We recommend the Homebrew package manager to install the missing development tools and libraries. For the tutorial in this book, we tested the package installations on macOS Catalina. On the terminal, we first have to execute the following command to get Homebrew:

```
$ /bin/bash -c "$(curl -fsSL https://raw.
    githubusercontent.com/Homebrew/install/HEAD/
    install.sh)"
```

The command is quite long, and it may look broken into several lines in this book or your terminal, but it is still a single line of command. You should ensure typing the command properly to the terminal as a unified entity, without pushing "Enter" key.

By having Homebrew in macOS, the prerequisites for the Quantum ESPRESSO installation can be obtained by another single line of command:

```
$ brew install git wget gcc gfortran fftw lapack
    openblas open-mpi eigen
```

Again, no matter how the above (long) command looks broken into separate lines, it is a single line of command.

2.2 Installation of Quantum ESPRESSO and its supporting software

With the various OSs that the readers have, we are ready to install Quantum ESPRESSO using the same approach on the terminal.

We will install Quantum ESPRESSO and an additional Wannier90 package, either in "real" Ubuntu Linux, WSL Ubuntu, or macOS.

For beginner users of Ubuntu Linux and WSL Ubuntu, the following commands are sufficient to install Quantum ESPRESSO and Wannier90:

```
$ sudo apt install quantum-espresso
$ sudo apt install wannier90
```

The above commands will install slightly earlier versions of Quantum ESPRESSO and Wannier90 which are already sufficient for running the tutorials given in Chapter 3. On the other hand, for readers who want to install the latest version of Quantum ESPRESSO and Wannier90 from source code, please follow the rest of this section.

We recommend the readers to create a new folder `opt` in the `$HOME` directory. Type this command on the terminal:

```
$ mkdir opt
```

Then, go into the `opt` folder:

```
$ cd opt
```

In the `opt` folder, we can download the latest source files of Quantum ESPRESSO using the Git utility as follows:

```
$ git clone https://github.com/QEF/q-e.git
```

After the source files are downloaded, we can enter the `q-e` folder:

```
$ cd ~/opt/q-e/
```

In this folder, we execute the configuration command:

```
$ ./configure
```

Quantum ESPRESSO will automatically determine the best setting for our OS. If no error occurs, we can continue with making the executable files:

```
$ make all
$ make w90
```

The purpose of the first line above is to make executable files for all standard Quantum ESPRESSO capabilities, such as electronic structure, phonon dispersion, and optical properties calculations. The second line is to make the additional Wannier90 package available in the binary folder. In addition to this compilation process, our OS must be "taught" how to find the executable files and include them in its variable path. For this purpose, we can execute two more commands to update the path:

```
$ echo 'export PATH=$PATH:$HOME/opt/q-e/bin' >> ~/.
    bashrc
$ source ~/.bashrc
```

The above command lines enable us to access all binary files of Quantum ESPRESSO without typing the complete path up to its `bin` folder.

To check that our Quantum ESPRESSO installation is done successfully, we can type the following command, which is the most basic command of Quantum ESPRESSO:

```
$ pw.x
```

We will learn how to use this command briefly in Sec. 2.4 and more extensively in Chapter 3. To exit the Quantum ESPRESSO environment generated from the above command line, push `CTRL+c` keystroke.

We can install some additional software to support our workflow when performing first-principles calculations with Quantum ESPRESSO. Executing the following commands line by line on the terminal will give us all necessary software:

```
$ sudo apt install xcrysden gnuplot
$ sudo apt install python3-dev python3-pip
$ pip3 install numpy scipy sympy
$ pip3 install matplotlib jupyterlab
$ echo 'export PATH="$HOME/.local/bin:$PATH"' >> ~/.
```

Some notes:

- XCrysden is useful for visualization of molecule and crystal structures.
- Gnuplot is a plotting software. We may also use other plotting software. In the above commands, we include Python numerical libraries and graphical tools such as Matplotlib and JupyterLab that will be mainly used in this book.
- GUI support in WSL is required to run XCrysden, Gnuplot, and Jupyter. Windows 11 with build number at least 22000 already natively support GUIs. If you have a lower Windows version, you need to install an additional "X server", e.g., VcXsrv, X410 App, or Kali App. See the following Microsoft Tech Community for the X server installation: `https://techcommunity.microsoft.com/t5/windows-dev-appconsult/running-wsl-gui-apps-on-windows-10/ba-p/1493242`
- Depending on the system, it may take around 15-30 minutes to complete all the installation in this section.
- We have created a simple Bash script (see Sec. 6.3.2 to learn more about Bash), which collects all the above commands to install Quantum ESPRESSO and its supporting software for Ubuntu Linux and WSL Ubuntu that can be downloaded from `https://github.com/nguyen-group/QE-SSP/blob/master/QEinstall.sh`. After downloading the script, the readers can execute the following command line in the terminal:

```
$ bash QEinstall.sh
```

2.3 VirtualBox approach

If you have already succeeded to obtain a working Quantum ESPRESSO environment from the previous sections, you do not need to follow this VirtualBox approach. VirtualBox is one of the virtual machines that can be used to run other guest OSs from a specific OS. In Windows, for example, we can install VirtualBox to run a Linux distribution that has already been configured to include Quantum ESPRESSO and other necessary software inside it. There are three recommended virtual machines for Quantum ESPRESSO:

1. Quantum Mobile:
   `https://quantum-mobile.readthedocs.io`
2. MateriApps LIVE!:
   `https://cmsi.github.io/MateriAppsLive/`
3. QE-2021: `http://qe2021.ijs.si`

Each virtual machine has its username and password, which may change over time. Therefore, it is better to check the documentation on one of the websites above, depending on which virtual machine you like. However, all of them use the same capability of VirtualBox to host a virtual machine. Here we just explain how to install Virtual Box and add the virtual machine. The steps to install and use the virtual machine is as follows:

1. Download VirtualBox from: `https://www.virtualbox.org/wiki/Downloads`. Choose the Windows host (or macOS if the macOS users decide to use VirtualBox, too). Then, install and open VirtualBox.
2. Download the extension pack from the same webpage and install it while opening VirtualBox.
3. Download a virtual machine, either Quantum Mobile, MateriApps LIVE!, or QE-2021 from their websites listed previously. It is huge, around 3–4 GB, so if your internet connection is slow, better download the file when you are going to sleep.
4. From the VirtualBox software, press `Machine` and then `Add`. Select one of the virtual machines you have downloaded.

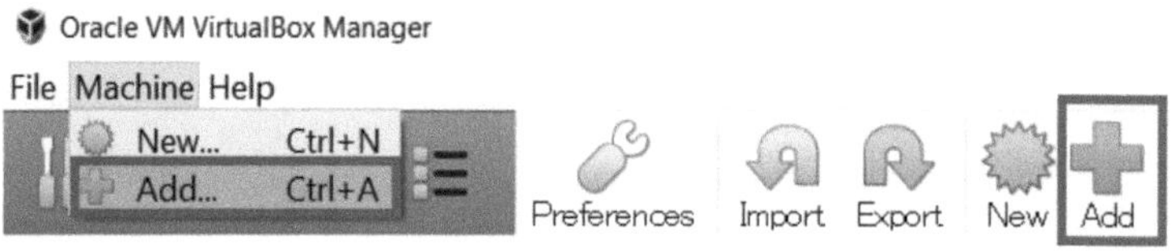

Figure 2.8 Adding a virtual machine in VirtualBox. Click the `Add` button, either from the `Machine` menu or from the toolbar.

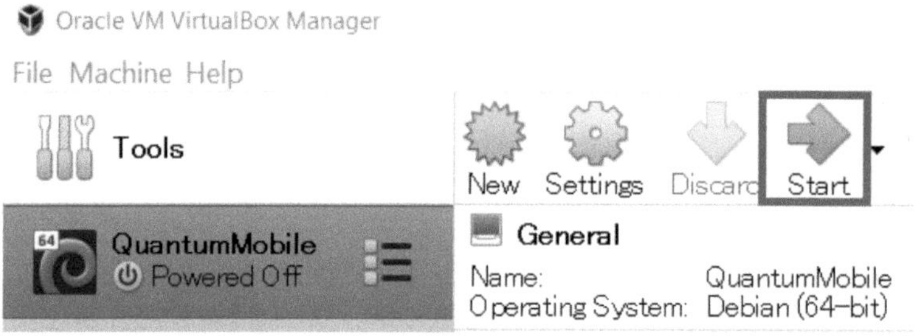

Figure 2.9 Starting the virtual machine.

Figure 2.10 Example desktop appearance of a virtual machine.

5. The virtual machine will be ready in a few minutes after extracting. Select the virtual machine from the left panel and press the "Start" button as shown in Fig. 2.9.
6. After starting up, we can see the desktop of the virtual machine as shown in Fig. 2.10.

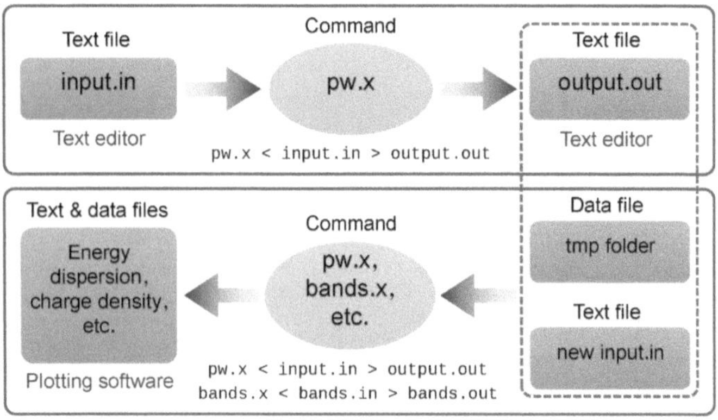

Figure 2.11 A typical Quantum ESPRESSO workflow.

2.4 Processing input and output files

With the Quantum ESPRESSO installed on our computer, we are now able to execute the simplest running instance of Quantum ESPRESSO, as shown in the diagram of a typical workflow of Quantum ESPRESSO in Fig. 2.11.

2.4.1 Basic execution of Quantum ESPRESSO commands

There are two main calculations in Quantum ESPRESSO. The first one is the self-consistent-field (SCF) calculation using the `pw.x` command. We will learn the details of the SCF calculation in Chapter 3, but we will need to prepare an input file that will be "read" by this command (for example: `input.in`), and then the calculation logs will be written into an output file (for example: `output.out`). The input file has some specifications for variables and parameters of the material simulation that should be prepared in advance using a text editor. We do not restrict the text editor that we use to edit the input file or check the output contents. Any text editor should be fine, for example, Notepad (or Notepad++) in Windows and `nano`, `vi`, or `emacs` in Linux. Suppose we already have the `input.in` file, the execution of the `pw.x` command is as follows:

```
$ pw.x < input.in > output.out
```

After the mandatory SCF calculation, the second calculation of a typical Quantum ESPRESSO simulation is the "real" stage for calculating the physical properties of the material under consideration. For example, in Fig. 2.11 we show another possibility of running pw.x, which is for the non-SCF (NSCF) calculation that includes the band structure and density of states (DOS). There will already be some data files produced from the NSCF calculations, but in most cases, we need to do "post-processing" before plotting and interpreting the data. Examples of the post-processing commands in Quantum ESPRESSO are bands.x, dos.x, epsilon.x, as shown in Table 2.1, which all will be learned in more detail in Chapter 3. The basic structure of the command execution is the same as the pw.x command, i.e.,

```
$ QEcommand < input.in > output.out
```

where QEcommand is any Quantum ESPRESSO command. For example, we may run the following in the case of bands.x command:

```
$ bands.x < band.in > band.out
```

We list some frequently-used Quantum ESPRESSO commands in Table 2.1. Note that besides the logs in the output.out file, the Quantum ESPRESSO commands also give other output (or data) files in a specified directory depending on the calculation type. The extension of the files is specified by the input file.

2.4.2 Choice of plotting software

In most cases, only after performing the post-processing commands we can obtain some meaningful data files related to the properties of the material and make some plots using our favorite plotting software. Although there are plenty of such softwares in the market, we decide to specifically suggest the readers of this book to

Table 2.1 Some frequently-used Quantum ESPRESSO commands.

Command	Purpose
`pw.x`	SCF and NSCF calculations
`bands.x`	band structure post-processing
`dos.x`	DOS postprocessing
`epsilon.x`	optical properties calculation
`ph.x`	phonon calculation

use Matplotlib libraries from the Python programming language. Therefore, for plotting the Quantum ESPRESSO calculation data, we will give the hands-on plotting examples either in `.py` or `.ipynb` extension, which is particularly comfortable to be edited and interacted with using the JupyterLab interface (`https://jupyter.org/`). The installation of Python, Matplotlib, JupyterLab, along with other important Python libraries, is already explained in Sec. 2.2. To open JupyterLab, we can type the following command in the terminal

```
$ jupyter-lab
```

which results in opening a web browser (e.g., Chrome/Firefox/Safari) with some menus shown in Fig. 2.12.

Following Fig. 2.12, there are three most important parts of JupyterLab we should understand:

(1) The address of JupyterLab instance: This address is shown in the web browser address locator. It will almost certainly start with `localhost`, followed by 4-digit number of the port used by JupyterLab. You need not worry about the number because JupyterLab will automatically determine it.

(2) List of files and folders at which we open JupyterLab: You will be able to see the files/folders in that panel if the working directory where we execute the `jupyter lab` command already consists of some files/folders. We can also browse through the files/folders by clicking the filename/folder name there.

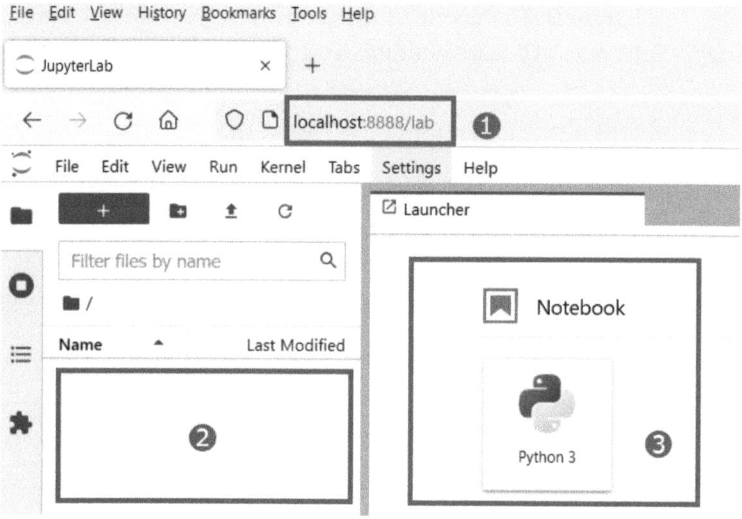

Figure 2.12 Jupyter lab interface.

(3) Jupyter Notebook: We should click the icon in order to open a Python 3 developing environment in the form of Jupyter Notebook. Then, we can save the file with the `.ipynb` or `.py` extension for the plotting purpose.

Note that some traditional Python users may prefer to directly access the Jupyter notebook from the command line by typing:

```
$ jupyter-notebook
```

2.4.3 Obtaining example files for hands-on tutorials

We provide example input/output files of Quantum ESPRESSO calculations, additional tips, and also errata of the book in the following public repository: `https://github.com/nguyen-group/QE-SSP`. The readers can use the input files from the repository to practice the Quantum ESPRESSO calculations directly on their computers. A step-by-step guide to run these files is given in

Chapter 3. To download the files from the terminal, we can use the one-line `git clone` utility:

```
$ git clone https://github.com/nguyen-group/QE-SSP.
    git
```

With all the information in this chapter, we are now ready to dive into hands-on calculations of Quantum ESPRESSO. It would be nice to read the remaining chapters and practice all the calculations accompanied by a cup of ESPRESSO coffee.

Chapter 3

Hands-On Tutorials of Quantum ESPRESSO

In this chapter, we show how to run Quantum ESPRESSO. We focus on several topics for obtaining the electronic, optical, transport, and mechanical properties of materials. Each section starts with an explanation of the background and purpose. Then, we show the command to run Quantum ESPRESSO by showing examples of input and output files. Before practicing with these tutorials, please ensure that Quantum ESPRESSO and all the necessary software packages have been installed on your computer (see Chapter 2 for installation). In Chapter 3, we only show how to run Quantum ESPRESSO. In order to understand how Quantum ESPRESSO works, the basic concepts of the density-functional theory are explained in Chapter 4. The concepts of solid-state physics behind these hands-on tutorials are briefly summarized in Chapter 5. Calculation job and plotting scripts are written in Bash and Python, which are explained in Chapter 6. We recommend using XCrySDen and VESTA for the visualization of input and output files.

We adopt two-dimensional graphene as an example of a material for these tutorials. Since graphene consists of two carbon atoms in a hexagonal unit-cell, we can quickly run Quantum ESPRESSO for

Quantum ESPRESSO Course for Solid-State Physics
Nguyen Tuan Hung, Ahmad R. T. Nugraha, and Riichiro Saito

ISBN 978-981-4968-37-9 (Hardcover), 978-981-4968-63-8 (Paperback), 978-1-003-29096-4 (eBook)
www.jennystanford.com

graphene. The calculation takes only a few minutes on desktop or notebook computers. Other materials such as MoS_2 and GeTe also appear in GitHub (`https://github.com/nguyen-group/QE-SSP`). It is noted that the format of the input files might be modified from version to version of Quantum ESPRESSO. In this book, we stick to the format of Quantum ESPRESSO version 7.0. However, alternative code lines will be provided in the input files whenever it is necessary to provide backward compatibility. The source files and scripts from this chapter are also available online in GitHub (`https://github.com/nguyen-group/QE-SSP`). The readers can download the whole repository, view them in a terminal, or read them at GitHub, where they are automatically rendered.

3.1 Basic parameters

When starting a new job of simulations with Quantum ESPRESSO, we first need to determine the parameters in the input file for each step of the calculation. In this section, we use graphene as an example material.

3.1.1 Total energy and self-consistent field calculations

❑ **Purpose:** By calculating the total energy with the self-consistent field (SCF) method, we get the ground-state properties, which will be used for electron and phonon dispersion calculations in Secs. 3.2 and 3.3, respectively.

❑ **Background:** The total energy and the SCF calculation are two concepts that we should know for this tutorial. Both concepts are given in Secs. 4.5 and 4.12.

❑ **How to run:** To run the SCF for the input file, the readers should type the following command lines:[1]

```
$ cd ~/QE-SSP/gr/scf/
$ mpirun -np 4 pw.x < scf.in > scf.out &
```

[1] If you do not find 'mpirun' command, you need to install libopenmpi-dev (see Chapter 2).

- Line 1: Go to the `scf` directory that includes the input files.
- Line 2: Run `pw.x` (`pw` = plane-wave, `x` = executable file) by using in parallel with 4 processors (`-np 4`) by the command `mpirun` (running a program with parallel processors) with the input file is `scf.in` and the output file is `scf.out`. The symbol `&` makes the command run in the background. Here, we select 4 processes for parallel calculations, but the readers can run with serial calculations (without `mpirun`, that is `pw.x < scf.in > scf.out &`) or any number of processes (e.g., `-np 8` for Intel Core i7 or Core i9), depending on your computer. For detailed commands for running in parallel, the readers can find on Sec. 6.2.3.

❑ **How to check:** To check whether or not the output file exists, the readers can use the `ls` command by typing:

```
$ ls
```

It will display a listing of all files in the `scf` directory as

```
scf.in   scf.out
```

If the readers see the `scf.out`, the readers can check the contents of `scf.out` by `tail` command as follows:

```
$ tail scf.out
```

The `tail` command prints the last few lines of the `scf.out` file. If the calculation normally finishes, a message `JOB DONE` is written at the end of this file as

```
=------------------------------------------------------=
   JOB DONE.
=------------------------------------------------------=
```

❑ **Input file:** Now, let us explain the details of the input file. To see the input file, the readers can use `vi` editor by typing:

```
$ vi scf.in
```

To quit the `vi` editor without saving any changes, the readers must press : (colon). Then the cursor should reappear in the lower-left corner of the screen beside a colon prompt. After that, the readers need to enter `q!` to quit the file without saving. Since `vi` editor is usually available in all Linux distributions, we would like to use `vi`. However, the readers can use any text editor, such as Emacs or Notepad, to open scf.in directly. When the readers open `scf.in` file, the readers will see the input variables as

QE-SSP/gr/scf/scf.in

```
1  &CONTROL
2  calculation    = 'scf'
3  pseudo_dir     = '../pseudo/'
4  outdir         = '../tmp/'
5  prefix         = 'gr'
6  /
7  &SYSTEM
8  ibrav          = 4
9  a              = 2.4623
10 c              = 10.0
11 nat            = 2
12 ntyp           = 1
13 occupations    = 'smearing'
14 smearing       = 'mv'
15 degauss        = 0.02
16 ecutwfc        = 60
17 /
18 &ELECTRONS
19 mixing_beta    = 0.7
20 conv_thr       = 1.0D-6
21 /
22 ATOMIC_SPECIES
23 C 12.0107 C.pbe-n-rrkjus_psl.0.1.UPF
24 ATOMIC_POSITIONS (crystal)
25 C  0.333333333  0.666666666  0.500000000
26 C  0.666666666  0.333333333  0.500000000
27 K_POINTS (automatic)
28 12 12 1 0 0 0
```

☞ **Explanation of scf.in:** There are three mandatory namelists (`CONTROL`, `SYSTEM`, and `ELECTRONS`), which start by &, and three mandatory input cards (`ATOMIC_SPECIES`, `ATOMIC_POSITIONS`, and `K_POINTS` in lines 22, 24, and 27, respectively). The SCF calculation is performed by selecting `calculation = 'scf'` in the namelist

Table 3.1 Meaning of input variables in `scf.in` file.

Line	Syntax	Meaning
1	`&CONTROL`	A *mandatory namelist* includes input variables that control the flux of the calculation and the amount of I/O on disk and on the screen.
2	`calculation`	A string describing the task to be calculated, in which `'scf'` is the SCF calculation.
3	`pseudo_dir`	Directory containing pseudopotential files.
4	`outdir`	Temporary folder to save output data.
5	`prefix`	Filenames of output data in tmp folder.
6	/	End of namelist `CONTROL`.
7	`&SYSTEM`	A *mandatory namelist* includes input variables that specify the system for the calculation.
8	`ibrav`	Bravais lattice index (see Table 3.2).
9, 10	`a, c`	Lattice constants in Angstrom unit.
11	`nat`	Number of atoms per unit cell.
12	`ntyp`	Number of types of atoms in unit cell.
13	`ecutwfc`	Cut-off energy (Ry) for pseudopotentials.
14	`occupations`	Gaussian smearing for the case of metal.
15	`smearing`	Smearing method, in which `'mv'` is the Marzari-Vanderbilt-DeVita-Payne cold smearing.
16	`degauss`	Value of the gaussian spreading (Ry) for Brillouin-zone integration in metal.
17	/	End of namelist `SYSTEM`.
18	`&ELECTRONS`	A *mandatory namelist* includes input variables that control the algorithms used to reach the SCF solution for the electrons.
19	`mixing_beta`	Mixing factor for self-consistency.
20	`conv_thr`	Convergence threshold for self-consistency.
21	/	End of namelist `ELECTRONS`.

Continued

Table 3.1 - *Continued*

Line	Syntax	Meaning
22, 23	ATOMIC_SPECIES	A *mandatory input card* includes name, mass and pseudopotential used for each atomic species present in the system.
24-26	ATOMIC_POSITIONS	A *mandatory input card* includes type and coordinates of each atom in the unit cell. The option 'crystal' indicates that atomic positions are in crystal coordinates.
27, 28	K_POINTS	A *mandatory input card* includes information of the ***k***-points used for Brillouin-zone integration.

CONTROL in the input file. In the namelist ELECTRONS, we need to set several variables that can be specified to control the SCF calculation. A short description of the input variables is given in Table 3.1. If the readers want to know the original information of input variables, the readers can visit the following web page: https://www.quantum-espresso.org/Doc/INPUT_PW.html.

☞ **Visualizing structure from scf.in:** To ensure that structure of the material is correctly represented by the input file, it is crucial to visualize the structure by reading the input file in XCrySDen software. XCrySDen is one of the visualization tools that we can use to check the structure directly. To run XCrySDen, we enter the command:

```
$ xcrysden &
```

One the window of XCrySDen is opened as shown in Fig. 3.1, we select tabs: **File → Open PWscf → Open PWscf Input File** and select scf.in. In the box **[PWSCF Input: "scf.in"]**, we click on the **OK** button, XCrySDen will automatically identify the structure of graphene from the input file (see Fig. 3.1).

The structure of graphene in Fig. 3.1 includes two concepts: "bravais lattice" and "atomic basis". A bravais lattice is specified by an integer number ibrav and six lattice constants a, b, c, cosAB, cosAC, and cosBC in the namelist SYSTEM. Here, cosAB = cosine of the angles between a and b (γ), cosAC = cosine of the angles between

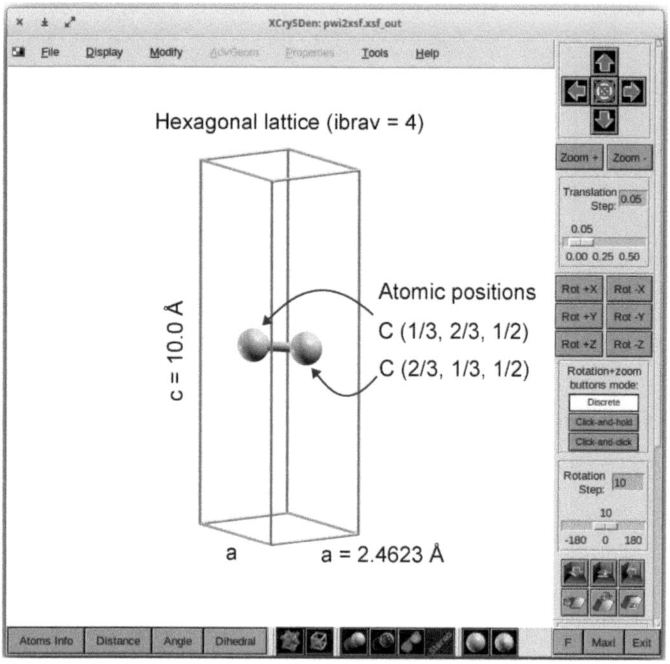

Figure 3.1 Visualizing structure of material from Quantum ESPRESSO input file by using XCrySDen.

a and c (β), and `cosBC` = cosine of the angles between b and c (α). The list of the bravais lattice indexes in Quantum ESPRESSO is shown in Table 3.2. Since graphene has a hexagonal lattice, we select `ibrav = 4` (hexagonal lattice) and the lattice constants $a = b = 2.4623$ Å and $c = 10$ Å as shown in Fig. 3.1. To confirm that our unit cell is hexagonal or not, let us increase the number of unit cells in XCrySDen. From XCrySDen in Fig. 3.1, we select **Modify → Number of Unit Drawn**. In the box **[Modify Number of Unit Drawn]**, we choose 3, 3, and 2 in the x, y, and z directions, respectively. Then we can see a $3 \times 3 \times 2$ supercell, as shown in Fig. 3.2. Since the periodic boundary conditions are always applied for any directions in Quantum ESPRESSO, for the "two-dimensional graphene", we set c = `10.0` Å as a sufficiently large value compared with distance of two graphene layers in graphite (3.35 Å) to avoid undesirable interactions in the z-direction. Nevertheless, too large c requires computational power.

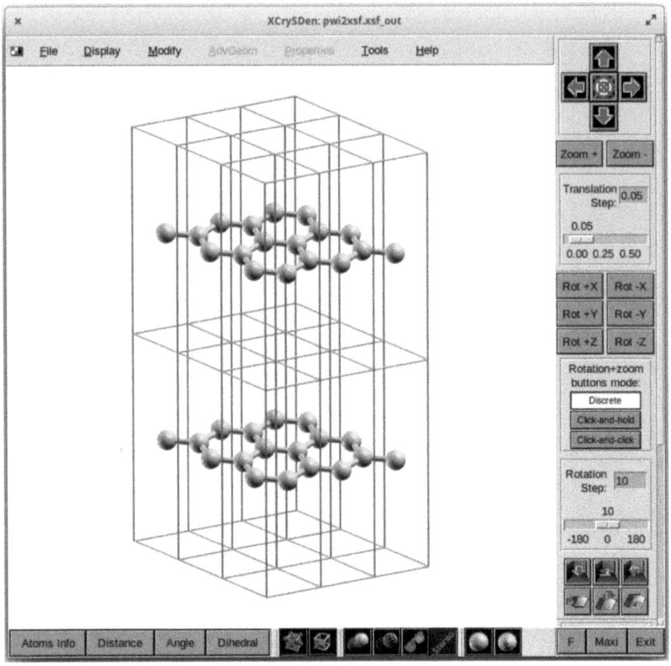

Figure 3.2 Visualizing a supercell 3 × 3 × 2 of graphene by using XCrySDen. We can see hexagonal lattice.

☞ **Note**: For the lattice constants, we can specify either [*a*, *b*, *c*, `cosAB`, `cosAC`, and `cosBC`] OR [`celldm(1)`- `celldm(6)`] but NOT both, in which `celldm(1) = a` (in a.u.), `celldm(2) =` *b*/*a* (in a.u.), `celldm(3) =` *c*/*a* (in a.u.), `celldm(4) = cosAB`, `celldm(5) = cosAC`, and `celldm(6) = cosBC`.

Since the bravais lattice is set, we now check the atomic basis. The atomic basis consists of the number, type, and positions of atoms in the unit cell, in which the number and type of atoms are specified by the integer numbers `nat` and `ntyp` in the namelist `SYSTEM`, respectively, while the positions of atoms are specified in the card `ATOMIC_POSITIONS` in the crystal coordinates `(crystal)`, i.e., the values of the positions are between 0.0 and 1.0 for each direction. For graphene, there are two C atoms in the unit cell. Thus, we set `nat = 2`, `ntyp = 1`, and the atomic positions are `(0.333333333 0.666666666 0.500000000)`

Table 3.2 Bravais lattice table in Quantum ESPRESSO.

ibrav	Structure	Lattice vectors
0	Free	Lattice vectors are given in card `CELL_PARAMETERS`
1	Cubic P	$v_1 = a(1,0,0)$, $v_2 = a(0,1,0)$, $v_3 = a(0,0,1)$
2	Cubic F	$v_1 = \frac{a}{2}(-1,0,1)$, $v_2 = \frac{a}{2}(0,1,1)$, $v_3 = \frac{a}{2}(-1,1,0)$
3	Cubic I	$v_1 = \frac{a}{2}(1,1,1)$, $v_2 = \frac{a}{2}(-1,1,1)$, $v_3 = \frac{a}{2}(-1,-1,1)$
−3	Cubic I	$v_1 = \frac{a}{2}(-1,1,1)$, $v_2 = \frac{a}{2}(1,-1,1)$, $v_3 = \frac{a}{2}(1,1,-1)$
4	Hexagonal and Trigonal P	$v_1 = a(1,0,0)$, $v_2 = a(-\frac{1}{2},\frac{\sqrt{3}}{2},0)$, $v_3 = a(0,0,\frac{c}{a})$
5	Trigonal R, 3-fold axis c	$v_1 = a(x,-y,z)$, $v_2 = a(0,2y,z)$, $v_3 = a(-x,-y,z)$, where $x = \sqrt{\frac{1-c}{2}}$, $y = \sqrt{\frac{1-c}{6}}$, $z = \sqrt{\frac{1+2c}{3}}$, $c = \cos\gamma$
−5	Trigonal R, 3-fold axis ⟨111⟩	$v_1 = \frac{a}{\sqrt{3}}(u,v,v)$, $v_2 = \frac{a}{\sqrt{3}}(u,v,v)$, $v_3 = \frac{a}{\sqrt{3}}(v,v,u)$, where $u = z - 2\sqrt{2}y$ and $v = z + \sqrt{(2)}y$ with y, z as for case `ibrav = 5`
6	Tetragonal P	$v_1 = a(1,0,0)$, $v_2 = a(0,1,0)$, $v_3 = a(0,0,\frac{c}{a})$
7	Tetragonal I	$v_1 = \frac{a}{2}(1,-1,\frac{c}{a})$, $v_2 = \frac{a}{2}(1,1,\frac{c}{a})$, $v_3 = \frac{a}{2}(-1,-1,\frac{c}{a})$
8	Orthorhombic P	$v_1 = (a,0,0)$, $v_2 = (0,b,0)$, $v_3 = (0,0,c)$
9	Orthorhombic	$v_1 = (\frac{a}{2},\frac{b}{2},0)$, $v_2 = (-\frac{a}{2},\frac{b}{2},0)$, $v_3 = (0,0,c)$
−9	Orthorhombic	$v_1 = (\frac{a}{2},-\frac{b}{2},0)$, $v_2 = (\frac{a}{2},\frac{b}{2},0)$, $v_3 = (0,0,c)$
91	Orthorhombic	$v_1 = (a,0,0)$, $v_2 = (0,\frac{b}{2},-\frac{c}{2})$, $v_3 = (0,\frac{b}{2},\frac{c}{2})$
10	Orthorhombic	$v_1 = (\frac{a}{2},0,\frac{c}{2})$, $v_2 = (\frac{a}{2},\frac{b}{2},0)$, $v_3 = (0,\frac{b}{2},\frac{c}{2})$
11	Orthorhombic	$v_1 = (\frac{a}{2},\frac{b}{2},\frac{c}{2})$, $v_2 = (-\frac{a}{2},\frac{b}{2},\frac{c}{2})$, $v_3 = (-\frac{a}{2},-\frac{b}{2},\frac{c}{2})$
12	Monoclinic P (unique axis c)	$v_1 = (a,0,0)$, $v_2 = (b\cos\gamma, b\sin\gamma, 0)$, $v_3 = (0,0,c)$

Continued

Table 3.2 - *Continued*

ibrav	Structure	Lattice vectors
−12	Monoclinic P (unique axis b)	$v_1 = (a, 0, 0), v_2 = (0, b, 0),$ $v_3 = (c\cos\beta, 0, c\sin\beta)$
13	Monoclinic base-centered (unique axis c)	$v_1 = (\frac{a}{2}, 0, -\frac{c}{2}), v_2 = (b\cos\gamma, b\sin\gamma, 0),$ $v_3 = (\frac{a}{2}, 0, \frac{c}{2})$
−13	Monoclinic base-centered (unique axis b)	$v_1 = (\frac{a}{2}, \frac{b}{2}, 0), v_2 = (-\frac{a}{2}, \frac{b}{2}, 0, 0),$ $v_3 = (c\cos\beta, 0, c\sin\beta)$
14	Triclinic	$v_1 = (a, 0, 0), v_2 = (b\cos\gamma, b\sin\gamma, 0),$ $v_3 = (c\cos\beta, c\frac{\cos\alpha - \cos\beta\cos\gamma}{\sin\gamma},$ $c\frac{\sqrt{1+2\cos\alpha\cos\beta\cos\gamma - \cos^2\alpha - \cos^2\beta - \cos^2\gamma}}{\sin\gamma})$

Note: α, β, and γ are angles between axis b and c, a and c, and a and b, respectively.

and (0.666666666 0.333333333 0.500000000) in the crystal coordinates.

☞ **Note**: In the card ATOMIC_POSITIONS, the positions of atoms can specify *in units of the lattice parameter* (either a or celldm(1)) with (alat) option. It is noted that (alat) is default option in the card ATOMIC_POSITIONS. For the (alat) option, the lines 24–26 in the scf.in file will be changed as follows:

```
ATOMIC_POSITIONS (alat)
C  0.0000000   0.5773503   2.0306218
C  0.5000000   0.2886751   2.0306218
```

❑ **Output file:** Now let us see the output file by command: vi scf.out. The total energy, as well as its decomposition into several terms, is obtained at the end of the scf.out as

QE-SSP/gr/scf/scf.out

```
!    total energy              =     -23.90991271 Ry
     Harris-Foulkes estimate   =     -23.90991328 Ry
     estimated scf accuracy    <       0.00000084 Ry
```

```
The total energy is the sum of the following terms:

one-electron contribution =     -90.80734321 Ry
hartree contribution      =      47.24141117 Ry
xc contribution           =      -8.30684749 Ry
ewald contribution        =      27.96304915 Ry
smearing contrib. (-TS)   =      -0.00018232 Ry

convergence has been achieved in  13 iterations
```

If the readers want to see only the total energy from `scf.out`, the readers can use the grep command to find the lines containing the symbol ! from `scf.out` as follows:

```
$ grep ! scf.out
```

The total energy is printed in the terminal as

```
$ !    total energy          =      -23.90991271 Ry
```

☞ **Explanation of scf.out:** This output file shows that the total energy of graphene is −23.90991271 Ry (1 Ry = 13.60569301 eV), and it is obtained in 13 SCF iterations. The total energy contains one-electron, Hartree, xc, ewald, and smearing contributions. These energy contributions are explained in Sec. 4.12. The smearing contribution is much small compared with other energy contributions. The magnitude of the total energy is not physically meaningful for a given cut-off energy and $\boldsymbol{k}$-points grid, but the convergence of the total energy with the related parameters is essential to determine the correct parameters. In the next tutorials, we discuss the total-energy value depending on the cut-off energy (Sec. 3.1.2) and the $\boldsymbol{k}$-points grid (Sec. 3.1.3).

Try It Yourself

1. Make `scf.in` file for the bulk Si with a face-centered-cubic structure (`ibrav = 2`) and visualize the Si structure by using XCrySDen.
2. Make `scf.in` file and calculate total energy of monolayer MoS_2.

3.1.2 Plane-wave cut-off energy

❑ **Purpose:** The controllable parameter to test the convergence is the cut-off energy related to how many plane waves are used in the calculation. The large numbers of plane waves give a better result through the memory and CPU time increase with increasing the cut-off energy. In this tutorial, we show how to select a suitable value of the cut-off energy.

❑ **Background:** The concept of plane-wave cut-off energy is given in Sec. 5.4. In Quantum ESPRESSO, the electronic wavefunction is represented by the linear combination of plane waves, where the maximum value of $\boldsymbol{G}_{\text{max}}$ is related to the cut-off energy $E_{\text{cut-off}} = \frac{\hbar^2}{2m}\boldsymbol{G}_{\text{max}}^2$ in Ry (1 Ry = 13.6 eV). We will change the value of the cut-off energy from 20 to 80 Ry to check the convergence of total energy in this tutorial.

❑ **How to run:** To run this tutorial, the readers should enter the following command lines:

```
1 $ cd ~/QE-SSP/gr/ecut/
2 $ ./run.sh &
```

- Line 1: Change directory to the `ecut` that includes the input files.
- Line 2: Run a bash script file `run.sh`, which generates and runs many jobs with changing the cut-off energies.

❑ **How to check:** To check whether or not the output file exists, the readers can use the `ls` command by typing:

```
$ ls
```

The `ls` command displays a listing of the many input and output files including `run.sh` in the `ecut` directory as

```
calc-ecut.dat   ecut.30.in    ecut.50.out   ecut.75.in
ecut.20.in      ecut.30.out   ecut.55.in    ecut.75.out
ecut.20.out     ecut.35.in    ecut.55.out   ecut.80.in
ecut.22.in      ecut.35.out   ecut.60.in    ecut.80.out
ecut.22.out     ecut.40.in    ecut.60.out   plot-ecut.ipynb
ecut.24.in      ecut.40.out   ecut.65.in    run.sh
ecut.24.out     ecut.45.in    ecut.65.out
ecut.26.in      ecut.45.out   ecut.70.in
ecut.26.out     ecut.50.in    ecut.70.out
```

QE-SSP/gr/ecut/run.sh

```
#!/bin/bash
# Convergence test of cut-off energy.
# Set a variable ecut from 20 to 80 Ry.
for ecut in 20 22 24 26 30 35 40 45 50 55 60 65 \
70 75 80; do
# Make input file for the SCF calculation.
# ecutwfc is assigned by variable ecut.
cat > ecut.$ecut.in << EOF
&CONTROL
calculation   = 'scf'
pseudo_dir    = '../pseudo/'
outdir        = '../tmp/'
prefix        = 'gr'
/
&SYSTEM
ibrav         = 4
a             = 2.4623
c             = 10.0
nat           = 2
ntyp          = 1
occupations   = 'smearing'
smearing      = 'mv'
degauss       = 0.02
ecutwfc       = ${ecut}
/
&ELECTRONS
mixing_beta   = 0.7
conv_thr      = 1.0D-6
/
ATOMIC_SPECIES
C 12.0107 C.pbe-n-rrkjus_psl.0.1.UPF
ATOMIC_POSITIONS (crystal)
C  0.333333333  0.666666666  0.500000000
C  0.666666666  0.333333333  0.500000000
K_POINTS (automatic)
12 12 1 0 0 0
EOF
# Run SCF calculation.
mpirun -np 4 pw.x <ecut.$ecut.in> ecut.$ecut.out
# Write cut-off and total energies in calc-ecut.dat.
awk '/!/ {printf"%d %s\n",'$ecut',$5}' ecut.$ecut.out >> calc-ecut.dat
# End of for loop
done
```

❑ **Output file:** The readers can open `calc-ecut.dat` generated by another pick-up-and-edit command `awk` in line 41 to make a plot of the total energy as a function of the cut-off energy.

```
$ vi calc-ecut.dat
```

QE-SSP/gr/ecut/calc-ecut.dat

```
20 -23.58222934
22 -23.74580589
24 -23.82861718
...
70 -23.91018738
75 -23.91023390
80 -23.91024577
```

- Column 1: Cut-off energy in units of Ry.
- Column 2: Total energy of graphene in units of Ry.

☞ **Plotting data from calc-ecut.dat:** To investigate how the total energy decreases with increasing the cut-off energy, we adopt Matplotlib, a plotting library for the Python programming language. Note that the readers can use any plotting software such as Gnuplot or Grace. Here, we would like to recommend the readers to install Matplotlib because it is synchronized for all examples in this book. For running Matplotlib, we make a JupyterLab `plot-ecut.ipynb` as an input file to automatically generate a plot from the extracted data in the output file. There is a detailed description of how to run a JupyterLab in Sec. 6.4. In order to run `plot-ecut.ipynb`, the readers can type as follows:

```
$ jupyter-lab plot-ecut.ipynb
```

QE-SSP/gr/ecut/plot-ecut.ipynb

```
# Import the necessary packages and modules
import matplotlib.pyplot as plt
plt.style.use('../../matplotlib/sci.mplstyle')
```

```
import numpy as np

# Open and read the file calc-ecut.dat
ecut, ener = np.loadtxt('calc-ecut.dat', delimiter=' ', unpack=True)

# Create figure object
plt.figure()
# Plot the data, using scatter plot
plt.scatter(ecut, ener, s=150)
# Plot a dashed line at 80 Ry
plt.axhline(ener[14], c='gray', ls='--')
# Add the x and y-axis labels
plt.xlabel('Cut-off energy (Ry)')
plt.ylabel('Total energy (Ry)')
# Set the axis limits
plt.xlim(20, 80)
plt.ylim(-23.94, -23.56)
# Save the figure
plt.savefig('plot-ecut.pdf')
# Show the figure
plt.show()
```

By running `jupyter-lab`, we obtain the total-energy plot for the plane-wave cut-off energy as shown in Fig. 3.3. The total energy rapidly converges around 30 Ry, and it shows a best-converged value at 40 Ry. In principle, the result of the lowest total energy is obtained for infinity cut-off energy by a variational principle. However, the higher the cut-off energy leads to more time-consuming the calculation. Therefore, in practice, 40 Ry is good enough for this tutorial. The total energy difference between 40 Ry and 80 Ry is only 1.15 meV, which is much smaller than $k_BT = 25$ meV (at $T = 300$ K) or optical phonon energy ($>$100 meV). Since most quantities that can be computed using Quantum ESPRESSO critically depend on the cut-off energy, it is essential to perform this test when running Quantum ESPRESSO calculations. Note that, sometimes, the pseudopotential files will suggest cut-off energy. By referring to this value, we can select the cut-off energy. The suggested minimum cut-off energy is usually 40–60 Ry for ultrasoft pseudopotentials, while it is 80–120 Ry for norm-conserving pseudopotentials. The cut-off energy dependence of the type of pseudopotential is explained in Sec. 5.4.

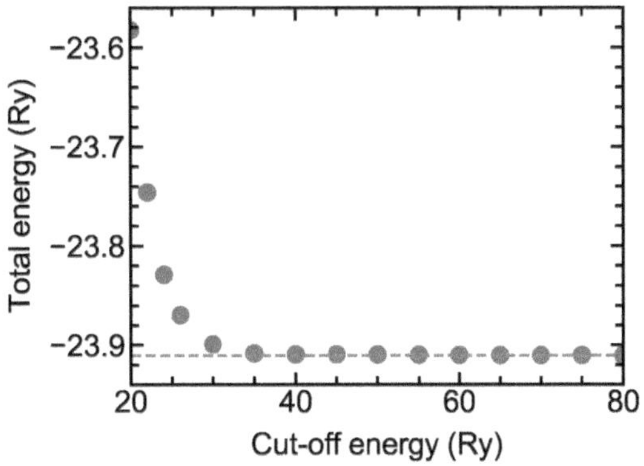

Figure 3.3 Total energy as a function of plane-wave cut-off energy. The dashed line is the total energy at 80 Ry. The cut-off energy shows a best-converged value at 40 Ry.

Try It Yourself

Check convergence value of cut-off energy for bulk Si and monolayer MoS_2.

3.1.3 *k*-points for Brillouin-zone integration

❑ **Purpose:** Other controllable parameter to test for convergence of the total energy is ***k***-points grid. The integration on ***k*** in the Brillouin zone (BZ) is approximated by summation of finite numbers of ***k***-points that is called ***k***-points grid. In this tutorial, we show how to select the ***k***-points grid.

❑ **Background:** The concept of ***k***-points in the BZ is given in Sec. 5.5. In Quantum ESPRESSO, there are six values to determine a ***k***-points grid, in which the first three values are set as the number of ***k***-points mesh of each axis, and the last three values are a parameter to move the lattice of each axis. The total number of ***k***-points is obtained by multiplying the first three values. For example, k_1, k_2, and k_3 are set to 10, 10, and 1, respectively, then the total number of ***k***-points is $10 \times 10 \times 1 = 100$. Table 3.3 shows the rule for selecting the ***k***-points grid based on the dimensions of the system.

Table 3.3 Selecting rule for ***k***-points grid based on dimension of system.

System	*k*-points grid
Three-dimensional (3D) system	$k_1 \times k_2 \times k_3$
Two-dimensional (2D) system in *xy*-plane	$k_1 \times k_2 \times 1$
One-dimensional (1D) system along *z*-direction	$1 \times 1 \times k_3$
Zero-dimensional (0D) system	$1 \times 1 \times 1$

Table 3.4 Selecting a rule for shifting (0 or 1) based on symmetry of system.

System	Shifting
Cubic, tetragonal, orthorhombic, and monoclinic systems	1
Hexagonal and trigonal systems	0
Rhombohedral and triclinic systems	1 or 0

The last three values should be set to either 0 or 1, where 0 indicates the default Gamma (Γ) position and 1 means that the ***k***-points grid is moved parallel, as shown in Fig. 3.4. Even though the ***k***-points are set with the same density by selecting the shift to move the ***k***-point mesh by selecting either 0 or 1, the number of inequivalent ***k***-points (red points in Fig. 3.4) can be reduced. The number of inequivalent ***k***-points depends on the crystal system of a cell. In Fig. 3.4 (a) and (b), we show a schematic diagram of the ***k***-point sampling in a cubic cell and hexagonal cell, respectively. In the cubic-cell case, the shifting decreases the number of inequivalent ***k***-points from 6 to 3 in the first BZ (see Fig. 3.4 (a)). However, in the case of a hexagonal cell, setting the shift breaks the hexagonal symmetry in the BZ. After the shift = 1 is set, symmetry operation is performed, and it needs more ***k***-points than the mesh with the case of shift = 0. Thus, the number of inequivalent ***k***-points increases due to the three-fold symmetry of the hexagonal cell, as shown in Fig. 3.4 (b). We thus should carefully select a shifting (0 or 1) depending on the symmetry of the system to calculate. Table 3.4 shows the selecting rule for shift based on the symmetry of the system.

Based on Tables 3.3 and 3.4, the ***k***-points grid for graphene is selected as "*k k* 1 0 0 0" with *k* is an integer from 4 to 14 in this tutorial.

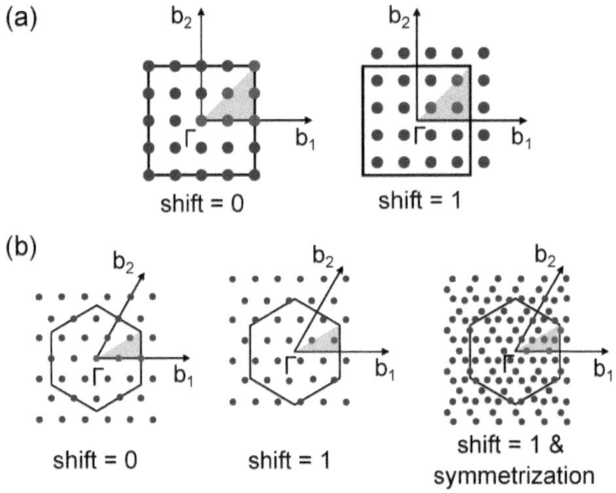

Figure 3.4 Schematic diagram of the ***k***-point sampling in a cubic cell (a) or a hexagonal cell (b). We also illustrate how to shift the ***k***-points by specifying "shift=1" and "symmetrization". Red points denote the inequivalent ***k***-points in the first Brillouin zone.

Next, we will learn how to perform the convergence test for this ***k***-points grid.

❑ **How to run:** To run this tutorial, the readers should type the following command lines:

```
1 $ cd ~/QE-SSP/gr/k-point/
2 $ ./run.sh &
```

- Line 1: Go to directory `k-point` that includes input files.
- Line 2: Run a bash script file `run.sh`, which consists of many jobs with changing the ***k***-points grid.

❑ **How to check:** To check whether or not the output file exists, the readers can use the `ls` command by typing:

```
$ ls
```

The `ls` command displays a listing of the many output files in the `k-point` directory as

```
calc-kpoint.dat  kpoint.14.in    kpoint.7.out
kpoint.10.in     kpoint.14.out   kpoint.8.in
kpoint.10.out    kpoint.4.in     kpoint.8.out
kpoint.11.in     kpoint.4.out    kpoint.9.in
kpoint.11.out    kpoint.5.in     kpoint.9.out
kpoint.12.in     kpoint.5.out    plot-kpoint.ipynb
kpoint.12.out    kpoint.6.in     run.sh
kpoint.13.in     kpoint.6.out
kpoint.13.out    kpoint.7.in
```

❑ **Input file:** All input files are generated automatically by a bash script file `run.sh` as

QE-SSP/gr/k-point/run.sh

```
#!/bin/bash
# Convergence test of k-points grid.
# Set a variable k-point from 4 to 14.
for k in 4 5 6 7 8 9 10 11 12 13 14; do

# Make input file for the SCF calculation.
# k-points grid is assigned by variable n.
cat > kpoint.$k.in << EOF
&CONTROL
calculation    = 'scf'
pseudo_dir     = '../pseudo/'
outdir         = '../tmp/'
prefix         = 'gr'
/
&SYSTEM
ibrav          = 4
a              = 2.4623
c              = 10.0
nat            = 2
ntyp           = 1
occupations    = 'smearing'
smearing       = 'mv'
degauss        = 0.02
ecutwfc        = 40
/
&ELECTRONS
mixing_beta    = 0.7
conv_thr       = 1.0D-6
/
ATOMIC_SPECIES
C 12.0107 C.pbe-n-rrkjus_psl.0.1.UPF
ATOMIC_POSITIONS (crystal)
C  0.333333333  0.666666666  0.500000000
C  0.666666666  0.333333333  0.500000000
```

```
35 K_POINTS (automatic)
36 ${k} ${k} 1 0 0 0
37 EOF
38
39 # Run pw.x for SCF calculation.
40 mpirun -np 4 pw.x <kpoint.$k.in>kpoint.$k.out
41 # Write the number of k-points (= k*k*1) and
42 # the total energy in calc-kpoint.dat
43 awk '/!/ {printf"%d %s\n",'$k*$k',$5}' kpoint.$k.out
       >> calc-kpoint.dat
44 # End of for loop.
45 done
```

☞ **Explanation of run.sh:** The ***k***-points grid is controlled with the card `K_POINTS` in line 35, wherein a ***k***-points grid needs to be set to determine how much density is necessary for the BZ integration. By selecting the `automatic` option in the card `K_POINTS`, it allows the ***k***-points grid through the Monkhorst-Pack method [Monkhorst and Pack (1976)].

❑ **Output file:** The number of ***k***-points ($= k \times k \times 1$) and the total energy are written in `calc-kpoint.dat`. The readers can open the `calc-kpoint.dat` file in the terminal by typing:

```
$ vi calc-kpoint.dat
```

QE-SSP/gr/ecut/calc-kpoint.dat

```
16 -23.91186965
25 -23.90766485
36 -23.90599780
...
144 -23.90913362
169 -23.90963897
196 -23.90953385
```

- Column 1: The number of ***k***-points.
- Column 2: Total energy of graphene in units of Ry.

☞ **Plotting data from calc-kpoint.dat:** The total energy as a function of the ***k***-point is plotted by running JupyterLab `plot-kpoint.ipynb`:

```
$ jupyter-lab plot-kpoint.ipynb
```

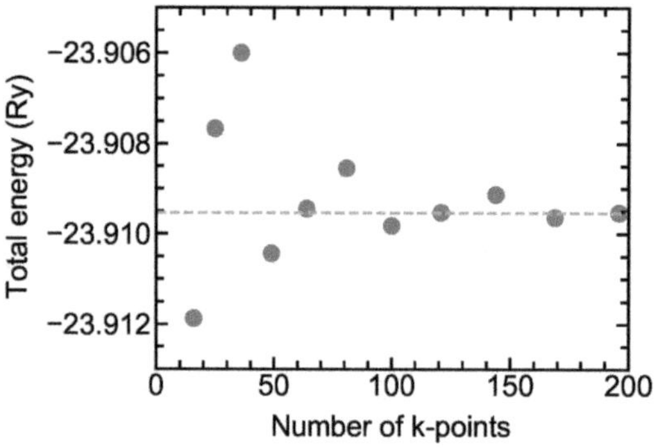

Figure 3.5 Total energy as a function of number of ***k***-points. It shows significant oscillations around the converged value. The dashed line is the total energy at $14 \times 14 \times 1 = 196$ ***k***-points. A ***k***-points grid of $10 \times 10 \times 1 = 100$ shows a good convergence.

QE-SSP/gr/k-point/plot-kpoint.ipynb

```
# Import the necessary packages and modules
import matplotlib.pyplot as plt
plt.style.use('../../matplotlib/sci.mplstyle')
import numpy as np

# Open and read the file calc-kpoint.dat
kp, ener = np.loadtxt('calc-kpoint.dat', delimiter=' ', unpack=True)

# Create figure object
plt.figure()
# Plot the data, using black color
plt.scatter(kp, ener, s=150)
# Plot a dashed line at 14x14x1
plt.axhline(ener[10], c='gray', ls='--')
# Add the x and y-axis labels
plt.xlabel('Number of $\bm k$-points')
plt.ylabel('Total energy (Ry)')
# Set the axis limits
plt.xlim(0, 200)
plt.ylim(-23.913, -23.905)
# Save the figure
plt.savefig('plot-kpoint.pdf')
# Show the figure
plt.show()
```

In Fig. 3.5, we plot the total energy as a function of the total number of **k**-points. Compared with the calculation of the cut-off energy convergence, the **k**-points convergence does not occur monotonically, but it may show some oscillations around the converged value. Since a convergence of 1 meV or less for the entire system is our goal here, as shown in Sec. 3.1.2, a **k**-points grid $k \times k \times 1 = 10 \times 10 \times 1 = 100$ is a reasonable choice.

Try It Yourself

1. Plot running-time, which can be found at the end of `scf.out` file, as a function of number of k-points for graphene.
2. Check convergence value of **k**-points grid for bulk Si and monolayer MoS_2.

3.1.4 Optimizing atomic positions

❑ **Purpose:** In this tutorial, we are going to learn how to optimize atomic positions of the system.

❑ **Background:** In Quantum ESPRESSO, the default algorithm for structural optimization is the **BFGS algorithm**, which was proposed by Broyden [Broyden (1970)], Fletcher [Fletcher (1991)], Goldfarb [Goldfarb (1970)], and Shanno [Shanno (1970)], independently. The BFGS algorithm is one of the most powerful methods to solve unconstrained optimization problem. Let us consider the unconstrained optimization problem as

$$\min f(x), \quad x \in \mathbb{R}^n, \tag{3.1}$$

where $f(x)$ is a general function that has continuous second derivatives. The goal of the optimization problem is to find a stationary point x^* such that $\nabla f(x^*) = 0$. Given a starting point x_0 and an initial estimate of $H_0^{-1} \approx \nabla^2 f(x_0)$, the $(k+1)$-th iteration of

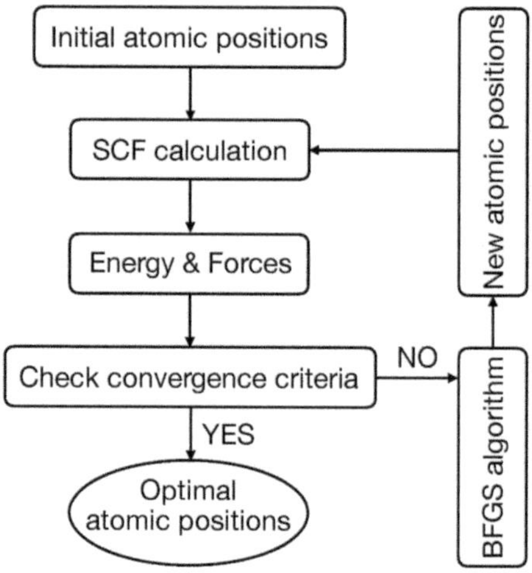

Figure 3.6 Flowchart of the optimizing atomic positions in Quantum ESPRESSO.

the BFGS algorithm is

$$\begin{aligned} x_{k+1} &= x_k - H_k^{-1}\nabla f(x_k), \\ s_k &= x_{k+1} - x_k, \\ y_k &= \nabla f(x_{k+1}) - \nabla f(x_k), \\ H_{k+1} &= H_k - \frac{H_k s_k s_k^T H_k}{s_k^T H_k s_k} + \frac{y_k y_k^T}{y_k^T s_k}, \text{ with } k = 0, 1, 2, \ldots \end{aligned} \tag{3.2}$$

The initial matrix H_0^{-1} is usually taken to be a positive multiple of the identity matrix, which means that the initial search direction will be the steepest descent direction. The advantage of the BFGS algorithm is that it does not need to compute $\nabla^2 f(x_k)$ for each iteration, which can be time-consuming.

In Fig. 3.6, we show the flowchart of the optimizing atomic positions in Quantum ESPRESSO. At each step of the BFGS algorithm, the new positions of the atoms are obtained. Then the total energy and the forces are calculated by using the SCF method and the Hellmann-Feynman theorem, respectively. When the change of the total energy and all components of all forces are smaller than the convergence criteria, which are given in the input file, the optimized atomic positions are obtained. In Fig. 3.6, we show the flowchart of the optimizing atomic positions in Quantum ESPRESSO. At each step of the BFGS algorithm, the new positions of the atoms are obtained. Then the total energy and the forces are calculated by using the SCF method and the Hellmann-Feynman theorem, respectively. When the change of the total energy and all components of all forces are smaller than the convergence criteria, which are given in the input file, the optimized atomic positions are obtained.

❑ **How to run:** To run this tutorial, the readers should type the following command lines:

```
1 $ cd ~/QE-SSP/gr/relax/
2 $ mpirun -np 4 pw.x <relax.in> relax.out &
```

- Line 1: Go to `relax` directory that includes input files.
- Line 2: Run `pw.x` by using in parallel with 4 processors (`-np 4`).

❑ **How to check:** The calculation will finish when `JOB DONE` is written at the end of the output file `relax.out`.

❑ **Input file:** The input file is showed in terminal by typing:

```
$ vi relax.in
```

QE-SSP/gr/relax/relax.in

```
&CONTROL
calculation    = 'relax'
pseudo_dir     = '../pseudo/'
outdir         = '../tmp/'
prefix         = 'gr'
etot_conv_thr  = 1.0D-5
forc_conv_thr  = 1.0D-4
/
&SYSTEM
ibrav          = 4
a              = 2.4623
c              = 10.0
nat            = 2
ntyp           = 1
occupations    = 'smearing'
smearing       = 'mv'
degauss        = 0.02
ecutwfc        = 60
/
```

```
20 &ELECTRONS
21 mixing_beta     = 0.7
22 conv_thr        = 1.0D-9
23 /
24 &IONS
25 ion_dynamics  = 'bfgs'
26 /
27 ATOMIC_SPECIES
28 C 12.0107 C.pbe-n-rrkjus_psl.0.1.UPF
29 ATOMIC_POSITIONS (crystal)
30 C  0.333333333  0.666666666  0.500000000  0  0  0
31 C  0.666666666  0.333333333  0.400000000
32 K_POINTS (automatic)
33 12 12 1 0 0 0
```

☞ **Explanation of relax.in:** The calculation of optimizing atomic position can be performed by selecting `calculation = 'relax'` (line 2) in the namelist `CONTROL`. The convergence criteria for the change of total energy and the forces are set by `etot_conv_thr` and `forc_conv_thr` (lines 6 and 7) in the namelist `CONTROL`, respectively. The BFGS algorithm is set in the namelist `IONS` as `ion_dynamics = 'bfgs'` (line 25). It is noted that the namelist `IONS` is the only mandatory addition. There are several variables that can be specified within this namelist, which control the algorithm used to find the optimized atomic positions. However, the readers can leave it empty if the readers are happy with selecting all defaults. The description of input variables is given in Tables 3.1, and 3.5. Here a.u. refers to the atomic unit (1 a.u. = 1 Hartree = 27.2 eV).

In order to observe the displacement of the carbon atoms in a unit cell of graphene, one carbon atom is fixed at a position `(0.333333333 0.666666666 0.500000000)` by adding `(0 0 0)`, and other carbon atom is set at position `(0.666666666 0.333333333 0.400000000)`. We expect that this carbon atom will move to the optimal position at `(0.333333333 0.666666666 0.500000000)`.

☞ **Note**: We can select the coordinates that will be fixed or not in relaxation by adding either `0` or `1`, respectively, in the atomic positions. For example, `(0 0 1)` means that the *x* and *y* coordinates are fixed, while *z* coordinate can be changed in structural optimization. If no option is specified, `(1 1 1)` is assumed.

Table 3.5 Meaning of input variables in `relax.in` file.

Line	Syntax	Meaning
6	`etot_conv_thr`	Convergence threshold on total energy (a.u.) for atomic minimization between two consecutive BFGS steps.
7	`forc_conv_thr`	Convergence threshold on forces (a.u.) for atomic minimization: the convergence criterion is satisfied when all components of all forces are smaller than this value.
24	`&IONS`	A namelist *must* be specified in the case of structural relaxation or molecular dynamics calculations.
25	`ion_dynamics`	Specify the type of atomic dynamics, in which `bfgs` (default) use BFGS quasi-newton algorithm.
26	`/`	End of namelist `IONS`.

❑ **Output file:** The optimized atomic positions are given in the output file `relax.out`. The readers can open output file in terminal by typing:

```
$ vi relax.out
```

To search forward for final coordinates, press `Esc` and then enter `/final`, vi editor will search the word `'final'` in `relax.out`.

QE-SSP/gr/relax/relax.out

```
  Forces acting on atoms (cartesian axes, Ry/au):

  atom    1 type  1   force =     0.00000000    0.00000000   -0.00001508
  atom    2 type  1   force =     0.00000000    0.00000000    0.00001508

  Total force =     0.000015     Total SCF correction =     0.000001

  bfgs converged in   8 scf cycles and   7 bfgs steps
  (criteria: energy <  1.0E-05 Ry, force <  1.0E-04 Ry/Bohr)

  End of BFGS Geometry Optimization

  Final energy   =     -23.9099132232 Ry
Begin final coordinates

ATOMIC_POSITIONS (crystal)
C          0.3333333330        0.6666666660        0.5000000000    0   0   0
C          0.6666666660        0.3333333330        0.4999977356
```

```
End final coordinates
```

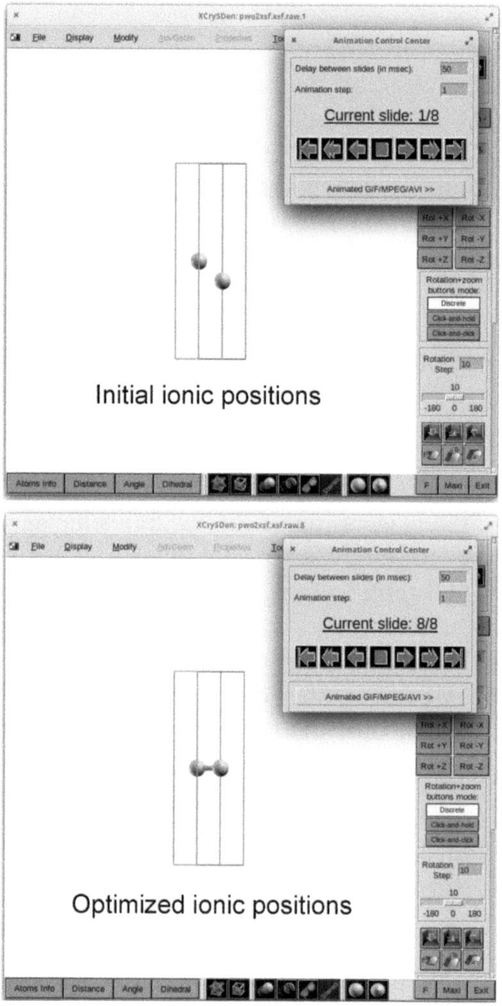

Figure 3.7 Visualizing optimized atomic positions from Quantum ESPRESSO output file by using XCrySDen. Top figure is the initial atomic positions (1/8 steps) and bottom figure is the optimized atomic positions (8/8 steps).

☞ **Explanation of relax.out:** This output file shows that the BFGS optimization is converged in 7 steps with the energy difference between two consecutive steps

$< 1.0 \times 10^{-5}$ Ry and the force $< 1.0 \times 10^{-4}$ Ry/Bohr. The second carbon atoms is moved from the initial position (`0.6666666660 0.3333333330 0.4000000000`) to the final position (`0.6666666660 0.3333333330 0.4999977356`).

☞ **Note**: The `Total force` in the output file is the square root of the sum of all the squared force components rather than the sum of the magnitudes of the individual forces on the atoms. If the `Total SCF correction` is large or comparable to the `Total force`, the output file will show a message "`SCF correction compared to forces is large: reduce conv_thr to get better values`". It usually means the readers need to try with a relatively smaller `conv_thr`.

☞ **Visualizing optimized ionic positions:** The readers can visualize the optimized ionic positions by using XCrySDen as

```
$ xcrysden &
```

Then the readers can go through the following steps: **File → Open PWscf → Open PWscf Output File** and select `relax.out`. In the box **[PWSCF Input: "relax.out"]**, we click on the **OK** button, then in the box **[Question]**, we select **Display All Coordinates as Animation** and click on the **Continue** button. The initial or optimized ionic positions can be showed by choosing **Current slide** in the box **[Animation Control Center]**, as shown in Fig. 3.7.

Try It Yourself

Run optimizing atomic positions for monolayer MoS_2 by changing position of only Mo atom.

3.1.5 Optimizing unit cell

❑ **Purpose:** In this tutorial, we show how to optimize the lattice vectors of the unit cell (or the variable-cell relaxation).

❑ **Background:** Similar to optimization the atomic positions, we can optimize the unit cell by using the BFGS algorithm.

❑ **How to run:** To run this tutorial, the readers should type as

```
1 $ cd ~/QE-SSP/gr/vc-relax/
2 $ mpirun -np 4 pw.x <vc-relax.in> vc-relax.out &
```

- Line 1: Go to `vc-relax` directory that includes input files.
- Line 2: Run `pw.x` by using in parallel calculation.

❑ **How to check:** The calculation will finish when `JOB DONE` is written at the end of the output file `vc-relax.out`.

❑ **Input file:** The readers can open the input file in terminal by typing:

```
$ vi vc-relax.in
```

QE-SSP/gr/vc-relax/vc-relax.in

```
&CONTROL
calculation    = 'vc-relax'
pseudo_dir     = '../pseudo/'
outdir         = '../tmp/'
prefix         = 'gr'
etot_conv_thr = 1.0D-5
forc_conv_thr = 1.0D-4
/
&SYSTEM
ibrav          = 4
a              = 2.5
c              = 15.0
nat            = 2
ntyp           = 1
occupations    = 'smearing'
smearing       = 'mv'
degauss        = 0.02
ecutwfc        = 60
/
&ELECTRONS
mixing_beta    = 0.7
conv_thr       = 1.0D-9
/
&IONS
ion_dynamics   = 'bfgs'
/
&CELL
cell_dynamics = 'bfgs'
press_conv_thr= 0.05
cell_dofree    = '2Dxy'
```

Table 3.6 Meaning of input variables in `vc-relax.in` file.

Line	Syntax	Meaning
27	`&CELL`	A namelist *must* be specified in the case of `calculation = 'vc-relax'`.
28	`cell_dynamics`	Specify the type of cell dynamics, in which `bfgs` (default) use BFGS quasi-newton algorithm.
29	`press_conv_thr`	Convergence threshold on the pressure for variable cell relaxation. Default value is 0.5 Kbar.
30	`cell_dofree`	Select the cell parameters to change, in which `'2Dxy'` is only x and y components are allowed to change.
31	/	End of namelist CELL.

```
31 /
32 ATOMIC_SPECIES
33 C 12.0107 C.pbe-n-rrkjus_psl.0.1.UPF
34 ATOMIC_POSITIONS (crystal)
35 C   0.333333333   0.666666666   0.500000000
36 C   0.666666666   0.333333333   0.500000000
37 K_POINTS (automatic)
38 12 12 1 0 0 0
```

☞ **Explanation of vc-relax.in:** The calculation of the optimizing unit cell can be performed by setting `calculation = 'vc-relax'` (line 2) in the namelist CONTROL. We need to specify both the namelist IONS, as shown in Sec. 3.1.4, and CELL. Note that the convergence criterion for optimizing unit cell is based on the pressure, which is set by `press_conv_thr` (line 29) in the namelist CELL. For the two-dimensional material as graphene, we also need to set `cell_dofree = '2Dxy'` (line 30) in the namelist CELL, that means that only *x* and *y* components are optimized. The description of input variables is given in Tables 3.1, 3.5, and 3.6.

❑ **Output file:** Both the optimized ionic positions and the lattice vectors are given in the output file `vc-relax.out`. The readers can open output file in terminal by typing:

```
$ vi vc-relax.out
```

To search forward for final coordinates, press Esc and then enter `/final`, vi editor will display as

QE-SSP/gr/vc-relax/vc-relax.out

```
     Computing stress (Cartesian axis) and pressure


     negative rho (up, down):  1.211E-04 0.000E+00
          total   stress  (Ry/bohr3)                (kbar)     P=       -0.13
   0.00000002   0.00000000  -0.00000000            0.00       0.00      -0.00
   0.00000000   0.00000002  -0.00000000            0.00       0.00      -0.00
   0.00000000   0.00000000  -0.00000259            0.00       0.00      -0.38


     bfgs converged in   5 scf cycles and   4 bfgs steps
     (criteria: energy<1.0E-05 Ry, force<1.0E-04Ry/Bohr, cell<5.0E-02kbar)

     End of BFGS Geometry Optimization

     Final enthalpy =     -23.9099689524 Ry
Begin final coordinates
     new unit-cell volume =      532.18989 a.u.^3 (     78.86240 Ang^3 )
     density =       0.50580 g/cm^3

CELL_PARAMETERS (alat=  4.72431533)
   0.985562233  -0.000000000   0.000000000
  -0.492781117   0.853521931   0.000000000
   0.000000000   0.000000000   6.000000000

ATOMIC_POSITIONS (crystal)
C             0.3333333330        0.6666666660        0.5000000000
C             0.6666666660        0.3333333330        0.5000000000
End final coordinates
```

☞ **Explanation of vc-relax.out:** This output file shows that the optimized structure (i.e., both ionic positions and lattice constants are optimized) with the convergence criteria of the energy difference $< 1.0 \times 10^{-5}$ Ry, the total forces of each atom $< 1.0 \times 10^{-4}$ Ry/Bohr and the in-plane pressures of unit cell $< 5.0 \times 10^{-2}$ kbar. Since we optimize the two-dimensional structure, the absolute value of the pressure in the z direction can be larger than the convergence threshold (5.0×10^{-2} kbar). To reduce the absolute value of the pressure in the z direction, the lattice constant in the z direction should be large enough. Here, we choose $c = 15.0$ Å as shown in the input file.

The optimized lattice vectors are given in `CELL_PARAMETERS` in the block of the `final coordinates`. Note that the lattice vectors are given explicitly in Bohr (1 Bohr = 0.529177211 Å) along with a scaling factor `alat = 4.72431533`, which is converted from a = `2.5` Å in the input file. The optimized lattice constant is a = `2.5×0.985562233` = `2.4639055825` Å, which is consistent with experimental value ($a \sim 2.46$ Å) [Ohta *et al.* (2006)]. Note that the $c = 15$ Å constant does not change because of setting `cell_dofree = '2Dxy'`.

Try It Yourself

Run optimizing unit cell for bulk Si and monolayer MoS_2. It is noted that `cell_dofree` is set to `all` for the bulk Si.

3.1.6 Selecting pseudopotential

❑ **Purpose:** In this tutorial, we learn how to select the pseudopotential in the input file of Quantum ESPRESSO.
❑ **Background:** As discussed in Sec. 5.4, the pseudopotential (PP) is an approximation of atomic potentials to minimize the number of plane wave basis, and PP contains various information for simulating a system. The PP in Quantum ESPRESSO uses a unified PP format (UPF).[2] Three types of PPs, including norm-conserving (NC) PPs, ultra-soft (US) PPs, and projector-augmented wave (PAW), are supported. Some calculations (e.g., Γ-point phonon, third-order energy derivatives, non-resonance Raman, anharmonic force constants) work only with the NC PPs because the NC PPs provide wavefunctions that we can use in the integration for calculating some physical properties. Thus, the readers should check what type of PP is supported and select a proper PP for each atom before calculating.

The main library of the PPs is PSlibrary, which includes the scalar relativistic and fully relativistic US PPs and PAW data sets for many elements. The readers can download the PPs from PSlibrary in this link: `http://pseudopotentials.quantum-espresso.org/legacy_tables/ps-library`. In Fig. 3.8, we show the naming convention for the PP file. A PP file includes not only the type of pseudopotential (NC, US, or PAW) but also the type of the exchange-correlation functional (LDA, GGA, or meta-GGA). The detail

[2] Unified pseudopotential format is the Extensible Markup Language (XML)-like structure to store pseudopotentials.

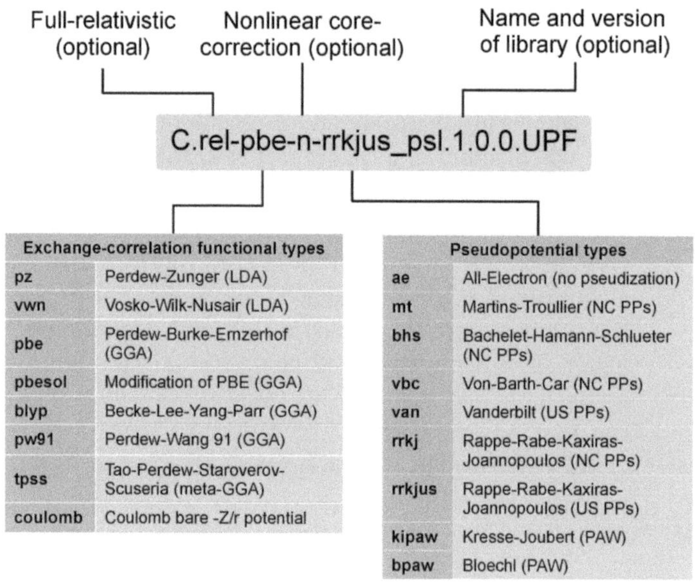

Figure 3.8 Naming convention for the pseudopotential from PSlibrary in unified pseudopotentials format (UPF).

of the exchange-correlation functional is given in Sec. 4.7. Besides PSlibrary, other libraries are also available as

- **Standard Solid-State PPs (SSSP)**: a collection of the best verified PPs, maintained by THEOS and MARVEL. Available on the Materials Cloud web site is `https://www.materialscloud.org/discover/sssp/table/efficiency`.
- **Pseudo Dojo**: a complete set of PPs and PAW sets. Available on the web site is `http://www.pseudo-dojo.org/index.html`.
- **GBRV high-throughput PPs**: a highly accurate and computationally inexpensive open-source US PP library. Available on the web site is `http://www.physics.rutgers.edu/gbrv`.
- **Optimized NC Vanderbilt PP (ONCVPSP)**: an accurate and inexpensive NC PPs by D. R. Hamann. Available on the web site is `http://www.quantum-simulation.org/potentials/sg15_oncv/upf/`.
- **Hartwigesen-Goedecker-Hutter PPs**: a collection of NC PPs from CPMD, by Matthias Krack. Available on the web site is

`http://pseudopotentials.quantum-espresso.org/legacy_tables/hartwigesen-goedecker-hutter-pp`.

A recommended way of selecting a good PP is based on testing and comparing the preliminary results to experimental data. Note that when comparing Quantum ESPRESSO calculation results, you should not directly compare the energy values by using different PPs. Since each PP has a different option or condition, using a different type of PP results in different energy values, which does not have any meaning. It is recommended to compare the properties such as lattice constant, energy band gap, etc., with experiments for observed materials. Since we showed how to obtain the lattice constant of graphene in Sec. 3.1.5, in this tutorial, we will compare the lattice constant for several PPs.

❑ **How to run:** To run this tutorial, the readers should type the following command lines:

```
1 $ cd ~/QE-SSP/gr/pps/
2 $ ./run.sh &
```

- Line 1: Go to pps directory that includes input files.
- Line 2: Run a bash script file `run.sh`.

❑ **How to check:** To check whether or not the output file exists, the readers can use the `ls` command by typing:

```
$ ls
```

The `ls` displays a listing of the all files in the pps folder as

```
run.sh
vc-relax.C.pbe-hgh.UPF.in
vc-relax.C.pbe-hgh.UPF.out
vc-relax.C.pbe-n-kjpaw_psl.1.0.0.UPF.in
vc-relax.C.pbe-n-kjpaw_psl.1.0.0.UPF.out
vc-relax.C.pbe-n-rrkjus_psl.0.1.UPF.in
vc-relax.C.pbe-n-rrkjus_psl.0.1.UPF.out
vc-relax.C.pz-hgh.UPF.in
vc-relax.C.pz-hgh.UPF.out
vc-relax.C.pz-n-kjpaw_psl.0.1.UPF.in
vc-relax.C.pz-n-kjpaw_psl.0.1.UPF.out
vc-relax.C.pz-n-rrkjus_psl.0.1.UPF.in
vc-relax.C.pz-n-rrkjus_psl.0.1.UPF.out
```

❑ **Input file:** All input files are generated automatically by a bash script file `run.sh` as

QE-SSP/gr/pps/run.sh

```
#!/bin/bash
# Set a variable pp for 6 PPs.
for pp in C.pbe-hgh.UPF \
C.pz-hgh.UPF \
C.pbe-n-rrkjus_psl.0.1.UPF \
C.pz-n-rrkjus_psl.0.1.UPF \
C.pbe-n-kjpaw_psl.1.0.0.UPF \
C.pz-n-kjpaw_psl.0.1.UPF; do
# Make input file for the vc-relax calculation.
cat > vc-relax.$pp.in << EOF
&CONTROL
calculation   = 'vc-relax'
pseudo_dir    = '../pseudo/'
outdir        = '../tmp/'
prefix        = 'gr'
etot_conv_thr = 1.0D-5
forc_conv_thr = 1.0D-4
/
&SYSTEM
ibrav         = 4
a             = 2.4639055825
c             = 15.0
nat           = 2
ntyp          = 1
occupations   = 'smearing'
smearing      = 'mv'
degauss       = 0.02
ecutwfc       = 80
/
&ELECTRONS
mixing_beta   = 0.7
conv_thr      = 1.0D-9
/
&IONS
ion_dynamics  = 'bfgs'
/
&CELL
cell_dynamics = 'bfgs'
press_conv_thr= 0.05
cell_dofree   = '2Dxy'
/
ATOMIC_SPECIES
C 12.0107 $pp
ATOMIC_POSITIONS (crystal)
```

Table 3.7 Calculated lattice constants (a) by 6 pseudopotentials.

Pseudopotential	Type	a (Å)
C.pbe-hgh.UPF	NC + GGA	2.4529
C.pz-hgh.UPF	NC + LDA	2.4321
C.pbe-n-rrkjus_psl.0.1.UPF	US + GGA	2.4643
C.pz-n-rrkjus_psl.0.1.UPF	US + LDA	2.4456
C.pbe-n-kjpaw_psl.1.0.0.UPF	PAW + GGA	2.4673
C.pz-n-kjpaw_psl.0.1.UPF	PAW + LDA	2.4456

```
45 C   0.333333333   0.666666666   0.500000000
46 C   0.666666666   0.333333333   0.500000000
47 K_POINTS (automatic)
48 12 12 1 0 0 0
49 EOF
50 # Run pw.x for vc-relax calculation.
51 mpirun -np 4 pw.x <vc-relax.$pp.in> vc-relax.$pp.out
52 # End of for loop.
53 done
```

☞ **Explanation of run.sh:** Six PP files are adopted in this tutorial. The folder path containing these PP files is set in syntax `pseudo_dir` (line 13) the namelist `CONTROL`, and the filename of PP is set as $pp in the namelist `ATOMIC_SPECIES` (line 43).

❑ **Output file:** The optimized lattice constants are obtained from the output files `vc-relax.*.out` (see Sec. 3.1.5) and listed in Table 3.7. Compared with the measured lattice constant of graphene (~ 2.46 Å [Ohta *et al.* (2006)]), the calculated lattice constant of `C.pbe-n-rrkjus_psl.0.1.UPF` (US PP + GGA) is in good agreement. The PP with GGA often overestimates the lattice constants of solids, while LDA almost always underestimates the lattice constants. In both cases, the typical errors amount to 1–2% of the lattice parameters.

Try It Yourself

Calculate lattice constants of bulk Si for several PP files, and compare the obtained results with the experimental value (5.431 Å).

3.1.7 Selecting smearing function and energy

❑ **Purpose:** Smearing is a concept for suppressing unstable electron density during the iteration in the calculation of metal. In this tutorial, we learn how to select the smearing function and energy.

❑ **Background:** Let us briefly explain the concept of smearing. In the DFT, the electron density $n(\boldsymbol{r})$ is calculated from the Kohn-Sham wavefuntion $\phi(\boldsymbol{r})$ as

$$n(\boldsymbol{r}) = \sum_i f_i |\phi_i(\boldsymbol{r})|^2, \tag{3.3}$$

where the index i runs over all states and f_i are the occupation numbers, which can be either 1 or 0 depending on that the i-th state is occupied or not, respectively, as shown in Fig. 3.9 (a). In Quantum ESPRESSO, the sum of Eq. (3.3) is replaced by an integration over the Brillouin zone (BZ) of the system as

$$n(\boldsymbol{r}) = \sum_i \int_{\mathrm{BZ}} f_{i\boldsymbol{k}} |\phi_{i\boldsymbol{k}}(\boldsymbol{r})|^2 d\boldsymbol{k}, \tag{3.4}$$

where $\boldsymbol{k}$ is the wavevector in the BZ. In practice, this integration is carried out numerically by summing over a finite set of $\boldsymbol{k}$-points. For the semiconductor or insulator, electrons in the valence band can not occupy the conduction band during the calculation since the energies of the valence and conduction bands are sufficiently separated. Therefore, the electron density and derived quantities (e.g., the total energy) converge quickly with the finite number of $\boldsymbol{k}$-points used in the integration. However, in the case of the metal or semimetal, the valence bands that cross the Fermi level are partially occupied. Thus, the electrons may occupy the unoccupied states during one iteration, making the algorithm unstable. In this case, it

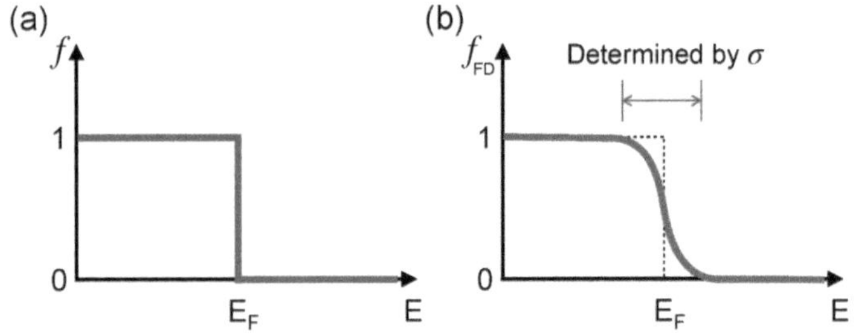

Figure 3.9 (a) The step function f and (b) the smooth Fermi-Dirac function f_{FD} are plotted as a function of the energy E. The width of the Fermi-Dirac function is determined by σ.

often needs a large number of $\boldsymbol{k}$-points in order to make calculations converge. To stabilize the algorithm and reduce the number of $\boldsymbol{k}$-points, we employ a smearing function, in which f_{ik} is replaced by a function that varies smoothly from 1 to 0 at near the Fermi level. By smoothly changing f_i as a function of energy, we can suppress the unstable change of electron density for each SCF iteration.

The Fermi-Dirac smearing: A simple smearing function is the Fermi-Dirac-like distribution function, which is defined by [Mermin (1965)]

$$f_{\mathrm{FD}} = \frac{1}{\exp(x)+1}, \text{ with } x = \frac{E_{ik} - E_{\mathrm{F}}}{\sigma}, \tag{3.5}$$

where E_{F} is the Fermi energy, and σ is the smearing energy, analogous to $k_{\mathrm{B}}T$ in the Fermi-Dirac statistics. Note that when $\sigma \rightarrow 0$, f approaches the step function. As shown in Fig. 3.9 (b), f is a smooth function. Thus, Eq. (3.4) does not cause problems during calculation. However, this approach has a problem. In order to calculate with a relatively small number of $\boldsymbol{k}$-points sampling, we need to use a very large σ value about $0.1 - 0.5$ eV, corresponding to thousands of degrees of the temperature, and the distribution has long tails. Thus, we need to calculate a relatively large number of unoccupied states compared with slowly decay to zero occupation.

The Gaussian smearing: Another smearing function is the Gaussian smearing, in which the step function is approximated by the (Gaussian) complementary error function (erfc), which is defined by [Fu and Ho (1983)]

$$f_G = \frac{1}{2}\text{erfc}(x), \tag{3.6}$$

where erfc(x) is defined as $\text{erfc}(x) = \frac{2}{\sqrt{\pi}} \int_x^\infty \exp(-t^2)dt$. As shown in Fig. 3.10, the width of the Gaussian smearing is smaller than that of the Fermi-Dirac smearing. This means that in order to get similar ***k***-points convergence as for the Fermi-Dirac smearing, the Gaussian smearing can use a smaller value of σ.

The Methfessel-Paxton smearing: Methfessel and Paxton proposed a more sophisticated approximation to the step function based on the Gaussian smearing [Methfessel and Paxton (1989)]. The first-order Methfessel-Paxton approximation is expressed as

$$f_{MP} = \frac{1}{2}\text{erfc}(x) - \frac{1}{2\sqrt{\pi}} x \exp\left(-x^2\right), \tag{3.7}$$

where the first term on the right-hand side corresponds to the Gaussian smearing, while the second term serves to correct the error introduced by the Gaussian smearing. Although only fewer ***k***-points are needed for the Methfessel-Paxton smearing, the occupation numbers become lower than zero, or larger than 1, as shown in Fig. 3.10. Note that while occupation numbers that are larger than 1 are less worrisome, negative occupation numbers can be conceptually problematic in the Methfessel-Paxton smearing.

The Marzari-Vanderbilt smearing: The Marzari-Vanderbilt approach has been proposed in order to avoid the negative occupancies introduced by the Methfessel-Paxton smearing. In this case, the step function is approximated by a Gaussian function multiplied by a first-order polynomial as [Marzari *et al.* (1999)]

$$f_{MV} = \frac{1}{2}\left[\sqrt{\frac{2}{\pi}} \exp(-x^2 - \sqrt{2}x - 1/2) + \text{erfc}\left(x + \frac{1}{\sqrt{2}}\right)\right]. \tag{3.8}$$

As shown in Fig. 3.10, the Marzari-Vanderbilt smearing is asymmetric but does not have negative values and thus the problem with negative electron density is safety avoided.

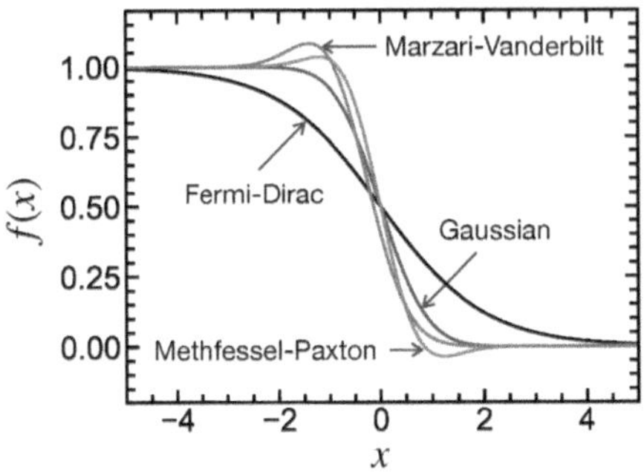

Figure 3.10 The several smearing functions $f(x)$ are plotted as a function of $x = (E_{ik} - E_{\mathrm{F}})/\sigma$.

In this tutorial, we plot the total energy as function of the smearing energy σ by using the Fermi-Dirac, Gaussian, Methfessel-Paxton, and Marzari-Vanderbilt smearing functions.

❑ **How to run:** To run this tutorial, the readers should type the following command lines:

```
1 $ cd ~/QE-SSP/gr/smearing/
2 $ ./run.sh &
```

- Line 1: Go to `smearing` directory that includes input files.
- Line 2: Run a bash script file `run.sh`, which consists of many jobs with changing both the smearing function and smearing energy.

❑ **How to check:** To check whether or not the output file exists, the readers can use the `ls` command by typing:

```
$ ls
```

The `ls` command displays a listing of the all files in the `smearing` folder as

```
calc-smearing.dat  gauss.0.020.in   mp.0.035.out
fd.0.005.in        gauss.0.020.out  mp.0.040.in
fd.0.005.out       gauss.0.025.in   mp.0.040.out
fd.0.010.in        gauss.0.025.out  mv.0.005.in
fd.0.010.out       gauss.0.030.in   mv.0.005.out
fd.0.015.in        gauss.0.030.out  mv.0.010.in
```

```
fd.0.015.out        gauss.0.035.in    mv.0.010.out
fd.0.020.in         gauss.0.035.out   mv.0.015.in
fd.0.020.out        gauss.0.040.in    mv.0.015.out
fd.0.025.in         gauss.0.040.out   mv.0.020.in
fd.0.025.out        mp.0.005.in       mv.0.020.out
fd.0.030.in         mp.0.005.out      mv.0.025.in
fd.0.030.out        mp.0.010.in       mv.0.025.out
fd.0.035.in         mp.0.010.out      mv.0.030.in
fd.0.035.out        mp.0.015.in       mv.0.030.out
fd.0.040.in         mp.0.015.out      mv.0.035.in
fd.0.040.out        mp.0.020.in       mv.0.035.out
gauss.0.005.in      mp.0.020.out      mv.0.040.in
gauss.0.005.out     mp.0.025.in       mv.0.040.out
gauss.0.010.in      mp.0.025.out      plot-smearing.ipynb
gauss.0.010.out     mp.0.030.in       run.sh
gauss.0.015.in      mp.0.030.out
gauss.0.015.out     mp.0.035.in
```

❑ **Input file:** All input files are generated automatically by a bash script file `run.sh` as

QE-SSP/gr/smearing/run.sh

```
#!/bin/bash
# Set a variable sf for the smearing functions.
for sf in fd gauss mp mv; do

# Set a variable se for the smearing energies.
for se in 0.005 0.010 0.015 0.020 0.025 0.030 \
0.035 0.040; do

# Make input file for the scf calculation.
cat > $sf.$se.in << EOF
&CONTROL
calculation    = 'scf'
pseudo_dir     = '../pseudo/'
outdir         = '../tmp/'
prefix         = 'gr'
/
&SYSTEM
ibrav          = 4
a              = 2.4639055825
c              = 15.0
nat            = 2
ntyp           = 1
occupations    = 'smearing'
smearing       = '$sf'
degauss        = $se
ecutwfc        = 40
/
&ELECTRONS
```

```
29 mixing_beta     = 0.7
30 conv_thr        = 1.0D-6
31 /
32 ATOMIC_SPECIES
33 C 12.0107 C.pbe-n-rrkjus_psl.0.1.UPF
34 ATOMIC_POSITIONS (crystal)
35 C   0.333333333   0.666666666   0.500000000
36 C   0.666666666   0.333333333   0.500000000
37 K_POINTS (automatic)
38 12 12 1 0 0 0
39 EOF
40
41 # Run pw.x for SCF calculation.
42 mpirun -np 4 pw.x <$sf.$se.in> $sf.$se.out
43
44 # Write the name of the smearing functions, the
       smearing energies, and total energies
45 awk -v var="$sf" '/!/ {printf"%-6s %1.3f %s\n",var,'
       $se',$5}' $sf.$se.out >> calc-smearing.dat
46 # End of for sf loop.
47 done
48 # End of for se loop.
49 done
```

☞ **Explanation of run.sh:** The smearing option is set by `occupations = 'smearing'` (line 23) in the namelist SYSTEM. The smearing function and energy σ in Ry are set in the syntaxes `occupations` (line 24) and `degauss` (line 25), respectively. The Fermi-Dirac, Gaussian, Methfessel-Paxton, and Marzari-Vanderbilt smearing can selected by setting `smearing = 'fd'`, `smearing = 'gauss'`, `smearing = 'mp'`, and `smearing = 'mv'`, respectively.

☞ **Note:** For the metal and semimetal materials such as graphene, the readers should select the smearing option, otherwise the program might be stopped with the error as

```
%%%%%%%%%%%%%%%%%%%%%%%%%%%%%%%%%%%%%%%%%%%%%%%%%%%%%%%%%%%%%%%%%%%%%%%%%%%%%%%%
     Error in routine electrons (1):
     charge is wrong: smearing is needed
%%%%%%%%%%%%%%%%%%%%%%%%%%%%%%%%%%%%%%%%%%%%%%%%%%%%%%%%%%%%%%%%%%%%%%%%%%%%%%%%
```

❑ **Output file:** The smearing functions, the smearing energies, and the total energy are written in `calc-smearing.dat`. The readers can open this file in terminal by typing:

```
$ vi calc-smearing.dat
```

QE-SSP/gr/smearing/calc-smearing.dat

```
fd      0.005 -23.90915383
fd      0.010 -23.90934651
fd      0.015 -23.90954805
...
mv      0.030 -23.90923457
mv      0.035 -23.90927836
mv      0.040 -23.90931451
```

- Column 1: The abbreviation of the smearing function.
- Column 2: The smearing energy in units of Ry.
- Column 3: The total energy in units of Ry.

☞ **Plotting data from calc-smearing.dat:** The total energy as a function of the smearing energy is plotted by running JupyterLab `plot-smearing.ipynb`:

```
$ jupyter-lab plot-smearing.ipynb
```

QE-SSP/gr/smearing/plot-smearing.ipynb

```
# Import the necessary packages and modules
import matplotlib.pyplot as plt
plt.style.use('../../matplotlib/sci.mplstyle')

# Open and read the file calc-smearing.dat
f = open('calc-smearing.dat', 'r')
f_smearing = [line for line in f.readlines() if line.strip()]
f.close()
nsf = 4 # Number of the smearing functions
nse = 8 # Number of the smearing energies
# Read the smearing energy (se) and total energy (ener) for each smearing function
se = []
ener = []
for i in range(nsf):
    se.append([])
    ener.append([])
    for j in range(nse):
        tmp1 = f_smearing[i*nse+j].split()[1]
        tmp2 = f_smearing[i*nse+j].split()[2]
```

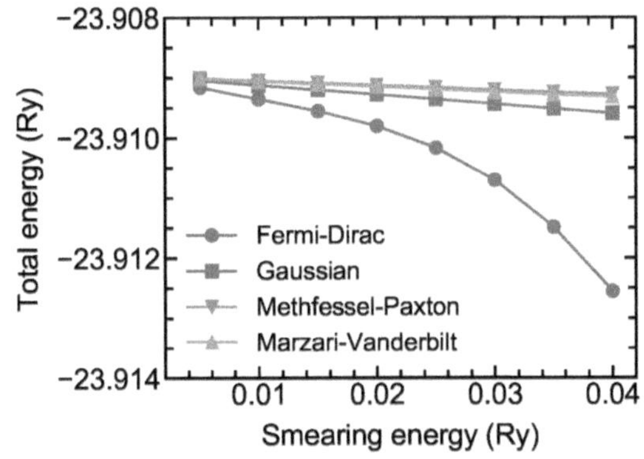

Figure 3.11 Total energy as a function of smearing energy by using the Fermi-Dirac, Gaussian, Methfessel-Paxton, and Marzari-Vanderbilt smearing functions.

```
            se[i].append(float(tmp1))
            ener[i].append(float(tmp2))

# Create figure object
plt.figure()
# Plot the data
plt.plot(se[0],ener[0],'o-', label='Fermi-Dirac')
plt.plot(se[1],ener[1],'s-', label='Gaussian')
plt.plot(se[2],ener[2],'v-', label='Methfessel-Paxton')
plt.plot(se[3],ener[3],'^-', label='Marzari-Vanderbilt')
# Add the legend
plt.legend(loc='lower left')
# Add the x and y-axis labels
plt.xlabel('Smearing energy (Ry)')
plt.ylabel('Total energy (Ry)')
# Set the axis limits
plt.xlim(0.003, 0.042)
plt.ylim(-23.914, -23.908)
# Save the figure.
plt.savefig('plot-smearing.pdf')
# Show the figure
plt.show()
```

In Fig. 3.11, we show the total energy of graphene as a function of smearing energy by using the Fermi-Dirac, Gaussian, Methfessel-Paxton, and Marzari-Vanderbilt smearing functions. We can see that the total energy decreases significantly as the smearing energy increases for the Fermi-Dirac function. In contrast, for the Methfessel-Paxton (MP) or Marzari-Vanderbilt (MV) functions, the decrease is not significant, and therefore it is possible to use relatively large smearing energy for the MP or MV functions. Note that the smearing energy is to treat the Fermi surface efficiently, and we should use small smearing energy from 0.01 to 0.02 Ry for the MP and MV functions. Thus the selection of the MV smearing is reasonable for graphene, and we will use this smearing for the next tutorials of graphene.

Try It Yourself

Plot total energy as a function of smearing energy with different smearing functions for monolayer MoS_2.

3.2 Electronic properties

In Sec. 3.1, we show how to converge our calculations with respect to the cut-off energy and the sampled $\boldsymbol{k}$-points grid. By using these parameters, the unit cell is optimized with a specific pseudopotential and smearing function. This section shows how to visualize and calculate charge density, band structure, and density of state of an optimized unit cell.

3.2.1 Charge density

❑ **Purpose:** In this tutorial we visually check the electron charge density in space.

❑ **Background:** By running `pw.x` in the SCF calculation as shown in Sec. 3.1.1, you can find `tmp/gr.save` directory containing the charge density file (`charge-density.dat`) and other output files. Now in order to do a post-process for converting the file format of `charge-density.dat` into another file format that we can visualize, we use the executable `pp.x` inside the Quantum ESPRESSO distribution. This code can read the output files produced by `pw.x` and convert them into files compatible with various visualization programs such as XCrySDen or VESTA.

❑ **How to run:** To run this tutorial, the readers should type the following command lines:

```
1 $ cd ~/QE-SSP/gr/rho/
2 $ mpirun -np 4 pw.x < scf.in > scf.out &
3 $ mpirun -np 4 pp.x < pp.in > pp.out &
```

- Line 1: Go to `rho` directory.
- Line 2: Run `pw.x` with the input file `scf.in` to obtain `scf.out`.
- Line 3: Run `pp.x` with the input file `pp.in` to obtain `pp.out`.

❑ **How to check:** When the calculations finish, a message `JOB DONE` is written at the end of both output files `scf.out` and `pp.out`.

❑ **Input file:** The detail of the `scf.in` file is given in Sec. 3.1.1. For other input file `pp.in`, the readers can use `vi` editor by typing:

```
$ vi pp.in
```

QE-SSP/gr/rho/pp.in

```
&INPUTPP
outdir        = '../tmp/'
prefix        = 'gr'
plot_num      = 0
/
&PLOT
iflag         = 3
output_format= 6
fileout       = 'gr_rho.cube'
nx            = 64
ny            = 64
nz            = 12
/
```

The description of input variables of `pp.in` is given in Table 3.8.

☞ **Note:** `outdir` and `prefix` must be the same as for `scf.in` for `pw.x` calculation.

☞ **Visualizing charge density:** To see the electron charge density, we launch VESTA and load the file `gr_rho.cube` by typing:

Table 3.8 Meaning of input variables in `pp.in` file.

Line	Syntax	Meaning
1	`&INPUTPP`	A *mandatory namelist* includes input variables, which read the output produced by `pw.x` and extract or calculate the desired quantities (rho, V, ...).
2	`outdir`	Directory includes input data, i.e. the same as in `pw.x`.
3	`prefix`	Prefix of files saved by program `pw.x`.
4	`plot_num`	Select the output quantities, in which 0 = electron charge density.
5	/	End of namelist `INPUTPP`.
6	`&PLOT`	A *mandatory namelist* includes input variables, which provide the desired quantity for outputting file in a suitable format for various plotting programs.
7	`iflag`	Type of plot, in which 3 = 3D plot.
8	`output_format`	Type of format, in which 6 = Gaussian cube file, which can be read by VESTA software.
9	`fileout`	Name of the file to which the plot is written.
10-12	`nx,ny,nz`	To determine the 3D grid for `iflag = 3`.
13	/	End of namelist `PLOT`.

```
$ VESTA gr_rho.cube
```

VESTA will automatically identify the 3D charge density of graphene as shown in Fig. 3.12 top. The readers might want to look at the charge density as a 2D contour plot or heat map. From VESTA, we go to **Utilities → 2D Data Display...**, then in the box **[2D Data Display - (gr_rho.cube)]**, we click on the **Slice** button to put the values for Miller indices, then click to **Ok** button. We will obtain the 2D plot of the charge density (see Fig. 3.12 bottom).

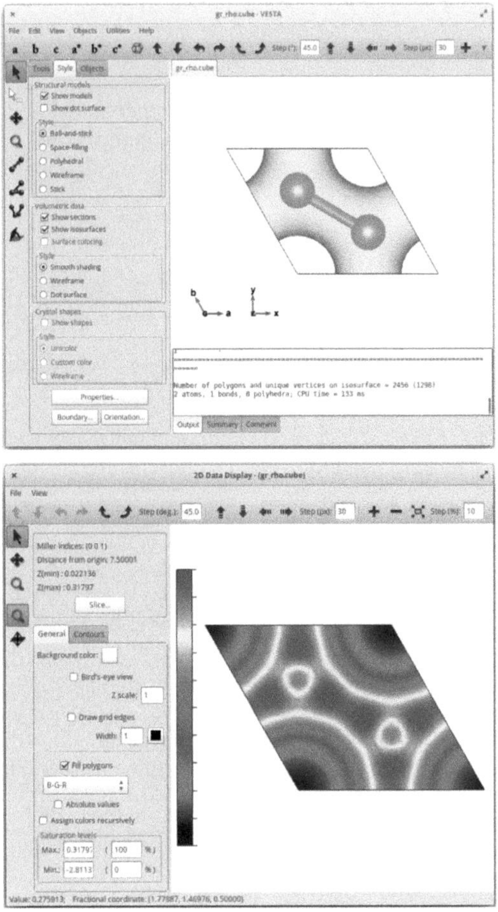

Figure 3.12 3D (top) and 2D (bottom) visualizing charge density of graphene by using VESTA.

Try It Yourself

Calculate the visualize the wavefunction $|\psi|^2$ of graphene. The readers need to set `plot_num = 7` and add two lines with `kpoint(1) = 1` and `kband(1) = 4` in the namelist `INPUTPP` of `pp.in`. This setting will calculate the wavefunction of the 4th band at the Gamma point.

3.2.2 Electronic energy dispersion

❑ **Purpose:** As discussed in Sec. 5.5, the electronic energy dispersion gives the electronic energy as a function of $\boldsymbol{k}$ wavevector, which we call energy band structure. The band structure can tell us useful information, such as a material is metallic or semiconductor. In this tutorial, we show how to calculate the energy dispersion of graphene.

❑ **Background:** In order to calculate the energy dispersion of graphene, we need three-step process as follows:

1. Calculate the Kohn-Sham states with a standard `scf` calculation. As discussed in Sec. 4.10.2, although the Kohn-Sham states are not strictly the real electronic states of the systems, they are often good first-approximations for the electronic states. Therefore, the band structure, which is plotted by the Kohn-Sham eigenvalues, is meaningful to understand the electronic property of the material.
2. Use the output data of the `scf` calculation to perform a non-self-consistent (non-SCF) calculation for many $\boldsymbol{k}$-points along the selected high-symmetry lines. The non-SCF calculates the Kohn-Sham eigenfunctions and eigenvalues without upgrading the Kohn-Sham Hamiltonian at every step.
3. Extract the energies from the non-SCF calculation and convert it to a dataset we can plot with Gnuplot or Python. For this step, we use the executable `bands.x` from Quantum ESPRESSO.

❑ **How to run:** To run this tutorial, the readers should type the following command lines:

```
1 $ cd ~/QE-SSP/gr/bands/
2 $ mpirun -np 4 pw.x < scf.in > scf.out &
3 $ mpirun -np 4 pw.x < nscf.in > nscf.out &
4 $ mpirun -np 4 bands.x < bands.in > bands.out &
```

- Line 1: Go to bands directory.
- Line 2: Run `pw.x` with the input file `scf.in` for step (1).
- Line 3: Run `pw.x` with the input file `nscf.in` for step (2).
- Line 4: Run `bands.x` with the input file `bands.in` for step (3).

❑ **How to check:** When the calculations finish, a message `JOB DONE` is written at the end of all output files `scf.out`, `nscf.out`, and `bands.out`.

❑ **Input file:** The detail of the `scf.in` file is given in Sec. 3.1.1. For input file `nscf.in`, the readers can use `vi` editor by typing:

```
$ vi nscf.in
```

QE-SSP/gr/bands/nscf.in

```
1 &CONTROL
2 calculation     = 'bands'
3 pseudo_dir      = '../pseudo/'
4 outdir          = '../tmp/'
5 prefix          = 'gr'
6 /
7 &SYSTEM
8 ibrav           = 4
9 a               = 2.4639055825
10 c              = 15.0
11 nat            = 2
12 ntyp           = 1
13 nbnd           = 16
14 occupations    = 'smearing'
15 smearing       = 'mv'
16 degauss        = 0.020
17 ecutwfc        = 40
18 /
19 &ELECTRONS
20 mixing_beta    = 0.7
21 conv_thr       = 1.0D-6
22 /
23 ATOMIC_SPECIES
24 C 12.0107 C.pbe-n-rrkjus_psl.0.1.UPF
25 ATOMIC_POSITIONS (crystal)
26 C  0.333333333  0.666666666  0.500000000
27 C  0.666666666  0.333333333  0.500000000
28 K_POINTS (crystal_b)
29 4
30 gG 40
31 K  20
32 M  30
33 gG 0
```

☞ **Explanation of nscf.in:** The non-SCF for the band structure calculation is performed by using the keyword `calculation = 'bands'` (line 2) in the namelist `CONTROL`. Here, we select the number of Kohn-Sham states with the keyword `nbnd = 16` (line 13) in the namelist

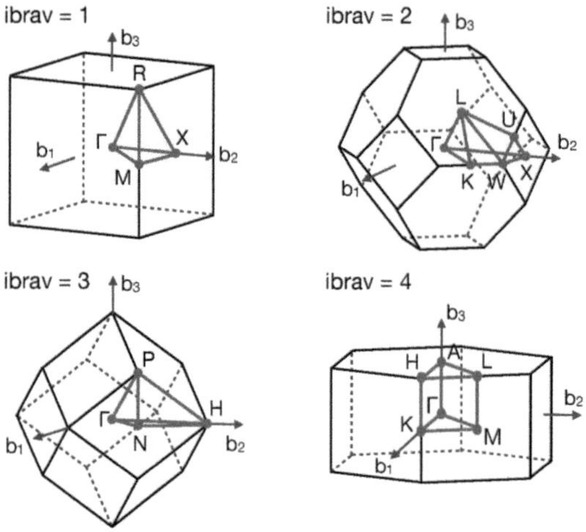

Figure 3.13 The labels of high-symmetry points in the Brillouin-zone of several structures. The Γ-point is denoted by gG in the input file.

SYSTEM, which means that 16 bands are calculated for the case of graphene.

We also need to specify the ***k***-points in the card `K_POINTS (crystal_b)` (line 28), in which `crystal_b` allows us to use the labels of high-symmetry points in the Brillouin-zone. In Fig. 3.13, we show the labels for several structures with `ibrav = 1, 2, 3,` and 4. Note that the Γ-point is denoted by gG (line 30) in the input file. The labels of all structures can be found in Ref. [Setyawan and Curtarolo (2010)]. It is noted that the ***k***-point path must be continuous in the Brillouin-zone.

To see the input file `bands.in`, the readers can type:

```
$ vi bands.in
```

QE-SSP/gr/bands/bands.in

```
&BANDS
outdir  = '../tmp/'
prefix  = 'gr'
filband = 'gr.bands'
/
```

☞ **Note:** `outdir` and `prefix` must be the same as for `scf.in` for `pw.x` calculation. By setting `lp = .true.` in the namelist &BANDS, the readers can obtain the square of the absolute value of the matrix elements of the momentum operator between valence and conduction bands in the output file `p_avg.dat`.

❑ **Output file:** The band structure of graphene is given by `gr.bands`, which is output of `bands.x`. The readers can use `vi` editor to see the `gr.bands` file as follows:

```
$ vi gr.bands
```

QE-SSP/gr/bands/gr.bands

```
1  &plot nbnd=  16, nks=     91 /
2           0.000000  0.000000  0.000000
3  -21.257 -9.357 -4.798 -4.798  1.157  1.900  2.126  4.139
4    4.937  6.590  6.590  8.097  9.243  9.721 10.839 13.470
5           0.016667  0.000000  0.000000
6  -21.252 -9.351 -4.819 -4.809  1.164  1.907  2.133  4.146
7    4.943  6.593  6.612  8.103  9.250  9.703 10.848 13.476
8           0.033333  0.000000  0.000000
9  -21.237 -9.334 -4.880 -4.842  1.184  1.927  2.153  4.166
10   4.963  6.603  6.676  8.123  9.270  9.651 10.874 13.494
11 ...
```

- Line 1: `nbnd` is the number of bands and `nks` is the number of ***k***-points.

- Line 2: The coordinates (k_x, k_y, k_z) of the first ***k***-point in crystal coordinates.

- Line 3 and 4: The list of 16 energies at the first ***k***-point in units of eV.

- Next lines: This format from line 2 to line 4 is repeated until the final ***k***-point.

Other output file is `gr.bands.gnu`, which has a simpler format than `gr.bands`. The `gr.bands.gnu` file contains two columns, the first column is the ***k***-points, and the second column is the energies (eV). We will use this file for plotting as below.

❑ **Plotting data from gr.bands.gnu:** The band structure is plotted by running JupyterLab `plot-bands.ipynb`, which reads the data from the `gr.bands.gnu` file:

```
$ jupyter-lab plot-bands.ipynb
```

QE-SSP/gr/bands/plot-bands.ipynb

```
# Import the necessary packages and modules
import matplotlib.pyplot as plt
plt.style.use('../../matplotlib/sci.mplstyle')
import numpy as np

# The Fermi energy, find it in scf.out
efermi = -1.6790

# Load data from gr.bands.gnu
data = np.loadtxt('gr.bands.gnu')
k = np.unique(data[:, 0])
bands = np.reshape(data[:, 1], (-1, len(k)))

# Set high-symmetry points from nscf.in
gG1 = k[0]; K = k[40]; M = k[60]; gG2 = k[90]

# Create figure object
plt.figure()
# Plot dotted line at Fermi energy
plt.axhline(0, c='gray', ls=':')
# Plot dotted lines at high-symmetry points
plt.axvline(K, c='gray')
plt.axvline(M, c='gray')

# Plot band structure
for band in range(len(bands)):
    plt.plot(k, bands[band, :]-efermi, c='b')

# Add the x and y-axis labels
plt.xlabel('')
plt.ylabel('Energy (eV)')
# Set the axis limits
plt.xlim(gG1, gG2)
plt.ylim (-20, 20)
# Add labels for high-symmetry points
plt.xticks([gG1, K, M, gG2], ['$\Gamma$', 'K', 'M',
    '$\Gamma$'])
# Hide x-axis minor ticks
plt.tick_params(axis='x', which='minor', bottom=
    False, top=False)
# Save the figure
plt.savefig('plot-bands.pdf')
# Show the figure
```

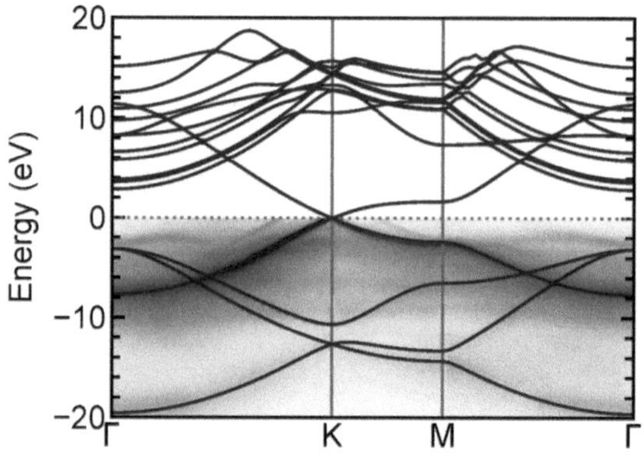

Figure 3.14 Electronic band structure of graphene. The blue lines plot the calculated electron dispersion. Dirac point is the transition between the valence and conduction bands at the K point and zero energy. The ARPES intensity of graphene appears on the electron dispersion of the valence band, which is reproduced with permission [Ohta *et al.* (2007)].

```
plt.show()
```

By running `plot-bands.ipynb`, we obtain the band structure, as shown in Fig. 3.14. It shows that the graphene is a semimetal with a zero band gap at $E = 0$ eV at the K point, which is called the Dirac point since the linear dispersion relation near the K point [Geim and Novoselov (2007)]. In order to compare with the experiment, the angle-resolved photoemission spectroscopy (ARPES) intensity of the electron dispersion of the valence band is plotted below the Dirac point (Fig. 3.14) [Ohta *et al.* (2007)]. The readers can see that the ARPES intensity is reproduced by the calculated electron dispersion (blue lines).

Try It Yourself

1. Calculate electronic energy dispersion of graphene by selecting a LDA functional such as `C.pz-hgh.UPF` and compare it with the GGA functional in Fig. 3.14.
2. Calculate electronic energy dispersion of bulk Si and monolayer MoS_2, and compare calculated energy band gaps with the experiment values (1.12 eV and 2.15 eV for bulk Si and monolayer MoS_2, respectively).

3.2.3 Electronic density of states

❑ **Purpose:** As discussed in Sec. 5.5, the electronic density of states (DOS) is a property on how many electronic states exist per volume per energy. In this tutorial, we will plot the DOS of graphene.

❑ **Background:** In a similar way to calculate the band structure in Sec. 3.2.2, we produce the DOS calculation in the following three steps:

1. Perform the SCF calculation as shown in Sec. 3.2.2.
2. Take the data in the previous step and use it to perform a non-SCF calculation on a more dense grid of ***k***-points. Since the DOS needs an integration over the Brillouin zone, we use a much denser grid for the non-SCF than that for the SCF. In this case, we use the **tetrahedron method** (see Sec. 5.5), in which the Brillouin zone is divided into non-overlapping small tetrahedra, and the integrated quantity is assumed to take a linear interpolation of energy for each tetrahedron. The obtained DOS is equivalent to a very large number of the ***k***-points grid.
3. Convert the state energies calculated on the dense grid of the ***k***-points to the DOS by using the executable `dos.x` from Quantum ESPRESSO. This step will produce `gr.dos` file, which can be plotted by JupyterLab.

❑ **How to run:** To run this tutorial, the readers should type the following command lines:

```
1 $ cd ~/QE-SSP/gr/edos/
2 $ mpirun -np 4 pw.x < scf.in > scf.out &
3 $ mpirun -np 4 pw.x < nscf.in > nscf.out &
4 $ mpirun -np 4 dos.x < dos.in > dos.out &
```

- Line 1: Go to edos directory.
- Line 2: Run `pw.x` with the input file `scf.in` for step (1).
- Line 3: Run `pw.x` with the input file `nscf.in` for step (2).
- Line 4: Run `dos.x` with the input file `dos.in` for step (3).

❑ **How to check:** When the calculations finish, a message `JOB DONE` is written at the end of all output files `scf.out`, `nscf.out`, and `dos.out`.

❑ **Input file:** The detail of the `scf.in` file is given in Sec. 3.1.1. For input file `nscf.in`, the readers can use `vi` editor by typing:

```
$ vi nscf.in
```

QE-SSP/gr/edos/nscf.in

```
1 &CONTROL
2 calculation    = 'nscf'
3 pseudo_dir     = '../pseudo/'
4 outdir         = '../tmp/'
5 prefix         = 'gr'
6 /
7 &SYSTEM
8 ibrav          = 4
9 a              = 2.4639055825
10 c              = 15.0
11 nat            = 2
12 ntyp           = 1
13 nbnd           = 16
14 occupations    = 'tetrahedra'
15 ecutwfc        = 40
16 /
17 &ELECTRONS
18 mixing_beta    = 0.7
19 conv_thr       = 1.0D-6
20 /
21 ATOMIC_SPECIES
22 C 12.0107 C.pbe-n-rrkjus_psl.0.1.UPF
23 ATOMIC_POSITIONS (crystal)
24 C   0.333333333   0.666666666   0.500000000
25 C   0.666666666   0.333333333   0.500000000
26 K_POINTS (automatic)
27 48 48 1 0 0 0
```

☞ **Explanation of nscf.in:** The non-SCF for the DOS is performed by using the keyword `calculation = 'nscf'` (line 2) in the namelist `CONTROL`, while the tetrahedron method is applied by using the keyword `occupations = 'tetrahedra'` (line 14) in the namelist `SYSTEM`. Here, a ***k***-points grid $48 \times 48 \times 1$ in the namelist `K_POINTS` is selected for the non-SCF calculation.

To see the input file `dos.in`, the readers can type:

```
$ vi dos.in
```

QE-SSP/gr/edos/dos.in

```
&DOS
outdir = '../tmp/'
prefix = 'gr'
fildos = 'gr.dos'
/
```

☞ **Note:** `outdir` and `prefix` must be the same as for `scf.in` for `pw.x` calculation.

❑ **Output file:** The output file for the DOS of graphene is given by `gr.dos`, which is output of step (3). The readers can use `vi` editor to see the `gr.dos` file as follows:

```
$ vi gr.dos
```

QE-SSP/gr/edos/gr.dos

```
 #  E (eV)    dos(E)      Int dos(E) EFermi =    -1.722 eV
 -21.257  0.0000E+00  0.0000E+00
 -21.247  0.4960E+00  0.2480E-02
 -21.237  0.2942E+00  0.5277E-02
...
```

- Column 1: The energies in units of eV.
- Column 2: The DOS in units of state/eV/unit-cell.
- Column 3: The integrated DOS in units of state/unit-cell.

☞ **Note:** The readers can find a difference between the Fermi energy value ($E_F = -1.6708$ eV) of the SCF calculation and $E_F = -1.722$ eV of the non-SCF calculation. $E_F = -1.6708$ eV can be found in `scf.out`, while $E_F = -1.722$ eV is given in the header row of `gr.dos`. This inconsistency might come from a small underestimation in the tetrahedron method.

☞ **Plotting data from gr.dos:** The DOS is plotted by running JupyterLab `plot-dos.ipynb` as follows:

```
$ jupyter-lab plot-dos.ipynb
```

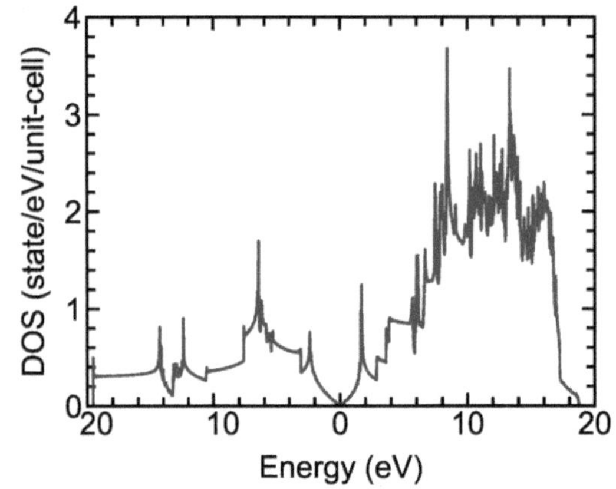

Figure 3.15 Electronic density of states (DOS) of graphene.

QE-SSP/gr/edos/plot-dos.ipynb

```
# Import the necessary packages and modules
import matplotlib.pyplot as plt
plt.style.use('../../matplotlib/sci.mplstyle')
import numpy as np

# The Fermi energy in header row of gr.dos
efermi = -1.724

# Open and read the file gr.dos
ener, dos, idos = np.loadtxt('gr.dos', unpack=True)

# Create figure object
plt.figure()
# Plot the DOS with the Fermi energy shifting
plt.plot(ener-efermi, dos, c='k')
# Add the x and y-axis labels
plt.xlabel('Energy (eV)')
plt.ylabel('DOS (state/eV/unit-cell)')
# Set the axis limits
plt.xlim(-20, 20)
plt.ylim(0, 2.5)
# Save the figure
plt.savefig('plot-dos.pdf')
# Show the figure
plt.show()
```

By running `plot-pdos.ipynb`, we obtain the DOS of graphene, as shown in Fig. 3.15. Since graphene is a zero-band-gap semimetal with bands intersecting at the Dirac point at $E = 0$ (see Fig. 3.14), the DOS is zero at the Dirac point.

Try It Yourself

1. Calculate the DOS of bulk Si and monolayer MoS_2.
2. Check if $\mathrm{DOS}(E) \propto \sqrt{E}$ for the 3D DOS, where E is measured from the bottom of the conduction band.

3.2.4 Partial density of states

❑ **Purpose:** Partial density of states (PDOS) is the relative contribution of a particular atom/orbital to the total DOS. In this tutorial, we will plot the PDOS of graphene.

❑ **Background:** For the SCF calculation in Sec. 3.2.2, we selected the pseudopotential `C.pbe-n-rrkjus_psl.0.1.UPF`, which gives us the information of the particular orbital for the C atom. The readers can find this information in `C.pbe-n-rrkjus_psl.0.1.UPF` as

```
Valence configuration:
nl pn  l   occ     Rcut   Rcut US    E pseu
2S  1  0  2.00    1.000    1.300  -1.010676
2P  2  1  2.00    1.000    1.450  -0.388489
```

The pseudopotential shows that the C atom contains the $2s$ and $2p$ orbitals. We can calculate the PDOS of the $2s$ and $2p$ orbitals for each C atom in the unit cell by using the executable `projwfc.x` from Quantum ESPRESSO.

❑ **How to run:** To run this tutorial, the readers should type the following command lines:

```
$ cd ~/QE-SSP/gr/pdos/
$ mpirun -np 4 pw.x < scf.in > scf.out &
$ mpirun -np 4 pw.x < nscf.in > nscf.out &
$ mpirun -np 4 projwfc.x < projwfc.in > projwfc.out &
```

- Line 1: Go to pdos directory.
- Line 2: Run SCF calculation by using `pw.x`.
- Line 3: Run non-SCF calculation by using `pw.x`.
- Line 4: Run `projwfc.x` with the input file `projwfc.in`.

☞ **Note:** The SCF and non-SCF calculations can be omitted if the readers have successfully finished the DOS calculation in Sec. 3.2.2.

❑ **How to check:** When the calculations finish, a message `JOB DONE` is written at the end of the all output files `scf.out`, `nscf.out`, and `projwfc.out`.

❑ **Input file:** The detail of the input files `scf.in` and `nscf.in` are given in Sec. 3.1.2. For input file `projwfc.in`, the readers can use `vi` editor by typing:

```
$ vi projwfc.in
```

QE-SSP/gr/pdos/projwfc.in

```
&projwfc
prefix   = 'gr'
outdir   = '../tmp/'
/
```

☞ **Note:** `outdir` and `prefix` in `projwfc.in` must be the same as for `scf.in` and `nscf.in`.

❑ **Output file:** The PDOS is written to output files `gr.pdos_atm#N(X)_wfc#M(l)`, where X is atom symbol (C), N is atom number (1, 2) in the unit cell, M is wavefunction number, and l is orbital (s, p, d, etc.). We can open the file `gr.pdos_atm#1(C)_wfc#1(s)` for the PDOS of 2*s* orbital of the first C atom in the unit cell as follows:

```
$ vi gr.pdos_atm#1\(C\)_wfc#1\(s\)
```

QE-SSP/gr/pdos/gr.pdos_atm#1(C)_wfc#1(s)

```
# E (eV)    ldos(E)     pdos(E)
  -21.259  0.000E+00   0.000E+00
  -21.249  0.243E+00   0.243E+00
  -21.239  0.144E+00   0.144E+00
   ...
```

```
  17.041  0.126E-03  0.126E-03
  17.051  0.557E-04  0.557E-04
  17.061  0.000E+00  0.000E+00
```

- Column 1: The energies (in eV).

- Column 2: The "total" PDOS of the 2*s* orbital (LDOS) in state/eV/unit-cell. It is noted that since we consider only wavefunction for *s* orbital, the LDOS is equal to the PDOS. In general, LDOS = PDOS_1 + PDOS_2 + ...

- Column 3: The PDOS of the 2*s* orbital with the first wavefunction of the first C atom (in state/eV/unit-cell).

☞ **Plotting data from gr.dos:** The PDOS of graphene is plotted by running JupyterLab `plot-pdos.ipynb` as follows:

```
$ jupyter-lab plot-dos.ipynb
```

QE-SSP/gr/edos/plot-dos.ipynb

```
# Import the necessary packages and modules
import matplotlib.pyplot as plt
plt.style.use('../../matplotlib/sci.mplstyle')
import numpy as np

# The Fermi energy, find it in nscf.out
efermi = -1.7241

# Define the function to read the data file
def r_dos(name):
    ener, dos = np.loadtxt(name, usecols=(0,1),
        unpack=True)
    return ener, dos

# Read the total PDOS
ener, dos = r_dos('gr.pdos_tot')
# Read the PDOS of C atom number 1
ener1, p1C1s = r_dos('gr.pdos_atm#1(C)_wfc#1(s)')
ener1, p1C2p = r_dos('gr.pdos_atm#1(C)_wfc#2(p)')
# Read the PDOS of C atom number 2
ener2, p2C1s = r_dos('gr.pdos_atm#2(C)_wfc#1(s)')
ener2, p2C2p = r_dos('gr.pdos_atm#2(C)_wfc#2(p)')

# Create figure object
plt.figure()
# Plot the DOS
```

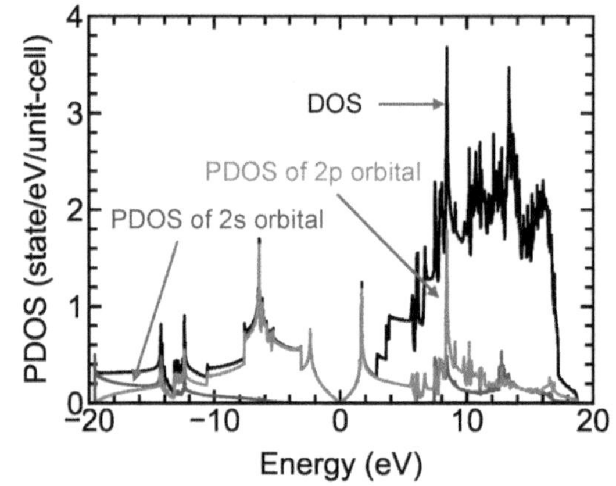

Figure 3.16 The DOS and PDOS of 2s and 2p orbital of graphene.

```
plt.plot(ener-efermi, dos, c='k')
# Plot the PDOS of s-orbital
plt.plot(ener1-efermi, p1C1s+p2C1s, c='b')
# Plot the PDOS of p-orbital
plt.plot(ener2-efermi, p1C2p+p2C2p, c='r')
# Add tho x and y-axis labels
plt.xlabel('Energy (eV)')
plt.ylabel('PDOS (state/eV/unit-cell)')
# Set the axis limits
plt.xlim(-20, 20)
plt.ylim(0, 4)
# Save the figure
plt.savefig('plot-pdos.pdf')
# Show the figure
plt.show()
```

In Fig. 3.16, we show the DOS and the PDOS of the 2s and 2p orbitals as functions of energy E. For the valence bands ($E < 0$), the DOS is equal to the total PDOS since all valence bands mainly come from the 2s and 2p orbitals. For the conduction bands, with $E > 2.5$ eV, the DOS is not equal to the sum of PDOS because the high energy bands include not only the 2s and 2p orbitals but also more delocalized wavefunctions far from the atoms, which are called interlayer-bands [Fretigny *et al.* (1989)].

Try It Yourself

Calculate the contribution of the PDOS of the S and Mo atoms to the total DOS of monolayer MoS_2.

3.3 Lattice oscillations

In this section, we learn how to calculate the frequencies of oscillations of solid. The oscillations of atoms in a crystal with a finite amplitude around the equilibrium positions, known as phonon modes (see Sec. 5.7), are essential to understand the dynamical properties of the material, such as thermal conductance or the Raman spectra.

3.3.1 Phonon dispersion

❑ **Purpose:** As discussed in Sec. 5.7, the phonon dispersion is the phonon frequency as a function of phonon wavevector.[3] The phonon dispersion, which is observed by neutron scattering, can tell us useful information of the material such as elastic constants or thermal conductivity. In this tutorial, we show how to plot the phonon dispersion of graphene.

❑ **Background:** The phonon angular frequencies $\omega_{\boldsymbol{q}}$ are determined by solving the simultaneous equation of motion for each $\boldsymbol{q}$ as

$$\sum_{J\beta} \tilde{D}_{I\alpha,J\beta}(\boldsymbol{q})\tilde{u}_{J\beta}(\boldsymbol{q}) = \omega_{\boldsymbol{q}}^2 \tilde{u}_{I\alpha}(\boldsymbol{q}) \text{ with } (I\alpha = 1, ..., 3N), \qquad (3.9)$$

where N is the number of atoms in the unit cell and α (β) corresponds to x, y, z components of oscillation. $\tilde{u}_{I\alpha}$ ($\tilde{u}_{J\beta}$) is a phonon eigenvector (or the amplitude of the phonon) of the I-th (J-th) atom. Summation on $J\beta$ is taken for neighbor atoms from a given I-th atom up to several nearest neighbors. $\tilde{D}_{I\alpha,J\beta}(\boldsymbol{q})$ is called the dynamical matrix, which is

[3] We use $\boldsymbol{k}$-points to refer to electron wavevectors and $\boldsymbol{q}$-points to refer to phonon wavevectors

defined by

$$\tilde{D}_{I\alpha,J\beta}(\boldsymbol{q}) = \frac{1}{\sqrt{M_I M_J}} \tilde{F}_{I\alpha,J\beta}(\boldsymbol{q}), \tag{3.10}$$

where M_I (M_J) is mass of the I-th (J-th) atom. $\tilde{F}_{I\alpha,J\beta}$ is second derivatives of the total energy E_{tot} with respect to the displacements $(\tilde{u}_{I\alpha}, \tilde{u}_{J\beta})$ of atomic positions, which is called the inter-atomic force constant (IFC), which is given by

$$\tilde{F}_{I\alpha,J\beta}(\boldsymbol{q}) = \frac{\partial^2 E_{\text{tot}}}{\partial \tilde{u}^*_{I\alpha}(\boldsymbol{q}) \partial \tilde{u}_{J\beta}(\boldsymbol{q})}. \tag{3.11}$$

$\tilde{D}_{I\alpha,J\beta}(\boldsymbol{q})$ in Eq. (3.10) is calculated by the executable `ph.x` in Quantum ESPRESSO within the density-functional perturbation theory[4] (DFPT) [Baroni *et al.* (2001)]. Then, $\omega_{\boldsymbol{q}}$ is obtained from Eq. (3.9) by diagonalizing $\tilde{D}_{I\alpha,J\beta}(\boldsymbol{q})$. By repeating this procedure for several $\boldsymbol{q}$, we obtain $\omega_{\boldsymbol{q}}$ as a function of $\boldsymbol{q}$ and plot the phonon dispersions. However, the DFPT calculation is time-consuming compared with the SCF calculation. Thus, we will calculate $\tilde{D}_{I\alpha,J\beta}(\boldsymbol{q})$ for a small set of $\boldsymbol{q}$ vectors and take an interpolation for $\omega_{\boldsymbol{q}}$ over the Brillouin zone.

In order to plot the $\omega_{\boldsymbol{q}}$ by interpolation, we use the executables `q2r.x` and `matdyn.x` in Quantum ESPRESSO. `q2r.x` reads $\tilde{D}_{I\alpha,J\beta}(\boldsymbol{q})$, which is calculated by `ph.x` for a uniform mesh of (q_1, q_2, q_3) vectors. Then, the IFC of a $q_1 \times q_2 \times q_3$ supercell in the real space, $F_{I\alpha,J\beta}(\boldsymbol{R})$, is calculated from $\tilde{F}_{I\alpha,J\beta}(\boldsymbol{q})$ by using the inverse Fourier transform as

$$F_{I\alpha,J\beta}(\boldsymbol{R}) = \frac{1}{N_q} \sum_{i=1}^{N_q} \tilde{F}_{I\alpha,J\beta}(\boldsymbol{q}_{ijk}) e^{i\boldsymbol{q}_{ijk}\cdot\boldsymbol{R}}, \tag{3.12}$$

where $N_q = q_1 \times q_2 \times q_3$, $\boldsymbol{R}$ is an atomic position in the $q_1 \times q_2 \times q_3$ supercell, and $\boldsymbol{q}_{ijk}$ is defined by

$$\boldsymbol{q}_{ijk} = \frac{i-1}{q_1}\boldsymbol{G}_1 + \frac{j-1}{q_2}\boldsymbol{G}_2 + \frac{k-1}{q_3}\boldsymbol{G}_3, \tag{3.13}$$

[4] The DFPT is a method that can directly calculate the second-order derivatives of the energy in Eq. (3.11) by a self-consistent procedure.

where $i = 1, \ldots, q_1$, $j = 1, \ldots, q_2$, $k = 1, \ldots, q_3$, and $\boldsymbol{G}_1$, $\boldsymbol{G}_2$, and $\boldsymbol{G}_3$ are the reciprocal lattice vectors. Then, `matdyn.x` reads the IFC that is calculated by `q2r.x` and calculates the IFC at an interpolated $\boldsymbol{q}'$ by using the Fourier transform as

$$\tilde{F}_{I\alpha J\beta}(\boldsymbol{q}') = \sum_{\boldsymbol{R}} F_{I\alpha J\beta}(\boldsymbol{R}) e^{i\boldsymbol{q}'\cdot\boldsymbol{R}}. \tag{3.14}$$

Eqs. (3.12) and (3.14) allow the interpolation of the dynamical matrix at arbitrary $\boldsymbol{q}'$, by a few IFCs. It is noted that the Fourier interpolation works well if the IFCs decay rapidly in the real space. Therefore, the Fourier interpolation might fail in some special cases when IFCs do not decay. For example, when the Kohn anomaly occurs in a metal, the dynamical matrix is not a smooth function of $\boldsymbol{q}$, and the IFCs decay slowly in the real space.

The calculation of the phonon dispersion of graphene has the following four steps:

1. Perform the SCF with `pw.x` calculation.
2. Calculate the dynamical matrix in Eq. (3.10) on a uniform mesh of $\boldsymbol{q}$-points by using `ph.x`.
3. Perform the inverse Fourier transform of the dynamical matrix in Eq. (3.12) to obtain a set of the IFCs in the real space by using `q2r.x`.
4. Perform the Fourier transform of the real space IFCs in Eq. (3.14) to obtain the dynamical matrix at any $\boldsymbol{q}'$ by using `matdyn.x`. This allows us to calculate the phonon frequencies for a set of points along lines between high-symmetry points in the Brillouin zone, which is similar to the electronic band structure in Sec. 3.2.2.

❑ **How to run:** The readers can run the following command lines:

```
1 $ cd ~/QE-SSP/gr/phonon/
2 $ mpirun -np 4 pw.x < scf.in > scf.out &
3 $ mpirun -np 4 ph.x < ph.in > ph.out &
4 $ mpirun -np 4 q2r.x < q2r.in > q2r.out &
5 $ mpirun -np 4 matdyn.x < matdyn.in > matdyn.out&
```

- Line 1: Go to phonon directory.
- Line 2: Run SCF calculation with `pw.x` for step (1).
- Line 3: Run phonon calculation with `ph.x` for step (2).
- Line 4: Run `q2r.x` for step (3).

- Line 5: Run `matdyn.x` for step (4).

❑ **How to check:** When the calculations finish, a message `JOB DONE` is written at the each end of the output files; `scf.out`, `ph.out`, `q2r.out`, and `matdyn.out`.

❑ **Input file:** The input file `scf.in` can be opened as follows:

```
$ vi scf.in
```

QE-SSP/gr/phonon/scf.in

```
1  &CONTROL
2  calculation      = 'scf'
3  pseudo_dir       = '../pseudo/'
4  outdir           = '../tmp/'
5  prefix           = 'gr'
6  /
7  &SYSTEM
8  ibrav            = 4
9  a                = 2.4639055825
10 c                = 15.0
11 nat              = 2
12 ntyp             = 1
13 occupations      = 'smearing'
14 smearing         = 'mv'
15 degauss          = 0.02
16 ecutwfc          = 80
17 assume_isolated = '2D'
18 /
19 &ELECTRONS
20 mixing_beta      = 0.7
21 conv_thr         = 1.0D-6
22 /
23 ATOMIC_SPECIES
24 C 12.0107 C.pbe-n-rrkjus_psl.0.1.UPF
25 ATOMIC_POSITIONS (crystal)
26 C  0.333333333  0.666666666  0.500000000
27 C  0.666666666  0.333333333  0.500000000
28 K_POINTS (automatic)
29 12 12 1 0 0 0
```

☞**Explanation of scf.in:** Compared with the `scf.in` file in the previous tutorials (see Sec. 3.2.2 or 3.2.3), a new syntax `assume_isolated = '2D'` (line 17) is added in the namelist SYSTEM. This syntax can help to avoid the "flexural" phonons (negative or imaginary acoustic frequencies at near Gamma point) in the 2D materials [Sohier *et al.* (2017)].

☞**Note:** For phonon calculation, the readers should use a larger value of `ecutwfc = 80` (line 16) compared with the SCF calculation (`ecutwfc = 40`). This is because that the phonon frequencies have the unit of cm^{-1} (~ 0.00012 eV), which requires a large cutoff energy.

To see the input file `ph.in`, the readers can type:

```
$ vi ph.in
```

QE-SSP/gr/phonon/ph.in

```
1  phonon calc.
2  &INPUTPH
3  outdir = '../tmp/'
4  prefix = 'gr'
5  tr2_ph = 1d-14
6  ldisp  = .true.
7  nq1    = 6
8  nq2    = 6
9  nq3    = 1
10 fildyn = 'gr.dyn'
11 /
```

☞**Explanation of ph.in:** In order to obtain the phonon dispersion, we need to set `ldisp = .true.` (line 6) and a grid of $\boldsymbol{q}$-points specified by `nq1, nq2, nq3`. In the case of the 2D materials, we can put `nq3 = 1`. Since the `ph.x` calculation is time-consuming, `nq1, nq2, nq3` are usually taken smaller than `nk1, nk2, nk3` of the `pw.x` calculation. The description of input variables of `ph.in` is given in Table 3.9.

To see the input file `q2r.in`, the readers can type:

```
$ vi q2r.in
```

QE-SSP/gr/phonon/q2r.in

```
&INPUT
fildyn = 'gr.dyn'
zasr   = 'crystal'
```

Table 3.9 Meaning of input variables in `ph.in` file.

Line	Syntax	Meaning
1	`title_line`	Title of the job, i.e., a line that is reprinted on output
4	`prefix`	Prefix of files saved by program `pw.x`.
5	`tr2_ph`	Threshold for self-consistency.
6	`ldisp`	The default is `ldisp = .false.`. If we set `ldisp = .true.`, the dynamical matrix is calculated for a grid of $\boldsymbol{q}$-points specified by `nq1, nq2, nq3`.
7-9	`nq1, nq2, nq3`	Parameters of the Monkhorst-Pack grid (no offset) used when `ldisp = .true.`. It is same meaning as for `nk1, nk2, nk3` in the input of `pw.x`.
10	`fildyn`	File name where the dynamical matrix is written.

```
4 flfrc    = 'gr.fc'
5 /
```

☞**Explanation of q2r.in:** The syntax `zasr` is used to impose the IFCs to adopt the acoustic sum rule (ASR).[5] There are several options for the ASR imposed. Here, we select `zasr = 'crystal'` (line 3), i.e., three translational ASR are imposed by optimized correction of the dynamical matrix. The file containing the IFCs in the real space is given by the `gr.fc` file.

☞ **Note:** `fildyn` in `q2r.in` must be the same as that for `ph.in` for `ph.x` calculation.

To see the input file `matdyn.in`, the readers can type:

```
$ vi matdyn_disp.in
```

[5] The ASR is a requirement for IFCs by translational and rotational invariance of the crystal. If we translate the whole solid by a uniform displacement or if we rotate the solid, the inter-atomic forces by the displacements on rotation should not appear. As a consequence of ASR, the frequencies of the acoustic modes become zero [Madelung (1978)].

QE-SSP/gr/phonon/matdyn.in

```
1  &INPUT
2  asr            = 'crystal'
3  flfrc          = 'gr.fc'
4  flfrq          = 'gr.freq'
5  flvec          = 'gr.modes'
6  loto_2d        = .true.
7  q_in_band_form = .true.
8  /
9  4
10 gG 40
11 K  20
12 M  30
13 gG 0
```

☞**Explanation of matdyn.in:** The syntax `loto_2d` is set to `.true.` (line 6) to activate two-dimensional treatment of LO-TO splitting,[6] which occurs in a polar 2D materials as h-BN or MoS_2. It is noted that LO-TO splitting does not occur in the case of graphene since it is not a polar material. The setting `q_in_band_form = .true` (line 7) allows us to use the labels of high-symmetry points in the Brillouin-zone (see Sec. 3.2.2) to plot the phonon dispersion.

☞ **Note:** By changing `asr = 'crystal'` (line 2) to `asr = 'simple'` in `matdyn.in`, the readers might obtain the negative phonon frequency at near the Γ point. The option `asr = 'crystal'` is usually better than `asr = 'simple'`.

❑ **Output file:** The output file for the phonon frequencies of graphene is given by `gr.freq`. The readers can open this file by

```
$ vi gr.freq
```

QE-SSP/gr/phonon/gr.freq

```
&plot nbnd=    6, nks=   91 /
          0.000000  0.000000  0.000000
-0.0000 -0.0000 -0.0000 876.7549 1488.7595 1488.7595
          0.016667  0.000000  0.000000
  7.4052 31.1526 49.6068 876.5584 1489.4337 1490.5041
  ...
```

[6] In the ionic crystal, we know the splitting between LO and TO as $\omega_{\mathrm{LO}}^2 = (\varepsilon_0/\varepsilon_\infty)\omega_{\mathrm{TO}}^2$ [Kittel (1976)].

- Line 1: nbnd is the number of phonon modes and nks is the number of $\boldsymbol{q}$-points.
- Line 2: The coordinates (q_x, q_y, q_z) of the first $\boldsymbol{q}$-point in crystal coordinates.
- Line 3: The list of 6 phonon frequencies at the first $\boldsymbol{q}$-point in units of cm^{-1}.
- Next lines: This format from line 2 to line 3 is repeated until the final $\boldsymbol{q}$-point.

The `matdyn.x` calculation in the step (4) also automatically outputs a file `gr.freq.gp`, which has a simpler format than `gr.freq`. The `gr.freq.gp` file contains 7 columns, in which the first column is the $\boldsymbol{q}$-points and other columns are phonon frequencies (cm^{-1}). We will use this file for plotting as below.

❑ **Plotting data from gr.freq.gp:** The phonon dispersion is plotted by running JupyterLab `plot-phonon.ipynb`, which reads the data from the `gr.freq.gp` file:

```
$ jupyter-lab plot-phonon.ipynb
```

QE-SSP/gr/phonon/plot-phonon.ipynb

```
# Import the necessary packages and modules
import matplotlib.pyplot as plt
plt.style.use('../../matplotlib/sci.mplstyle')
import numpy as np
# Number of phonon modes and q-points gr.freq
nbnd = 6; nks = 91
# Open gr.freq.gp file
qs, *ph = np.loadtxt('gr.freq.gp', unpack=True)
# Read phonon at each phonon index
ph0 = []
for iq in range(nks):
    ph0.append([])
    for ib in range(nbnd):
        tmp = ph[ib][iq]
        ph0[iq].append(float(tmp))
# Set high-symmetry points from matdyn.in
G1 = qs[0]; K = qs[40]; M = qs[60]; G2 = qs[90]
# Create figure object
plt.figure()
# Plot dotted lines at high-symmetry points
plt.axvline(K, c='gray')
plt.axvline(M, c='gray')
```

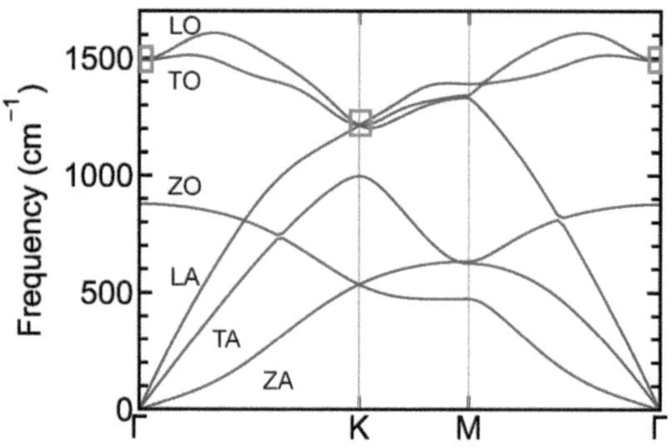

Figure 3.17 Phonon dispersion of graphene along the high symmetry-lines of hexagonal Brillouin zone. The red boxes show the Kohn anomalies at the Γ and the K points of the Brillouin zone.

```
# Plot phonon dispersion
plt.plot(qs, ph0, c='b')
# Add the x and y-axis labels
plt.xlabel('')
plt.ylabel('Frequency (cm$^{-1}$)')
plt.xlim(G1, G2)
plt.ylim(0, 1700)
# Add labels for high-symmetry points
plt.xticks([G1, K, M, G2], ['$\Gamma$', 'K', 'M', '$
    \Gamma$'])
# Hide x-axis minor ticks
plt.tick_params(axis='x', which='minor', bottom=
    False, top=False)
# Save the figure
plt.savefig('plot-phonon.pdf')
# Show the figure
plt.show()
```

By running `plot-phonon.ipynb`, we obtain the phonon dispersion of graphene, as shown in Fig. 3.17. There are 6 phonon modes including three acoustic branches (ZA, TA, and LA) and three optical branches (ZO, TO, and LO). The red boxes in Fig. 3.17 show the Kohn anomaly[7] [Kohn (1959)] for certain phonon modes at the Γ

[7] The Kohn anomaly is an anomaly in the phonon dispersion of a metal. For a specific wavevector, the frequency of the associated phonon is softened, and there is a discontinuity in the derivative of frequency.

and the K points of the Brillouin zone, arising from perturbation of a phonon by electron-phonon interaction. Specifically, the Kohn anomaly is manifested by a non-zero slope of LO and TO at the Γ point and a non-zero slope of TO at the K point. This effect was observed experimentally by the Raman spectra of the G band, due to the LO/TO phonon, at different charge doping level [Lazzeri and Mauri (2006)].
☞ **Visualizing phonon dispersion:** The readers can also visualize the phonon dispersion calculated with a display graphically the phonon modes of the lattice vibrations. First, the readers need to upload the input file `scf.in` and the output files `scf.out` and `gr.modes` to the website: `https://www.materialscloud.org/work/tools/interactivephonon`, as shown in the red box of Fig. 3.18 (top). Then, the readers click on the button "Calculate phonon dispersion". A new window will appear, as shown in Fig. 3.18 (bottom). In the phonon section, the readers can click on any point in the phonon dispersion and see an animation of how the atoms vibrate according to this model.

Try It Yourself

1. Calculate phonon dispersion of graphene without `assume_isolated = '2D'` option in `scf.in`, and check if the negative phonon frequency appears for the ZA mode.
2. Calculate phonon dispersion of monolayer MoS_2.

3.3.2 Phonon density of states

❑ **Purpose:** Phonon density of states (DOS) is defined in exactly the same way as the electronic density of states (see Sec. 3.2.3), i.e., the number of phonon modes per volume per frequency. In this tutorial, we will plot the phonon DOS of graphene.
❑ **Background:** The phonon DOS can be obtained by choosing some input options for `matdyn.x` in step (4) of the phonon calculation (see Sec. 3.3.1). Therefore, the readers are asked to finish the phonon calculation in Sec. 3.3.1 before this tutorial.
❑ **How to run:** To run this tutorial, the readers should type the following command lines:

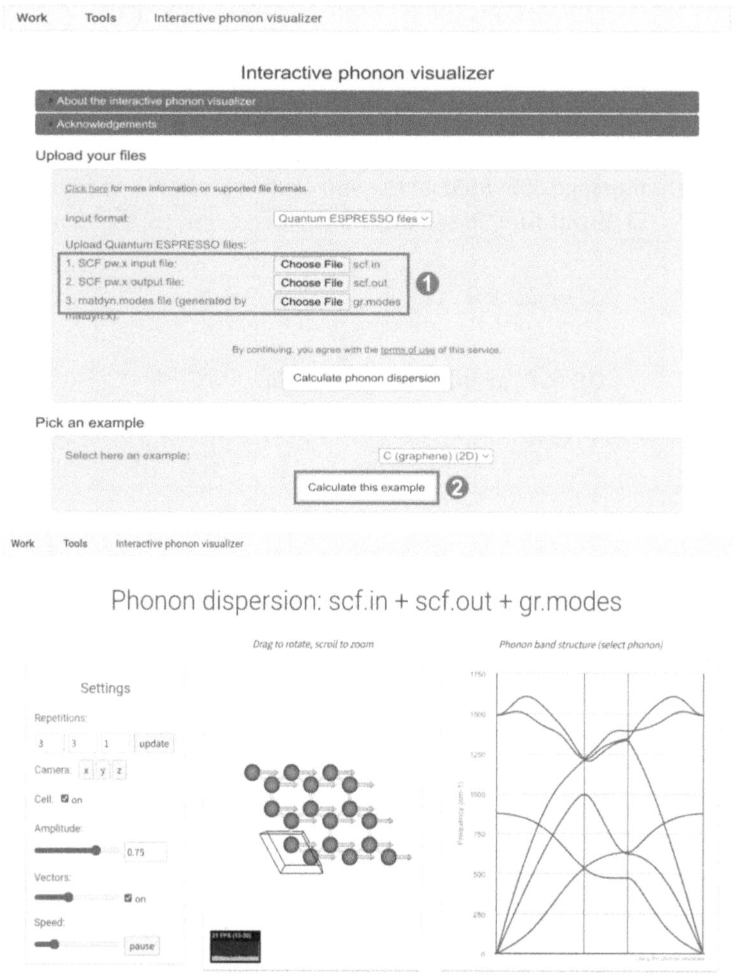

Figure 3.18 Visualizing phonon dispersion by using free-online-tool "Interactive phonon visualizer".

```
1 $ cd ~/QE-SSP/gr/phdos/
2 $ cp ../phonon/gr.fc ./
3 $ mpirun -np 4 matdyn.x < matdyn.in> matdyn.out &
```

- Line 1: Go to phdos directory.

- Line 2: Copy the IFCs file `gr.fc` from Sec 3.3.1.

- Line 3: Run `matdyn.x` with the input file `matdyn.in` for the phonon DOS calculation. It is noted that this `matdyn.in` file is different with the `matdyn.in` file in Sec 3.3.1.

❑ **How to check:** When the calculations finish, you can find a message `JOB DONE` at the end of the output file `matdyn.out`.

❑ **Input file:** To see the input file `matdyn.in`, the readers can type:

```
$ vi matdyn.in
```

QE-SSP/gr/phdos/matdyn.in

```
1  &INPUT
2  asr        = 'crystal'
3  flfrc      = 'gr.fc'
4  flfrq      = 'gr.freq'
5  flvec      = 'gr.modes'
6  loto_2d    = .true.
7  fldos      = 'gr.dos'
8  dos        = .true.
9  nk1        = 48
10 nk2        = 48
11 nk3        = 1
12 /
```

☞**Explanation of matdyn.in:** The syntax dos is set to `.true.` (line 8) to obtain the phonon DOS. This is similar to the electronic density of states in Sec. 3.2.3, we need to select a dense $\boldsymbol{q}$-points grid to calculate the phonon DOS compared with $\boldsymbol{q}$-points of the `ph.x` calculation. Here, we select $48 \times 48 \times 1$ for the case of graphene.

❑ **Output file:** The output file for the phonon DOS of graphene is given by `gr.dos`. The readers can use `vi` editor to see the `gr.dos` file as follows:

```
$ vi gr.dos
```

QE-SSP/gr/phdos/gr.dos

```
# Frequency[cm^-1] DOS PDOS
-2.4978196418E-05 0.0000000000E+00 0.0000E+00 0.0000E+00
 9.9997502180E-01 2.4555277523E-05 1.2278E-05 1.2278E-05
```

```
1.9999750218E+00 4.9110570407E-05 2.4555E-05 2.4555E-05
...
```

- Column 1: The phonon frequencies in units of cm^{-1}. We can neglect a tiny negative value `-2.4978196418E-05` and assumed to be zero.

- Column 2: The phonon DOS in units of state/cm^{-1}/unit-cell.

- Columns 3 and 4: The partial phonon density of states (PDOS) for the C atoms 1 and 2 in the unit cell (in state/cm^{-1}/unit-cell).

☞ **Plotting data from gr.dos:** The phonon DOS is plotted by running JupyterLab `plot-phdos.ipynb`, which reads the data from the `gr.dos` file:

```
$ jupyter-lab plot-phdos.ipynb
```

QE-SSP/gr/edos/plot-dos.ipynb

```
# Import the necessary packages and modules
import matplotlib.pyplot as plt
plt.style.use('../../matplotlib/sci.mplstyle')
import numpy as np
# Load data from gr.dos
omega, ph_tot, ph_c1, ph_c2 = np.loadtxt('gr.dos',
    unpack=True)
# Create figure object
plt.figure()
# Plot the phonon DOS
plt.plot(omega, ph_tot, c='b', label='Total')
plt.plot(omega, ph_c1, c='r', ls='dashed', label='C
    atom')
# Add the x and y-axis labels
plt.xlabel('Frequency (cm$^{-1}$)')
plt.ylabel('Phonon DOS (state/cm$^{-1}$/unit-cell)')
# Add the legend
plt.legend(loc='upper left')
# Set the axis limits
plt.xlim(0, 1700)
plt.ylim(0, 0.025)
# Save the figure
plt.savefig('plot-phdos.pdf')
# Show the figure
plt.show()
```

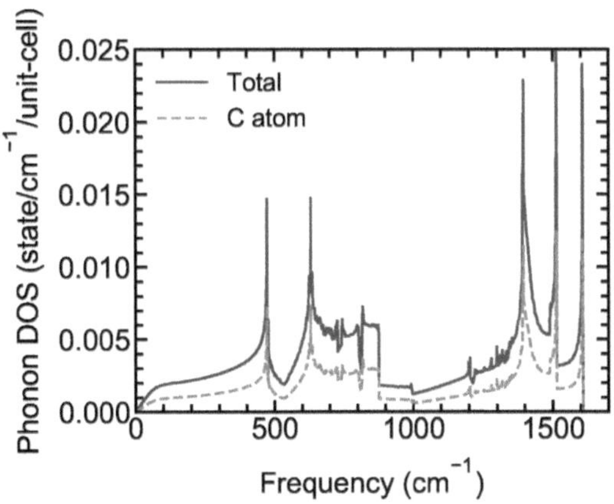

Figure 3.19 Phonon density of states (solid line) and phonon PDOS of a C atom (dashed line) of graphene.

By running `plot-phdos.ipynb`, we obtain the phonon DOS of graphene, as shown in Fig. 3.19. The total phonon DOS (solid line) of graphene is twice as large as the PDOS of a single C atom (dashed line) since there are two C atoms in the unit cell of graphene.

Try It Yourself

Calculate contributions of the phonon DOS of the Mo and S atoms to total phonon DOS of monolayer MoS_2.

3.3.3 Electron-phonon interaction

❑ **Purpose:** In solids, the electron-phonon interaction plays an important role in a variety of physical phenomena, such as temperature-dependent energy band, the Kohn anomaly, electrical conductivity, or superconductivity. As discussed in Sec. 5.8, the origin of electron-phonon interaction is the oscillation of atomic potential by lattice oscillation. In this tutorial, we calculate the electron-phonon interaction of graphene.

❑ **Background:** In Quantum ESPRESSO, the electron-phonon matrix elements, $g_{\nu\boldsymbol{q}}$, for the phonon mode ν at a wavevector $\boldsymbol{q}$ are defind as

$$g_{\nu\boldsymbol{q}}(\boldsymbol{k}, i, j) = \left(\frac{\hbar}{2M\omega_{\nu\boldsymbol{q}}}\right)^{1/2} \langle\psi_{i,\boldsymbol{k}}|\partial_{\nu\boldsymbol{q}}\mathcal{V}_{\mathrm{SCF}}|\psi_{j,\boldsymbol{k}+\boldsymbol{q}}\rangle, \tag{3.15}$$

where M is the total mass of the atoms in the unit cell, $\omega_{\nu\boldsymbol{q}}$ is the phonon frequency, $\partial_{\nu\boldsymbol{q}}\mathcal{V}_{\mathrm{SCF}}$ is the derivative of the self-consistent potential $\mathcal{V}_{\mathrm{SCF}}$ by ionic displacement by phonon mode ν at $\boldsymbol{q}$, and $\psi_{i,\boldsymbol{k}}$ and $\psi_{j,\boldsymbol{k}+\boldsymbol{q}}$ are the wavefunctions at initial state i and final state j, respectively.

The phonon linewidth $\gamma_{\nu\boldsymbol{q}}$, which is the imaginary part of the phonon self-energy, is defined by $g_{\nu\boldsymbol{q}}$ in Eq. (3.15) as

$$\gamma_{\nu\boldsymbol{q}} = 2\pi\omega_{\nu\boldsymbol{q}} \sum_{ij} \int_{\mathrm{BZ}} \frac{\mathrm{d}\boldsymbol{k}}{\Omega_{\mathrm{BZ}}} |g_{\nu\boldsymbol{q}}(\boldsymbol{k}, i, j)|^2 \delta(\epsilon_{i,\boldsymbol{k}} - \epsilon_F)\delta(\epsilon_{j,\boldsymbol{k}+\boldsymbol{q}} - \epsilon_F), \tag{3.16}$$

where Ω_{BZ} is the volume of the Brillouin zone (BZ), ϵ_F is the Fermi energy, and $\epsilon_{i,\boldsymbol{k}}$ and $\epsilon_{j,\boldsymbol{k}+\boldsymbol{q}}$ are the energies of the initial state i and final state j, respectively.

The electron-phonon coupling constant $\lambda_{\nu\boldsymbol{q}}$ is defined by $\gamma_{\nu\boldsymbol{q}}$ in Eq. (3.16) as

$$\lambda_{\nu\boldsymbol{q}} = \frac{\gamma_{\nu\boldsymbol{q}}}{\pi\hbar D(\epsilon_F)\omega_{\nu\boldsymbol{q}}^2}, \tag{3.17}$$

where $D(\epsilon_F)$ is the density of states (DOS) at the Fermi energy.

The calculation of the electron-phonon interaction of graphene consists the following 5 steps:

1. Perform the SCF for a dense $\boldsymbol{k}$-points grid with `pw.x` calculation. The dense grid must contain all $\boldsymbol{k}$ and $\boldsymbol{k}+\boldsymbol{q}$ points, which is used in the electron-phonon calculation and must be dense enough to produce the phonon linewidth accurately.
2. Perform other SCF for a grid of $\boldsymbol{k}$-points that is suitable for the phonon calculation by using `pw.x`.
3. Perform the phonon calculation based on the SCF in step (2) by using `ph.x`.

4. Perform the inverse Fourier transform of dynamical matrix and the phonon linewidth, which are obtained in step (3), in real space by using `q2r.x`.
5. Perform the Fourier transform to obtain the dynamical matrix and the phonon linewidth for a set of points along lines between high-symmetry points in the Brillouin zone by using `matdyn.x`.

❑ **How to run:** To run this tutorial, the readers should type the following command lines:

```
1 $ cd ~/QE-SSP/gr/elph/
2 $ mpirun -np 4 pw.x < scf-dense.in > scf-dense.out &
3 $ mpirun -np 4 pw.x < scf.in > scf.out &
4 $ mpirun -np 4 ph.x < ph.in > ph.out &
5 $ mpirun -np 4 q2r.x < q2r.in > q2r.out &
6 $ mpirun -np 4 matdyn.x < matdyn.in> matdyn.out &
```

- Line 1: Go to `elph` directory.
- Line 2: Run `pw.x` for the SCF calculation with a dense ***k***-points grid.
- Line 3: Run `pw.x` for the SCF calculation with a suitable ***k***-points grid for phonon calculation.
- Line 4: Run `ph.x` with the input file `ph.in` for the electron-phonon calculation.
- Line 5: Run `q2r.x` with the input file `q2r.in`.
- Line 6: Run `matdyn.x` with the input file `matdyn.in`.

❑ **How to check:** When the calculations finish, you can find a message `JOB DONE` at the end of each output file.

❑ **Input file:** The readers can open the input file `scf-dense.in` as follows:

```
$ vi scf-dense.in
```

QE-SSP/gr/elph/scf-dense.in

```
&CONTROL
calculation      = 'scf'
pseudo_dir       = '../pseudo/'
outdir           = '../tmp/'
prefix           = 'gr'
/
```

```
7 &SYSTEM
8 ibrav            = 4
9 a                = 2.4639055825
10 c               = 15.0
11 nat             = 2
12 ntyp            = 1
13 occupations     = 'smearing'
14 smearing        = 'mv'
15 degauss         = 0.020
16 ecutwfc         = 80
17 assume_isolated = '2D'
18 la2F            = .true.
19 /
20 &ELECTRONS
21 mixing_beta     = 0.7
22 conv_thr        = 1.0D-6
23 /
24 ATOMIC_SPECIES
25 C 12.0107 C.pbe-n-rrkjus_psl.0.1.UPF
26 ATOMIC_POSITIONS (crystal)
27 C  0.333333333  0.666666666  0.500000000
28 C  0.666666666  0.333333333  0.500000000
29 K_POINTS (automatic)
30 36 36 1 0 0 0
```

☞**Explanation of scf-dense.in:** The option `la2F = .true.` (line 18) in the namelist SYSTEM instructs the `pw.x` to save the eigenvalues on the dense $\boldsymbol{k}$-points grid into a file `gr.a2Fsave`, which can be found in `outdir` (line 4). Here, we select a dense $\boldsymbol{k}$-points grid is $36 \times 36 \times 1$ (line 30) in the namelist `K_POINTS`. The remaining part is same as Sec. 3.3.1.

☞**Note:** The dense $\boldsymbol{k}$-points grid must be unshifted (line 30) because the grid should include the Γ point (see Sec. 3.1.3). Since the dense grid must contain all $\boldsymbol{k}$ and $\boldsymbol{k}+\boldsymbol{q}$ points, it should be a multiple of $\boldsymbol{q}$-point grid in `ph.in` for phonon calculations, otherwise the `ph.x` calculation will be stopped by showing the following error:

```
%%%%%%%%%%%%%%%%%%%%%%%%%%%%%%%%%%%%%%%%%%%%%%%%%%%%%%%%%%%%%%%%%%%
 task #         3
 from elphsum : error #         2
 q is not a vector in the dense grid
%%%%%%%%%%%%%%%%%%%%%%%%%%%%%%%%%%%%%%%%%%%%%%%%%%%%%%%%%%%%%%%%%%%
```

The input file `scf.in` is same as Sec. 3.3.1. For the input file `ph.in`, the readers can open by typing:

```
$ vi ph.in
```

QE-SSP/gr/elph/ph.in

```
1  phonon calc.
2  &INPUTPH
3  outdir           = '../tmp/'
4  prefix           = 'gr'
5  tr2_ph           = 1d-14
6  ldisp            = .true.
7  nq1              = 6
8  nq2              = 6
9  nq3              = 1
10 fildyn           = 'gr.dyn'
11 fildvscf         = 'grdv'
12 electron_phonon  = 'interpolated'
13 el_ph_nsigma     = 10
14 el_ph_sigma      = 0.02
15 /
```

☞**Explanation of ph.in:** The electron-phonon calculation is performed by setting `electron_phonon = 'interpolated'` (line 12). The option `fildvscf = 'grdv'` (line 11) is the file name of derivative of the self-consistent potential $\partial_{\nu \boldsymbol{q}} \mathcal{V}_{\mathrm{SCF}}$ in Eq. (3.15), which is stored at `_ph0` in `outdir` (line 3) for each $\boldsymbol{q}$ point. `el_ph_nsigma` (line 13) is the number of the Gaussian broadening values, which is used for delta functions in Eq. (3.16). `el_ph_sigma` (line 14) is the spacing (in Ry) between the Gaussian broadening values. The remaining part is same as Sec. 3.3.1.

☞**Explanation of q2r.in and matdyn.in:** The input files `q2r.in` and `matdyn.in` are the same as Sec. 3.3.1, but the option `la2F = .true.` should be added in the namelist `INPUT` of these input files to obtain the electron-phonon coupling along lines between high-symmetry points in the Brillouin zone.

❑ **Output file:** The output files of both the phonon linewidth and the electron-phonon coupling are stored in the files `elph_dir/elph.inp_lambda.*` for each $\boldsymbol{q}$ point. For example, the readers can open the file `elph_dir/elph.inp_lambda.1` for the Γ point as

```
$ vi elph_dir/elph.inp_lambda.1
```

QE-SSP/gr/elph/elph_dir/elph.inp_lambda.1

```
1     0.000000       0.000000       0.000000     10     6
2 0.119733E-06  0.119733E-06  0.173580E-06  0.638932E-04
     0.183390E-03  0.183390E-03
3   Gaussian Broadening:    0.020 Ry, ngauss=    0
4   DOS =   0.122860 states/spin/Ry/Unit Cell at Ef= -4.238911
       eV
5   lambda(     1)=  0.3002    gamma=     0.05 GHz
6   lambda(     2)=  0.1533    gamma=     0.02 GHz
7   lambda(     3)=  0.0001    gamma=     0.00 GHz
8   lambda(     4)=  0.0000    gamma=     0.00 GHz
9   lambda(     5)=  1.3810    gamma=   321.60 GHz
10  lambda(     6)=  1.3816    gamma=   321.72 GHz
11  Gaussian Broadening:    0.040 Ry, ngauss=    0
12  DOS =   0.242754 states/spin/Ry/Unit Cell at Ef= -4.244465
       eV
13  lambda(     1)=  2.5752    gamma=     0.77 GHz
14 ...
```

- Line 1: The coordinates (q_x, q_y, q_z) of $\boldsymbol{q}$, `el_ph_nsigma`, and the number of phonon modes `3*nat`.

- Line 2: $\omega^2_{\nu\boldsymbol{q}}$ (in Ry^2) for each phonon mode ν from 1 to 6.

- Line 3: The Gaussian broadening value (in Ry).

- Line 4: The DOS at the Fermi energy (in states/spin/Ry/unit-cell).

- Lines 5 to 10: $\lambda_{\nu\boldsymbol{q}}$ and $\gamma_{\nu\boldsymbol{q}}$ (in GHz) for each ν.

- From line 11: The DOS, $\lambda_{\nu\boldsymbol{q}}$, and $\gamma_{\nu\boldsymbol{q}}$ for next Gaussian broadening value.

It is noted that only $\gamma_{\nu\boldsymbol{q}}$ is interpolated by `matdyn.x` calculation for each Gaussian broadening value. The name of this file is `elph.gamma.*`. For example, the readers can open the file `elph.gamma.1`, which corresponds to the Gaussian broadening = 0.02 Ry, as follows:

```
$ vi elph.gamma.1
```

QE-SSP/gr/elph/elph.gamma.1

```
  &plot nbnd=    6, nks=   91 /
  0.000000  0.000000  0.000000
0.0000    -0.0000     0.0000    -0.0000   321.6265   321.6265
```

```
4|   0.016667  0.000000  0.000000
5| -0.0000    -0.0715    -0.1440    -0.0000  316.9246  316.9963
6| ...
```

- Line 1: nbnd is the number of phonon modes and nks is the number of $\boldsymbol{q}$-points.
- Line 2: The crystal coordinates (q_x, q_y, q_z) of the first $\boldsymbol{q}$-point.
- Line 3: The values of $\gamma_{\nu\boldsymbol{q}}$ for 6 phonon modes.
- Next lines: This format from line 2 to line 3 is repeated for 91 $\boldsymbol{q}$-points.

☞ **Plotting data from elph.gamma.*:** We will select the `elph.gamma.5` file to plot the phonon linewidth $\gamma_{\nu\boldsymbol{q}}$. Other files can be plotted similarly. First, the readers run the file `plotband.in` as

```
$ plotband.x < plotband.in
```

QE-SSP/gr/elph/plotband.in

```
elph.gamma.5
-30 300
elph.gamma.5.gnu
elph.gamma.5.ps
0.0
0.0 0.0
```

The `plotband.x` calculation will read the `elph.gamma.5` file, and output the `elph.gamma.5.gnu` and `elph.gamma.5.ps` files. Next, we will use the `elph.gamma.5.gnu` file to plot with the JupyterLab `gamma.ipynb` as

```
$ jupyter-lab gamma.ipynb
```

QE-SSP/gr/elph/gamma.ipynb

```
# Import the necessary packages and modules
import matplotlib.pyplot as plt
plt.style.use('../../matplotlib/sci.mplstyle')
import numpy as np
# Convert from GHz to meV
```

```
ghz2mev = 4.13567587265e-3
# Load elph.gamma.5.gnu file
data = np.loadtxt('elph.gamma.5.gnu')
k = np.unique(data[:, 0])
gamma = np.reshape(data[:, 1], (-1, len(k)))
# Set high-symmetry points from matdyn.in
gG1 = k[0]; K = k[40]; M = k[60]; gG2 = k[90]
# Create figure object
plt.figure()
plt.axhline(0, c='gray', ls=':')
# Plot dotted lines at high-symmetry points
plt.axvline(K, c='gray')
plt.axvline(M, c='gray')
# Plot phonon linewidth
for i in range(len(gamma)):
    plt.plot(k, gamma[i, :]*ghz2mev, c='r', ls='None', marker='o', markersize=7)
# Add the x and y-axis labels
plt.xlabel('')
plt.ylabel('Phonon linewidth (meV)')
# Set the axis limits
plt.xlim(gG1, gG2)
plt.ylim (-0.1, 1.1)
# Add labels for high-symmetry points
plt.xticks([gG1, K, M, gG2], ['$\Gamma$', 'K', 'M', '$\Gamma$'])
# Hide the x-axis minor ticks
plt.tick_params(axis='x', which='minor', bottom=False, top=False)
# Save the figure
plt.savefig('plot-gamma.pdf')
# Show the figure
plt.show()
```

In Fig. 3.20, we plot the phonon linewidth $\gamma_{\nu q}$ of graphene. At the Γ point, both the TO and LO modes show the maximum values about $\gamma_{\nu q} = 1.0$ meV, while at the K point, only the TO shows the maximum value about $\gamma_{\nu q} = 0.75$ meV. The maximum values of the phonon linewidth at the Γ and K points are consistent with the observed Kohn anomaly at the Γ point (TO and LO) and the K point (TO) of graphene, as discussed in Fig. 3.17 in Sec. 3.3.1.

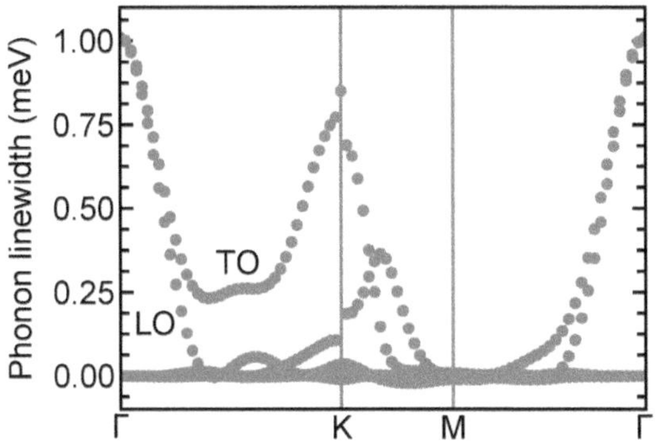

Figure 3.20 Phonon linewidth of graphene along the high symmetry-lines of the Brillouin zone.

Try It Yourself

1. Plot the phonon linewidth $\gamma_{\nu q}$ with the Gaussian broadening about 0.02, 0.08, 0.14, and 0.2.
2. Calculate the electron-phonon coupling $\lambda_{\nu q}$ based on $\gamma_{\nu q}$ from the `elph.gamma.5` file by using Eq. (3.17), and plot $\lambda_{\nu q}$ along lines between high-symmetry points in the Brillouin zone.

3.3.4 Eliashberg spectral function

❑ **Purpose:** The Eliashberg spectral function $\alpha^2F(\omega)$ is a sum over the contributions from scattering processes, in which the electrons are scattered by the phonons on the Fermi surface [Eliashberg (1960)]. $\alpha^2F(\omega)$ is an important parameter to determine the critical temperature of superconductors [McMillan (1968)]. In this tutorial, we will calculate $\alpha^2F(\omega)$ of graphene.

❑ **Background:** From Eq. (3.17) in Sec. 3.3.3, the isotropic Eliashberg spectral function [Eliashberg (1960)] can be obtained by taking an average over the Brillouin zone (BZ) as

$$\alpha^2F(\omega) = \frac{1}{2}\sum_{\nu}\int_{\mathrm{BZ}}\frac{\mathrm{d}\boldsymbol{q}}{\Omega_{\mathrm{BZ}}}\omega_{\nu\boldsymbol{q}}\lambda_{\nu\boldsymbol{q}}\delta(\omega-\omega_{\nu\boldsymbol{q}}). \tag{3.18}$$

The parameter λ, which is a dimensionless measure of the strength of $\alpha^2F(\omega)$, can also be defined as

$$\lambda = 2\int\frac{\alpha^2F(\omega)}{\omega}\mathrm{d}\omega = \sum_{\nu\boldsymbol{q}}\lambda_{\nu\boldsymbol{q}}. \tag{3.19}$$

The superconducting critical temperature T_c using the McMillan formula as a function of λ [McMillan (1968)] is given by

$$T_c = \frac{\omega_{\ln}}{1.2}\exp\left[\frac{-1.04(1+\lambda)}{\lambda(1-0.62\mu^*)-\mu^*}\right], \tag{3.20}$$

where $\omega_{\ln}$ is defined by [Dynes (1972)]

$$\omega_{\ln} = \exp\left[\frac{2}{\lambda}\int\frac{\mathrm{d}\omega}{\omega}\alpha^2F(\omega)\log(\omega)\right], \tag{3.21}$$

and μ^* in Eq. (3.20) is an empirical parameter that describes the Coulomb screening. μ^* has typical values between 0.1 and 0.16.

Since $\alpha^2F(\omega)$ is obtained by using $\lambda_{\nu\boldsymbol{q}}$ in Eq. (3.17), the readers are asked to finish the electron-phonon calculation in Sec. 3.3.3 before this tutorial.

❑ **How to run:** To run this tutorial, the readers should type the following command lines:

```
1 $ cd ~/QE-SSP/gr/alpha/
2 $ cp -rf ../elph/elph_dir/ ../elph/gr.fc ./
3 $ mpirun -np 4 matdyn.x < matdyn.in> matdyn.out &
```

- Line 1: Go to `alpha` directory.
- Line 2: Copy the `elph_dir` directory, which contains $\gamma_{\nu\boldsymbol{q}}$ and $\lambda_{\nu\boldsymbol{q}}$ in Sec. 3.3.3, and `gr.fc`, which contains the inter-atomic force constants in the real space, to the current direction (`./`).
- Line 3: Run `matdyn.x` with the input file `matdyn.in` to obtain $\alpha^2F(\omega)$ and λ.

❑ **How to check:** When the calculations finish, you can find a message `JOB DONE` at the end of each output file.

❑ **Input file:** The input file `matdyn.in` is same as Sec. 3.3.2. We need to add the option `la2F = .true.` and select `ndos = 100` (number of energy steps for DOS calculation) in the namelist `INPUT`.

❑ **Output file:** The Eliashberg spectral function $\alpha^2F(\omega)$ is given by the output files `a2F.dos*` for each Gaussian broadening. For example, the readers can open the file `a2F.dos1` for the Γ point as follows:

```
$ vi a2F.dos1
```

QE-SSP/gr/alpha/a2F.dos1

```
1 # Eliashberg function a2F (per both spin)
2 #  frequencies in Rydberg
3 # DOS normalized to E in Rydberg: a2F_total, a2F(mode)
4
5 0.739762E-04  -0.382895E-04   0.000000E+00  -0.291145E-04  -0.917506E-05   0.000000E+00   0.000000E+00   0.000000E+00
6 0.221929E-03  -0.344606E-03  -0.105875E-29  -0.262031E-03  -0.825757E-04   0.000000E+00   0.000000E+00   0.000000E+00
7 ...
```

- Lines 1–3: The head of the a2F.dos1 file.
- Lines 5–6: The 1st column is ω in Ry unit, the 2nd column is the total of $\alpha^2F(\omega)$, the 3rd–7th columns are the $\alpha^2F(\omega)$ for the each phonon mode ν from 1 to 6.
- Next lines: The format from lines 5 and 6 is repeated until the last value of ω.

☞ **Plotting data from a2F.dos*:** The readers can plot the file `a2F.dos*` as follows:

```
$ jupyter-lab alpha.ipynb
```

QE-SSP/gr/alpha/alpha.ipynb

```
# Import the necessary packages and modules
import matplotlib.pyplot as plt
plt.style.use('../../matplotlib/sci.mplstyle')
import numpy as np
```

```

# Convert from Ry to cm^-1
ry2cm = 109737.07176

# Load data for the broadening of 0.1 Ry
omega1, aF1 = np.genfromtxt('a2F.dos5', usecols=(0,1), skip_footer=1, unpack=True)
# Load data for the broadening of 0.2 Ry
omega2, aF2 = np.genfromtxt('a2F.dos10', usecols=(0,1), skip_footer=1, unpack=True)

# Create figure object
plt.figure()
# Plot the JDOS
plt.plot(omega1*ry2cm, aF1, c='b', label='Broadening 0.1 Ry')
plt.plot(omega2*ry2cm, aF2, c='r',ls='--', label='Broadening 0.2 Ry')
# Add the x and y-axis labels
plt.xlabel(r'$\omega$ (cm$^{-1}$)')
plt.ylabel(r'$\alpha^2 F(\omega)$')
# Add the the legend
plt.legend(loc='upper left')
# Set the axis limits
plt.xlim(0, 1700)
plt.ylim(0, 0.45)
# Save a figure to the pdf file
plt.savefig('plot-aF.pdf')
# Show plot
plt.show()
```

In Fig. 3.21, we plot the total $\alpha^2F(\omega)$ as a function of the phonon frequency ω of graphene with the Gaussian broadening of 0.1 Ry (solid line) and 0.2 Ry (dashed line). As shown in Fig. 3.20, the TO and LO modes show the highest electron-phonon coupling at the Γ and the K points, which correspond to $\omega = 1500$ cm^{-1} and 1400 cm^{-1}, respectively (see Fig. 3.17). Moreover, the phonon DOS in Fig. 3.19 also shows highest value around $\omega = 1500$ cm^{-1}. Therefore, $\alpha^2F(\omega)$ becomes a highest value around $\omega = 1500$ cm^{-1}, as shown in Fig. 3.21. It is noted that the calculated result of $\alpha^2F(\omega)$ might not be converged since $\alpha^2F(\omega)$ converges very slowly with increasing size of the ***k***-point and ***q***-point grids. The readers should check the convergence by increasing ***k***-point and ***q***-point grids, and it is time-consuming for electron-phonon calculation.

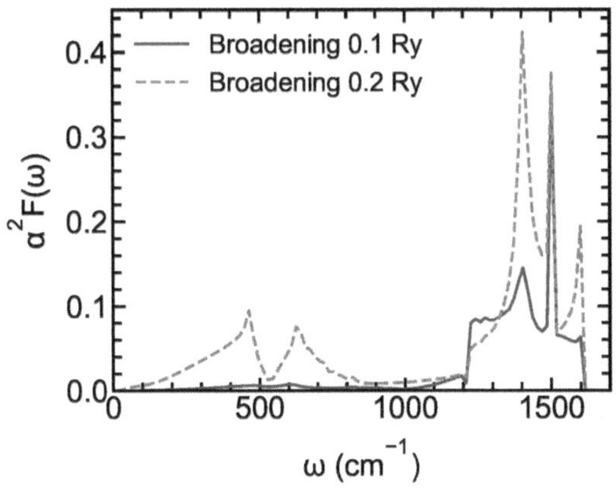

Figure 3.21 Electron-phonon Eliashberg spectral function $\alpha^2F(\omega)$ of graphene with the Gaussian broadening of 0.1 Ry (solid line) and 0.2 Ry (dashed line).

For the parameters λ and ω_{ln}, they are given by the output file `lambda` for each Gaussian broadening. The readers can open `lambda` as follows:

```
$ vi lambda
```

QE-SSP/gr/alpha/lambda

```
Electron-phonon coupling constant, lambda

Broadening   0.0200 lambda    0.1516 dos(Ef)   0.1229
     omega_ln [K]   3612.2615
Broadening   0.0400 lambda    0.0507 dos(Ef)   0.2428
     omega_ln [K]   3600.9008
...
```

This output file shows that $\lambda = 0.1517$ and $\omega_{\text{ln}} = 3611.6988$ K at the broadening of 0.02 Ry. By inserting these values and the constant $\mu^* = 0.1$ into Eq. (3.20), we obtain the critical temperature $T_c \sim 0$ for the graphene. This is because that graphene has the DOS $=$ 0 at the Fermi energy. Thus, it is not good for superconductivity. Nevertheless, graphene can become a superconductor by heavily electron-doped or twisted bilayer graphene. For example, Ca-intercalated graphite (CaC6) is a superconductor with $T_c = 11.5$

K [Emery *et al.* (2009)]. Cao *et al.* [Cao *et al.* (2018)] showed that bilayer graphene with a twist angle of about 1.1° shows T_c of up to 1.7.

Try It Yourself

1. Calculate the $\alpha^2 F(\omega)$ and T_c for the 3D Pb. The 3D Pb has the structure of face centered cubic and $T_c = 7.2$ K.
2. Explain why is T_c of the 3D Pb higher than that of graphene?

3.4 Optical properties

In this section, we learn how to calculate the optical properties of solids, such as optical absorption spectra and non-resonant Raman spectra. We select monolayer MoS_2 as an example for this section since it is a semiconductor. It is noted that some optical calculations in Quantum ESPRESSO work only for semiconductors or insulators. This is because the metallic system has an infinite dielectric constant at $\omega = 0$, which is required in some calculations.

3.4.1 Dielectric function and absorption spectra

❑ **Purpose:** In this tutorial, we show how to obtain the optical absorption of the monolayer MoS_2.
❑ **Background:** The optical absorption spectra $\alpha(\omega)$ can be calculated by using the real and imaginary parts of the dielectric function $\varepsilon(\omega)$, as shown in Eq. (5.32) in Sec. 5.9. $\varepsilon(\omega)$ is calculated by using the the executable `epsilon.x` in Quantum ESPRESSO. `epsilon.x` is based on the single-particle approximation.[8]
☞**Note:** `epsilon.x` supports for only the norm-conserving pseudopotentials.
❑ **How to run:** To run this tutorial, the readers should type the following command lines:

[8] In optical absorption, we have either single-particle excitation of an electron or collective excitations of electrons. In single-particle approximation, we calculate a sum of independent excitation of an electron from occupied state to empty states.

```
1 $ cd ~/QE-SSP/mos2/optic/
2 $ mpirun -np 4 pw.x < scf.in > scf.out &
3 $ mpirun -np 4 pw.x < nscf.in > nscf.out &
4 $ mpirun -np 4 epsilon.x < epsilon.in > epsilon.out
     &
```

- Line 1: Go to `optic` directory.
- Line 2: Run SCF by using `pw.x` with the input file `scf.in`.
- Line 3: Run non-SCF by using `pw.x` with the input file `nscf.in`. It is noted that the non-SCF calculation might take about 20 minutes.
- Line 4: Run `epsilon.x` with the input file `epsilon.in`.

❑ **How to check:** When calculation finishes, a message `JOB DONE` is written at the end of each output file `scf.out`, `nscf.out`, and `epsilon.out`.

❑ **Input file:** The readers can use `vi` editor to open the input file `scf.in` as follows:

```
$ vi scf.in
```

QE-SSP/mos2/optic/scf.in

```
&CONTROL
calculation      = 'scf'
pseudo_dir       = '../pseudo/'
outdir           = '../tmp/'
prefix           = 'mos2'
/
&SYSTEM
ibrav            = 4
a                = 3.1378055413
c                = 20.0
nat              = 3
ntyp             = 2
ecutwfc          = 80.0
/
&ELECTRONS
mixing_beta      = 0.7
conv_thr         = 1.0d-10
/
ATOMIC_SPECIES
Mo   95.94    Mo.pz-hgh.UPF
S    32.065   S.pz-hgh.UPF
ATOMIC_POSITIONS (crystal)
```

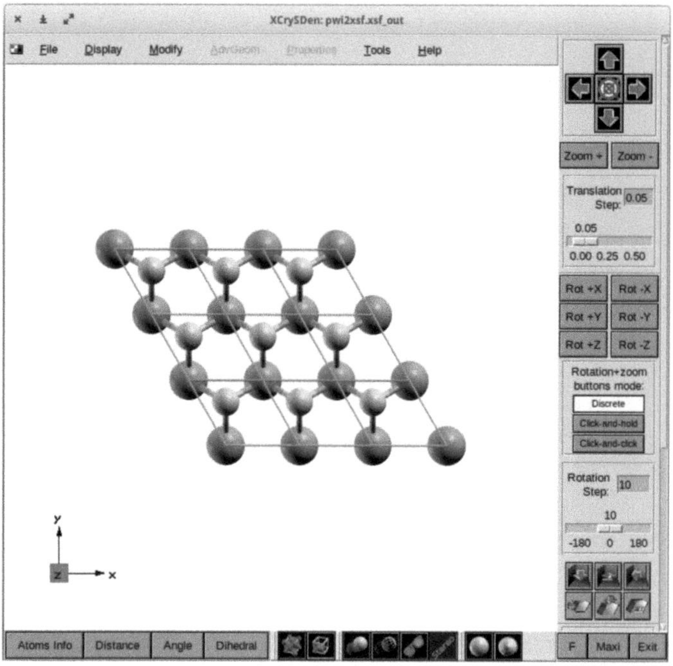

Figure 3.22 Atomic structure of monolayer MoS_2 in a 3 × 3 × 1 supercell.

```
Mo   -0.0000000000   -0.0000000000    0.5000000000
S     0.3333333333    0.6666666667    0.4218571774
S     0.3333333333    0.6666666667    0.5781427066
K_POINTS (automatic)
6 6 1 0 0 0
```

☞ **Visualizing structure from scf.in:** By using the XCrySDen software to open the `scf.in` file, the readers can see the structure of the monolayer MoS_2, as shown in Fig. 3.22.

For the input file `nscf.in`, the readers can see it by typing as follows:

```
$ vi nscf.in
```

QE-SSP/mos2/optic/nscf.in

```
1  &CONTROL
2  calculation      = 'nscf'
3  pseudo_dir       = '../pseudo/'
4  outdir           = '../tmp/'
5  prefix           = 'mos2'
6  /
7  &SYSTEM
8  ibrav            = 4
9  a                = 3.1378055413
10 c                = 20.0
11 nat              = 3
12 ntyp             = 2
13 nbnd             = 40
14 ecutwfc          = 80.0
15 nosym            = .true.
16 noinv            = .true.
17 /
18 &ELECTRONS
19 mixing_beta      = 0.7
20 conv_thr         = 1.0d-10
21 /
22 ATOMIC_SPECIES
23 Mo   95.94    Mo.pz-hgh.UPF
24 S    32.065   S.pz-hgh.UPF
25 ATOMIC_POSITIONS (crystal)
26 Mo   -0.0000000000   -0.0000000000    0.5000000000
27 S     0.3333333333    0.6666666667    0.4218571774
28 S     0.3333333333    0.6666666667    0.5781427066
29 K_POINTS (automatic)
30 24 24 1 0 0 0
```

☞**Explanation of nscf.in:** It is noted that `epsilon.x` does not support the reduction of the ***k***-points grid into the irreducible Brillouin zone.[9] Therefore, the non-SCF calculation must be performed with a uniform ***k***-points grid, and all ***k***-points weights must be equal to each other. Since the auto-symmetrization of ***k***-points grid in the namelist `K_POINTS (automatic)` can produce a non-uniform distribution of ***k***-points weights, we need to set `nosym = .true.` and `noinv = .true.` (lines 16 and 17), respectively, in the namelist `SYSTEM` to disable the auto-symmetrization.

[9] Irreducible Brillouin zone is the first Brillouin zone, which is reduced by all symmetries in the point group of the lattice.

☞**Note:** For the non-SCF calculations, the ***k***-points grid should be checked carefully to obtain the convergence of the dielectric function. In particular, for graphene, an extremely high density of ***k***-points is needed to obtain accuracy near the Dirac point (for low transition energies) because the Fermi surface is reduced to a dot. A ***k***-points grid of $600 \times 600 \times 1$ might be possible to get a good result. It is important to note that the default maximum number of the ***k***-points in Quantum ESPRESSO is 40,000. In order to run with the number of ***k***-points $>40,000$, the readers need to change the variable "npk" in `Modules/parameters.f90` and recompile the Quantum ESPRESSO package.

For the input file `epsilon.in`, the readers can open as follows:

```
$ vi epsilon.in
```

QE-SSP/mos2/optic/epsilon.in

```
1  &INPUTPP
2  outdir      = '../tmp/'
3  prefix      = 'mos2'
4  calculation = 'eps'
5  /
6  &ENERGY_GRID
7  smeartype   = 'gauss'
8  intersmear  = 0.15
9  intrasmear  = 0.0
10 wmin        = 0.0
11 wmax        = 10.0
12 nw          = 500
13 shift       = 0.0
14 /
```

☞**Explanation of epsilon.in:** The dielectric function $\varepsilon(\omega)$ is calculated by using the keyword `calculation = 'eps'` (line 4) in the namelist `INPUTPP`. The type of broadening of syntax `smeartype` (line 7) can be `gauss` or `lorentz` for a Gaussian or Lorentzian broadening, respectively. `intersmear` and `intrasmear` (lines 8 and 9), respectively, are the broadening parameters in eV for the interband and intraband transitions, respectively. It is noted that we do not use `intrasmear` for the semiconductor, but it is mandatory for a metal, like graphene. The readers should try several values for `intersmear` and `intrasmear` to find a best-converged value. A good value should be between 0.1 and 0.2 eV. [`wmin`, `wmax`] is the range of values for the photon energy (in units of eV), which is given by $\hbar\omega$, where $\hbar$ is the reduced Planck constant and ω is the angular frequency of light. `nw` (line 12) is the number of points of the photon energy mesh. Finally, `shift` (line 13) is an optional rigid shift (in units of eV) of the imaginary part of the dielectric function. `shift` is used when we want to correct the calculated energy band gap with an experimental value.

☞**Note:** `outdir` and `prefix` for `epsilon.in` must be the same as that for `scf.in` and `nscf.in`.

❑ **Output file:** There are four output files `epsr_mos2.dat`, `epsi_mos2.dat`, `eels_mos2.dat`, and `ieps_mos2.dat`. The first and second files contain the real $\text{Re}(\varepsilon(\omega))$ and imaginary $\text{Im}(\varepsilon(\omega))$ parts of diagonal part of the dielectric tensor, respectively, as a function of photon energy $\hbar\omega$ in eV. The third file contains the electron energy loss spectra, $\text{Im}(1/\varepsilon(\omega))$, calculated from the diagonal elements of the dielectric tensor. The last file contains the diagonal components of the dielectric tensor, $\varepsilon(i\omega)$, which is calculated on the imaginary axis of frequency. Let us open the file `epsr_mos2.dat` as follows: ❑ **Output file:** There are four output files `epsr_mos2.dat`, `epsi_mos2.dat`, `eels_mos2.dat`, and `ieps_mos2.dat`. The first and second files two contain the real $\text{Re}(\varepsilon(\omega))$ and imaginary $\text{Im}(\varepsilon(\omega))$ parts of diagonal part of the dielectric tensor, respectively, as a function of photon energy $\hbar\omega$ in eV. The third file contains the electron energy loss spectra, $\text{Im}(1/\varepsilon(\omega))$, calculated from the diagonal elements of the dielectric tensor. The last file contains the diagonal components of the dielectric tensor, $\varepsilon(i\omega)$, which is calculated on the imaginary axis of frequency. Let us open the file `epsr_mos2.dat` as follows:

```
$ vi epsr_mos2.dat
```

QE-SSP/gr/optic/epsr_mos2.dat

```
# energy grid [eV]     epsr_x   epsr_y   epsr_z
#
0.000000000  4.485200865  4.485200862  2.930873960
0.020040080  4.485316552  4.485316549  2.930898420
0.040080160  4.485663685  4.485663682  2.930971808
```

```
...
```

- Column 1: The photon energies, $\hbar\omega$, in units of eV.
- Columns 2 to 4: The real parts of the dielectric function, $\mathrm{Re}(\varepsilon_{xx}(\omega))$, $\mathrm{Re}(\varepsilon_{yy}(\omega))$, and $\mathrm{Re}(\varepsilon_{zz}(\omega))$ in the x-, y-, and z-directions, respectively.

Other output files can be opened in a similar way.

☞ **Note:** For the 2D material, we must reduce the value of $\mathrm{Re}(\varepsilon(\omega))$ and $\mathrm{Im}(\varepsilon(\omega))$ by a dimensionless c/L, where c is the height of the unit cell including the vacuum region and L is the real thickness of the 2D material. For the monolayer MoS_2, c is 20 Å in `scf.in` and $L = 6.5$ Å is given by the experiment [Eda *et al.* (2011)].

☞ **Plotting data from epsr_mos2.dat and epsi_mos2.dat:** In order to plot the dielectric function and the absorption coefficient, the readers can run the JupyterLab `eps.ipynb`, which reads both real and imaginary parts of the dielectric function:

```
$ jupyter-lab eps.ipynb
```

QE-SSP/mos2/optic/eps.ipynb

```
# Import the necessary packages and modules
import matplotlib.pyplot as plt
plt.style.use('../../matplotlib/sci.mplstyle')
import numpy as np
# Set parameter
c = 299792458 # velocity of light (m/s)
hbar = 6.582119569e-16 # reduced Planck constant (eV
    .s)
l = 2.0/0.65 # to reduce thinkness the unit cell (2
    nm) to real thickness (0.65 nm)
# Load the real part of dielectric tensor
ener, repsx, repsy, repsz = np.loadtxt('epsr_mos2.
    dat', unpack=True)
# Load the imaginary part of dielectric tensor
ener, iepsx, iepsy, iepsz = np.loadtxt('epsi_mos2.
    dat', unpack=True)
# Absorption coefficient in x-, y-, z-directions
alphax = 2*(ener/hbar)*np.sqrt((np.sqrt((l*repsx)
    **2+(l*iepsx)**2)-l*repsx)/2)/c
alphay = 2*(ener/hbar)*np.sqrt((np.sqrt((l*repsy)
    **2+(l*iepsy)**2)-l*repsy)/2)/c
```

```
alphaz = 2*(ener/hbar)*np.sqrt((np.sqrt((1*repsz)
     **2+(1*iepsz)**2)-1*repsz)/2)/c
# Create figure object
fig, (ax1, ax3) = plt.subplots(1, 2,
     constrained_layout=True, figsize=(10, 4))
ax2 = ax1.twinx()
# Plot the epsilon
ax1.plot(ener, 1*repsx, 'b-')
ax2.plot(ener, 1*iepsx, 'r--')
# Add the x and y-axis labels
ax1.set_xlabel('Photon energy (eV)')
ax1.set_ylabel(r'Re$(\varepsilon_{xx})$', color='b')
ax1.tick_params(axis='y', labelcolor='b')
ax2.set_ylabel(r'Im$(\varepsilon_{xx})$', color='r')
ax2.tick_params(axis='y', labelcolor='r')
# Set the axis limits
ax1.set_xlim(0, 6)
ax1.set_ylim(0, 36)
ax2.set_ylim(0, 36)
# Plot the abosorption coefficient
ax3.plot(ener, alphax/10**8, c='k')
# Add the x and y-axis labels
ax3.set_xlabel('Photon energy (eV)')
ax3.set_ylabel(r'$\alpha_{xx}\times 10^8$ (1/m)')
# Set the axis limits
ax3.set_xlim(0, 6)
ax3.set_ylim(0, 1.4)
# Save the figure
plt.savefig('plot-optic.pdf')
# Show the figure
plt.show()
```

By running eps.ipynb, we obtain the in-plane optical properties of monolayer MoS_2, as shown in Fig. 3.23. In Fig. 3.23 (a), we plot the real part and imaginary part of isotropic dielectric function in the in-plane with $\varepsilon_{xx} = \varepsilon_{yy}$, in which x- and y-directions correspond to zigzag and armchair directions, as shown in Fig. 3.22. The dielectric function shows two peaks A and B (see Fig. 3.23 (a)) around 2 eV, which correspond to transition energies at the K point of the Brillouin zone [Li *et al.* (2014)].[10] We highly recommend readers to calculate the electronic energy dispersion of the monolayer MoS_2 (see the

[10] Quantum ESPRESSO does not support the calculation of exciton energies. Nevertheless, peaks A and B show excitation energy. This is because the correction to the transition energy from the exciton binding energy is largely offset by many-body corrections to the band gap in the DFT. Therefore, the predicted transition energy within a single-particle calculation is closer to the experiment than might be expected [Li *et al.* (2014)].

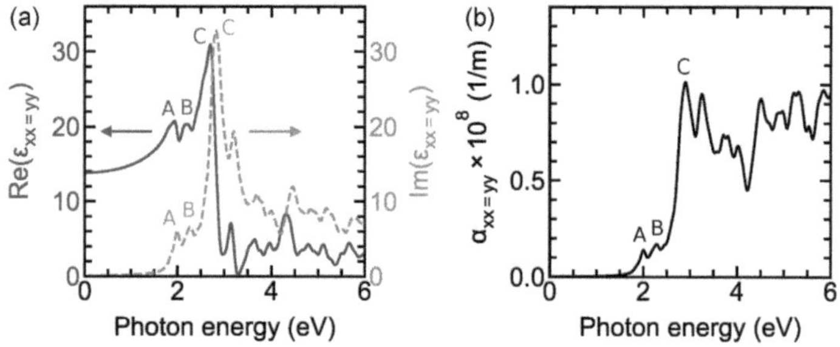

Figure 3.23 In-plane optical properties of monolayer MoS_2. (a) Real part $\text{Re}(\varepsilon_{xx})$ and imaginary part $\text{Im}(\varepsilon_{xx})$ of the dielectric tensor are plotted by the solid and dashed lines, respectively. (b) Absorption coefficient in x-direction, $\alpha_{xx=yy}$, of the monolayer MoS_2. It is noted that $\varepsilon_{xx} = \varepsilon_{yy}$ and $\alpha_{xx} = \alpha_{yy}$. The peaks labeled A and B correspond to transition energies at the K point of the Brillouin zone, and C peak corresponds to higher-lying interband transitions, including the transitions between the K and Γ points.

tutorial in Sec. 3.2.2). By checking the energy dispersion, the readers will see that the transitions occur at the K point for the photon energy of 2 eV. For the peak C in Fig. 3.23 (a), the transitions occur between the K and Γ points in the Brillouin zone. Based on $\text{Re}(\varepsilon_{xx})$ and $\text{Im}(\varepsilon_{xx})$, we can obtain the in-plane absorption coefficient α_{xx} by using the Eq. 5.32, as shown in Fig. 3.23 (b). It is noted that $\varepsilon_{xx} = \varepsilon_{yy}$ and $\alpha_{xx} = \alpha_{yy}$. The calculated α_{xx} is consistent with the experiment, in which $\alpha_{xx} \sim 0.25 \times 10^8$ 1/m for the peaks A and B, and $\alpha_{xx} \sim 1.0 \times 10^8$ 1/m for the peak C [Liu *et al.* (2020)].

Try It Yourself

1. Change the values of `nbnd` and the number of ***k***-points grid in `nscf.in` to check the convergence of the dielectric function of monolayer MoS_2.
2. Change the values of `intersmear` and check the convergence of the dielectric function of monolayer MoS_2.
3. Calculate absorption coefficient of graphene. It is noted that, for graphene, since the Fermi surface is reduced to a dot at low transition energies, an extremely high density of ***k***-points in `nscf.in` have to be used to achieve accurate results near the Fermi level.

3.4.2 Joint density of states

❑ **Purpose:** In this tutorial, we calculate the joint density of states (JDOS) to understand the peaks of the absorption spectra in Sec. 3.4.1.

❑ **Background:** As discussed in Sec. 5.9, the JDOS is the number of states for a pair of the initial and final states by given energy. Therefore, a large peak in the JDOS corresponds to the peak in the absorption spectra.[11] The JDOS is defined by Eq. (5.34), and the JDOS is obtained by using the executable `epsilon.x`.

❑ **How to run:** To run this tutorial, the readers should type the following command lines:

```
1 $ cd ~/QE-SSP/mos2/optic/
2 $ mpirun -np 4 pw.x < scf.in > scf.out &
3 $ mpirun -np 4 pw.x < nscf.in > nscf.out &
4 $ mpirun -np 4 epsilon.x < epsilon-jdos.in > epsilon
      -jdos.out &
```

- Line 1: Go to `optic` directory.
- Line 2: Run SCF calculation by using `pw.x`.
- Line 3: Run non-SCF calculation by using `pw.x`.
- Line 4: Run JDOS calculation by using `epsilon.x`

☞ **Note:** The SCF and non-SCF calculations can be omitted if the readers have successfully finished the tutorial in Sec. 3.4.1.

❑ **How to check:** When the calculations finish, a message `JOB DONE` is written at the end of each output file `scf.out`, `nscf.out`, and `epsilon-jdos.out`.

❑ **Input file:** The input files `scf.in` and `nscf.in` are shown in Sec. 3.4.1. For the input file `epsilon-jdos.in`, the readers can see by typing as follows:

```
$ vi epsilon-jdos.in
```

[11] The optical absorption intensity is proportional to the product of JDOS and the square of the matrix element of the electron-photon matrix element.

QE-SSP/mos2/optic/epsilon-jdos.in

```
1  &INPUTPP
2  outdir      = '../tmp/'
3  prefix      = 'mos2'
4  calculation = 'jdos'
5  /
6  &ENERGY_GRID
7  smeartype   = 'gauss'
8  intersmear  = 0.15
9  intrasmear  = 0.0
10 wmin        = 0.0
11 wmax        = 10.0
12 nw          = 500
13 shift       = 0.0
14 /
```

☞**Explanation of epsilon-jdos.in:** The JDOS calculation is performed by using the keyword `calculation = 'jdos'` (line 4) in the namelist `INPUTPP`. The remaining part is same as Sec. 3.4.1.

❑ **Output file:** The executable `epsilon.x` produces the file `jdos_mos2.dat`, which contains the photon energy in the first column in eV and the JDOS in the second column in states/eV/unit-cell.

☞ **Plotting data from jdos_mos2.dat:** In order to plot the JDOS of the monolayer MoS_2, the readers can run the JupyterLab `jdos.ipynb` as follows:

```
$ jupyter-lab jdos.ipynb
```

QE-SSP/mos2/optic/jdos.ipynb

```
# Import the necessary packages and modules
import matplotlib.pyplot as plt
plt.style.use('../../matplotlib/sci.mplstyle')
import numpy as np
# Load data
ener, jdos = np.loadtxt('jdos_mos2.dat', unpack=True
    )
# Create figure object
plt.figure()
# Plot the JDOS
plt.plot(ener, jdos, c='b')
```

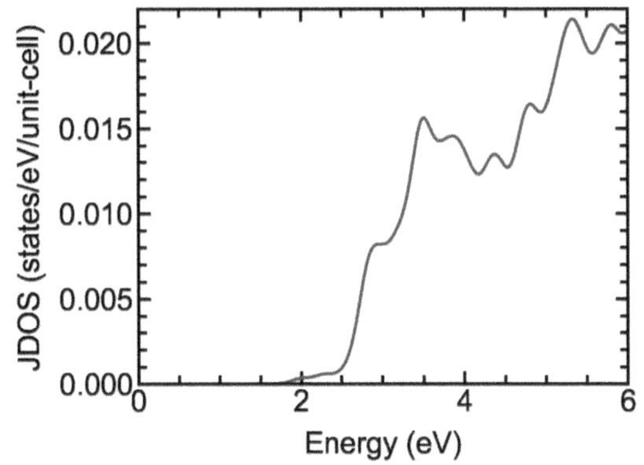

Figure 3.24 Joint density of states (JDOS) of monolayer MoS_2.

```
# Add the x and y-axis labels
plt.xlabel('Energy (eV)')
plt.ylabel('JDOS (1/eV/unit-cell)')
# Set the axis limits
plt.xlim(0, 6)
plt.ylim(0, 0.022)
# Save the figure
plt.savefig('plot-jdos.pdf')
# Show the figure
plt.show()
```

In Fig. 3.24, we show the JDOS of the monolayer MoS_2. Since the peaks A and B of the absorption coefficient in Fig. 3.23 (b) are relatively weak, we can not see the corresponding peaks in the JDOS. Nevertheless, we can see the peak in the JDOS around 3 eV, which corresponds to the peak C in Fig. 3.23 (b).

Try It Yourself

Change the values of `nbnd` and the number of ***k***-points grid in `nscf.in` to check the convergence of the JDOS of monolayer MoS_2.

3.4.3 Non-resonant Raman spectra

❑ **Purpose:** The Raman spectroscopy is a common spectroscopic technique, which is used to determine phonon modes of a solid. As discussed in Sec. 5.13, the peak positions of Raman spectra correspond to the optical phonon frequency of the Raman active mode. In this tutorial, we show how to plot the non-resonant Raman spectra of monolayer MoS_2.

❑ **Background:** The Raman calculation in Quantum ESPRESSO is implemented based on second-order response of DFT developed by Lazzeri and Mauri [Lazzeri and Mauri (2003)]. The non-resonant Raman intensities are given by Eq. (5.53) in Sec. 5.13, which can be obtained by using the executable `dynmat.x` in Quantum ESPRESSO. It is noted that the calculation is only implemented for norm-conserving pseudopotentials and LDA functional. If the material is a metal, the dielectric function will become infinite, which is not suitable for non-resonant Raman calculation, in which we use the dielectric function. Thus, the non-resonant Raman calculation works only for the semiconductor and insulator systems.

❑ **How to run:** To run this tutorial, the readers should type the following command lines:

```
1 $ cd ~/QE-SSP/mos2/raman/
2 $ mpirun -np 4 pw.x < scf.in > scf.out &
3 $ mpirun -np 4 ph.x < ph.in > ph.out &
4 $ mpirun -np 4 dynmat.x < dynmat.in> dynmat.out &
```

- Line 1: Go to `raman` directory.
- Line 2: Run SCF by using `pw.x` with the input file `scf.in`.
- Line 3: Run phonon by using `ph.x` with the input file `ph.in`.
- Line 4: Run `dynmat.x` with the input file `dynmat.in`.

❑ **How to check:** When the calculations finish, a message `JOB DONE` is written at the end of each output file `scf.out`, `ph.out`, and `dynmat.out`.

❑ **Input file:** The input file `scf.in` is shown in Sec. 3.4.1. It is noted that the option `assume_isolated = '2D'` (not shown here) is added to `scf.in` in the namelist `SYSTEM` for the phonon calculation of the 2D materials (see Sec. 3.3.1). For the input file `ph.in`, the readers can open with `vi` editor as follows:

```
$ vi ph.in
```

QE-SSP/mos2/raman/ph.in

```
1  phonon calc.
2  &INPUTPH
3  outdir     = '../tmp/'
4  prefix     = 'mos2'
5  fildyn     = 'mos2.dmat'
6  tr2_ph     = 1d-14
7  lraman     = .true.
8  epsil      = .true.
9  trans      = .true.
10 asr        = .true.
11 /
12 0.0 0.0 0.0
```

☞**Explanation of ph.in:** The non-resonant Raman calculation is performed by setting `lraman = .true.` (line 7) in the namelist `&INPUTPH`. We also have to set `epsil = .true.` (line 8) and `trans = .true.` (line 9) to calculate the effective charges. We apply the acoustic sum rule (see Sec. 3.3.1) for the dynamical matrix by setting `asr = .true.` (line 10). Since we only consider the first-order Raman scattering, only the Γ point (`0.0 0.0 0.0`) (line 12) is calculated.

For the input file `dynmat.in`, the readers can see by typing as follows:

```
$ vi dynmat.in
```

QE-SSP/mos2/raman/dynmat.in

```
&INPUT
fildyn = 'mos2.dmat'
asr    = 'crystal'
/
```

❑ **Output file:** The output file containing the Raman intensities is `dynmat.out`. The readers can use `vi` editor to open the output as follows:

```
$ vi dynmat.out
```

QE-SSP/mos2/raman/dynmat.out

```
   IR activities are in (D/A)^2/amu units
   Raman activities are in A^4/amu units
   multiply Raman by 0.256128 for Clausius-Mossotti
        correction

# mode [cm-1]   [THz]       IR        Raman   depol.fact
1     -0.00   -0.0000   0.0000     0.0018     0.7500
2     -0.00   -0.0000   0.0000     0.2344     0.7500
3      0.00    0.0000   0.0000     0.2361     0.7500
4    283.62    8.5028   0.0000     0.0008     0.7500
5    283.62    8.5028   0.0000     0.0008     0.7500
6    384.72   11.5335   1.2891   178.7551     0.7500
7    384.72   11.5335   1.2891   178.7551     0.7500
8    403.67   12.1018   0.0000   176.8187     0.1061
9    464.86   13.9363   0.0045     0.0000     0.7500
```

For lines 6 to 14:

- Column 1: The ordering numbers of photon modes.
- Column 2: The phonon frequencies in units of cm^{-1}.
- Column 3: The infrared (IR) intensities.
- Column 4: The Raman intensities.
- Column 5: The depolarization ratio factor.[12]

☞ **Plotting data from dynmat.out:** The Raman spectra $I(\omega)$ can be plotted as a smoothed function of the frequency ω by fitting the Gaussian function, which is defined by

$$I(\omega) = I_0 \exp\left[-\left(\frac{\omega - \omega_0}{\Gamma}\right)^2\right], \tag{3.22}$$

where ω_0 and I_0 are the phonon frequencies and Raman intensities, respectively, which are obtained from the `dynmat.out` file. Γ is a broadening parameter, which can be obtained by the full width at half maximum (FWHM) in the experiment.[13] We run the JupyterLab `raman.ipynb` to plot the Raman spectra as follows:

[12] Depolarization ratio is the intensity ratio of the perpendicular component to the parallel component for the Raman peak intensity.

[13] FWHM is inversely proportional to phonon lifetime. A typical value of the FWHM is 10 cm^{-1}, whose phonon lifetime is 2 ps.

```
$ jupyter-lab raman.ipynb
```

QE-SSP/mos2/raman/raman.ipynb

```
# Import the necessary packages and modules
import matplotlib.pyplot as plt
plt.style.use('../../matplotlib/sci.mplstyle')
import numpy as np
# Define the Gaussian function
def gaussian(w, G, w0, I0):
    return I0*np.exp(-((w-w0)/G)**2)
# Peaks list of non-resonant Raman in dynmat.out
peak =[(384.72,178.7551),(403.67,176.8187)]
# Fitting with the Gaussian function
def fit(w, G):
    fit = gaussian(w, G, 0, 0)
    for w0, I0 in peak:
        fit = fit + gaussian(w, G, w0, I0)
    return fit
w = np.linspace(300, 500, 500)
# Create figure object
plt.figure()
# Plot the non-resonant Raman
plt.plot(w, fit(w,2), c='b')
# Add the x and y-axis labels
plt.xlabel('Raman shift (cm$^{-1}$)')
plt.ylabel('Intensity (a.u.)')
# Hide y-axis minor ticks
plt.tick_params(axis='y', which='both', right=False,
      left=False, labelleft=False)
plt.tick_params(axis='x', which='both',top=False)
# Set the axis limits
plt.xlim(370, 420)
# Save the figure
plt.savefig('plot-raman.pdf')
# Show the figure
plt.show()
```

In Fig. 3.25, we plot the non-resonant Raman spectra of monolayer MoS_2. The calculated result shows the two peaks at 385 cm^{-1} and 404 cm^{-1}, which correspond to the E^1_{2g} and A_{1g} Raman active modes [Li *et al.* (2012)]. These peaks have the same intensity and it is consistent with the experimentally observed Raman spectra ($\sim$ 384 cm^{-1} and $\sim$ 403 cm^{-1} for the E^1_{2g} and A_{1g} modes, respectively [Li *et al.* (2012)]).

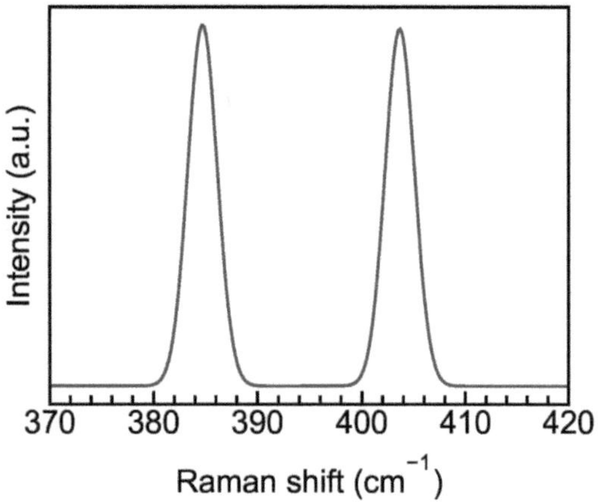

Figure 3.25 Non-resonant Raman spectra of monolayer MoS_2.

Try It Yourself

1. Change lattice constant a in the `scf.in` of monolayer MoS_2, and plot the Raman spectra as a function of the lattice constant.
2. Calculate non-resonant Raman spectra of bulk Si.

3.5 Subjects for two-dimensional materials

This section will calculate three important tutorials for two-dimensional (2D) materials: the spin-orbit coupling in the monolayer MoS_2, the van der Waals interaction of bilayer graphene, and an external electric field to bilayer graphene. These calculations can modify the electronic energy dispersion and the optimized structure of the materials.

3.5.1 Spin-orbit coupling

❑ **Purpose:** In Sec. 3.4, the monolayer MoS_2 is calculated without spin-orbit coupling[14] (SOC). However, it is observed that the SOC leads to the band splitting in the valence bands at the K points [Roch *et al.* (2019)] for the monolayer MoS_2. In this tutorial, we show how to calculate the band splitting of the monolayer MoS_2 in the presence of the SOC.

❑ **Background:** Since the SOC is a relativistic effect, we need to use the fully relativistic pseudopotentials (PPs) for SOC calculation. As shown in Sec. 3.1.6, the fully relativistic PPs can be obtained from the PSlibrary (`https://www.quantum-espresso.org/pseudopotentials/ps-library`). In this tutorial, we select `Mo.rel-pbe-spn-rrkjus_psl.1.0.0.UPF` and `S.rel-pbe-nl-rrkjus_psl.1.0.0.UPF` for the Mo and S atoms, respectively.

☞ **Note**: Some calculation does not support the fully relativistic PPs, for example, the optimizing structure (`relax` or `vc-relax`). Therefore, the readers should optimize the structure by using scalar relativistic PPs, such as `Mo.pbe-spn-rrkjus_psl.1.0.0.UPF` and `S.pbe-nl-rrkjus_psl.1.0.0.UPF` for the Mo and S atoms, respectively. Then, replacing the scalar relativistic PPs with the fully relativistic PPs for the SOC calculation.

❑ **How to run:** To run this tutorial, the readers should type the following command lines:

```
1 $ cd ~/QE-SSP/mos2/soc/
2 $ mpirun -np 4 pw.x < scf.in > scf.out &
3 $ mpirun -np 4 pw.x < nscf.in > nscf.out &
4 $ mpirun -np 4 bands.x < bands.in> bands.out &
```

- Line 1: Go to `soc` directory.
- Line 2: Run SCF calculation by using `pw.x`.
- Line 3: Run non-SCF calculation by using `pw.x`.
- Line 4: Run `bands.x` with the input file `bands.in`.

---

14 The spin-orbit coupling is a relativistic interaction between a spin and an orbital angular momentum of an electron.

❑ **How to check:** When the calculations finish, a message `JOB DONE` is written at the end of each output file `scf.out`, `nscf.out`, and `bands.out`.

❑ **Input file:** For the input file `scf.in`, the readers can open with `vi` editor as follows:

```
$ vi scf.in
```

QE-SSP/mos2/soc/scf.in

```
1  &CONTROL
2  calculation      = 'scf'
3  pseudo_dir       = '../pseudo/'
4  outdir           = '../tmp/'
5  prefix           = 'mos2'
6  /
7  &SYSTEM
8  ibrav            = 4
9  a                = 3.1825188839
10 c                = 20.0
11 nat              = 3
12 ntyp             = 2
13 ecutwfc          = 60.0
14 noncolin         = .true.
15 lspinorb         = .true.
16 /
17 &ELECTRONS
18 mixing_beta      = 0.7
19 conv_thr         = 1.0d-6
20 /
21 ATOMIC_SPECIES
22 Mo   95.94    Mo.rel-pbe-spn-rrkjus_psl.1.0.0.UPF
23 S    32.065   S.rel-pbe-nl-rrkjus_psl.1.0.0.UPF
24 ATOMIC_POSITIONS (crystal)
25 Mo    0.0000000000    0.0000000000    0.5000000000
26 S     0.3333333333    0.6666666667    0.4217548051
27 S     0.3333333333    0.6666666667    0.5782450789
28 K_POINTS (automatic)
29 6 6 1 0 0 0
```

☞**Explanation of scf.in:** The SOC calculation is performed by setting `noncolin = .true.` and `lspinorb = .true.` (lines 14 and 15), respectively. The first setting allows a non-collinear calculation, and the second one allows using a PP with spin-orbit such as fully relativistic PP. In collinear calculation, up- or down-spin is taken into account on the calculation. In non-collinear calculation, the spin direction can be taken in any direction. If all PPs are scalar relativistic, the calculation is non-collinear, but there is no SOC.

For the input file `nscf.in`, the readers can open as follows:

```
$ vi nscf.in
```

QE-SSP/mos2/soc/nscf.in

```
&CONTROL
calculation      = 'bands'
pseudo_dir       = '../pseudo/'
outdir           = '../tmp/'
prefix           = 'mos2'
/
&SYSTEM
ibrav            = 4
a                = 3.1825188839
c                = 20.0
nat              = 3
ntyp             = 2
nbnd             = 60
ecutwfc          = 60.0
noncolin         = .true.
lspinorb         = .true.
/
&ELECTRONS
mixing_beta      = 0.7
conv_thr         = 1.0d-6
/
ATOMIC_SPECIES
Mo   95.94    Mo.rel-pbe-spn-rrkjus_psl.1.0.0.UPF
S    32.065   S.rel-pbe-nl-rrkjus_psl.1.0.0.UPF
ATOMIC_POSITIONS (crystal)
Mo     0.0000000000    0.0000000000    0.5000000000
S      0.3333333333    0.6666666667    0.4217548051
S      0.3333333333    0.6666666667    0.5782450789
K_POINTS (crystal_b)
4
gG 40
K  20
M  30
gG 0
```

☞**Note for nscf.in:** Since we consider both spin-up and spin-down of an electron, the number of Kohn-Sham states (`nbnd`) should be

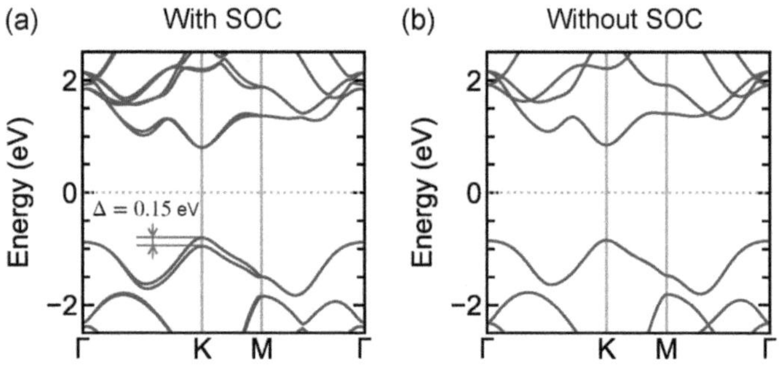

Figure 3.26 Electronic band structure of monolayer MoS_2 (a) with SOC and (b) without SOC.

selected more than twice the number of electrons. Here, we set `nbnd = 60` (line 13) since there are 26 electrons in the monolayer MoS_2, which can be found in the `scf.out` file.

For the input file `bands.in`, the readers can open as follows:

```
$ vi bands.in
```

QE-SSP/mos2/soc/bands.in

```
&BANDS
 outdir  = '../tmp/'
 prefix  = 'mos2'
 filband = 'mos2.bands'
/
```

❑ **Plotting band structure with SOC:** The band structure including the SOC is given by output file `mos2.bands.gnu`. To plot this output file, the readers can run JupyterLab `plot-bands.ipynb` as follows:

```
$ jupyter-lab plot-bands.ipynb
```

It is noted that `plot-bands.ipynb` is obtained from the Python script in page 77 (Sec. 3.2.2) by changing the filename to `mos2.bands.gnu` and the value of the Fermi energy to `-0.0925`. By running `plot-bands.ipynb`, we obtain the band structure of the monolayer MoS_2 with the SOC, as shown in Fig. 3.26 (a). Compared with the band structure without SOC in Fig. 3.26 (b), we can see that the electronic bands split by the spin-orbit interaction that is generally larger in the valence band than in the conduction band. The split bands have the energy difference, Δ, because the two electronic states with spin-up and spin-down are separated. As discussed above, we can not separate spin-up and spin-down in the band structure for non-collinear calculation. The largest value $\Delta = 0.15$ eV is found at the K point for the valence bands, which reproduces the experimental value for bulk MoS_2 ($\Delta = 0.17$ eV [Latzke *et al.* (2015)]).

Try It Yourself

1. Calculate band structure of monolayer MoS_2 without SOC by selecting `noncolin = .false.` and `lspinorb = .false.` with the scalar relativistic PPs. The band structure will be obtained as shown in Fig. 3.26 (b).
2. Calculate the spin-orbit splitting (Δ) of monolayer WSe_2, and discuss the reason why Δ of monolayer WSe_2 is larger than that of monolayer MoS_2?

3.5.2 Van der Waals interaction

❑ **Purpose:** Van der Waals (vdW) interaction is a weak attractive interaction between induced dipole moments that occurs even where there is no overlap between electron densities. Therefore, the vdW interaction is important for stacking layered materials. In this tutorial, we check the effect of the vdW interaction by calculating the interlayer distance of the bilayer graphene. We use the bilayer graphene with the AB-stacking (Bernal) structure, as shown in Fig. 3.27.

❑ **Background:** The vdW energy should be included in the total energy calculated by the exchange-correlation functional. However, the exchange-correlation functionals, such as LDA and GGA, do not describe the vdW energy since the LDA and GGA assume overlapping

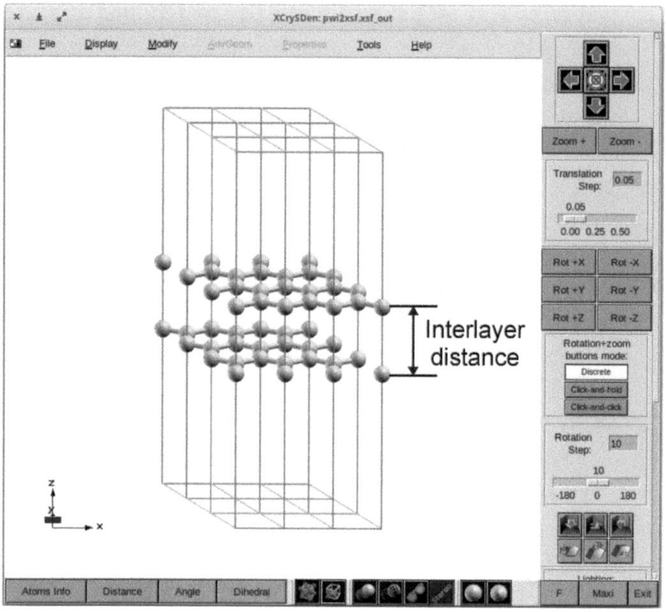

Figure 3.27 Atomic structure of bilayer graphene with the AB-stracking in a $3 \times 3 \times 1$ supercell.

electron densities (see Sec. 3.1.1). Therefore, several methods have been developed to solve this problem. Each method has a different way of calculating vdW energy from others. Details of each method are listed as follows:

DFT-D: It is empirically known that the vdW energy is described by damped inter-atomic potential, which is proportional to $1/R^6$, where R is the inter-atomic distance [Grimme (2004)]. Thus, a simple method is that we add an energy term with $1/R^6$ into the total energy as

$$E_{\text{DFT-D}} = E_{\text{DFT}} + E_{\text{vdW}}, \tag{3.23}$$

where E_{vdW} denotes the vdW energy. E_{vdW} is given by

$$E_{\text{vdW}} = -\frac{s_6}{2} \sum_{I \neq J} \frac{C_6^{IJ}}{R_{IJ}^6} f_d(R_{IJ}), \tag{3.24}$$

where s_6 is a global scaling factor depending on the specific GGA and C_6^{IJ} denotes the dispersion coefficient for atom pair IJ. The C_6^{IJ} coefficients are taken by a least-square fitting procedure from the work of Wu and Yang [Wu and Yang (2002)]. $f_d(R_{IJ})$ is a damping function, which is given by [Grimme (2004)]:

$$f_d(R_{IJ}) = \frac{1}{1 + e^{-d(R_{IJ}/R_r - 1)}}, \tag{3.25}$$

where the parameters d and R_r are fitted to experimental or accurate theoretical data. The DFT-D method is an inexpensive and straightforward calculation, but it is not a fully first-principles approach.

DFT-D3: This method is a refined model of the DFT-D method with more parameters and more interaction terms [Grimme *et al.* (2010)].

Tkatchenko-Scheffler (TS): For the TS method, the C_6^{IJ} coefficients in Eq. (3.24) are computed from the ground-state electron density and reference values for the free atoms [Tkatchenko and Scheffler (2009)].

Exchange-dipole model (EDM): The EDM shows another way to obtain the C_6^{IJ} coefficients in Eq. (3.24) based on second-order perturbation theory [Becke and Johnson (2007), Otero-De-La-Roza and Johnson (2012)], which is referred to as exchange-dipole model.

Nonlocal vdW functionals: This method replaces the vdW energy term $E_{\rm vdW}$ in Eq. (3.23) by a nonlocal energy functional of the electron density $n(\boldsymbol{r})$, which is given by a six-dimensional integral as

$$E_{\rm vdW}^{\rm nl}[n(\boldsymbol{r})] = \frac{1}{2} \iint n(\boldsymbol{r})\Phi(\boldsymbol{r},\boldsymbol{r}')n(\boldsymbol{r}')\mathrm{d}\boldsymbol{r}\mathrm{d}\boldsymbol{r}', \tag{3.26}$$

where $\Phi(\boldsymbol{r},\boldsymbol{r}')$ is a function depending on $\boldsymbol{r} - \boldsymbol{r}'$ and the densities n in the vicinity of $\boldsymbol{r}$ and $\boldsymbol{r}'$ [Dion *et al.* (2004)]. Several nonlocal vdW functionals have been proposed, such as vdW-DF [Dion *et al.* (2004), Thonhauser *et al.* (2015)], vdW-DF2 [Lee *et al.* (2010)], vdW-DF3-opt1 [Chakraborty *et al.* (2020)], vdW-DF3-opt2 [Chakraborty *et al.* (2020)], vdW-DF-C6 [Berland *et al.* (2019)], or rVV10 [Sabatini *et al.* (2013)]. A full list of the nonlocal vdW functionals that are supported by Quantum ESPRESSO can be found in Modules/funct.f90.

Since the nonlocal vdW functionals implemented in Quantum ESPRESSO are sufficiently fast and can be calculated at almost the same cost as LDA and GGA, we recommend using the nonlocal vdW functionals (vdW-DF, rVV10, etc.) for calculating the vdW interaction. Therefore, in this tutorial, we consider only the nonlocal vdW functionals to optimize the structure of bilayer graphene.

❑ **How to run:** To run this tutorial, the readers should type the following command lines:

```
1 $ cd ~/QE-SSP/bi-gr/vdw/
2 $ ./run.sh &
```

- Line 1: Go to vdw directory.
- Line 2: Run a bash script file `run.sh`, which generates and runs many jobs with changing the vdW functionals in `scf.in`.

❑ **How to check:** To check whether or not the output file exists, the readers can use the `ls` command by typing:

```
$ ls
```

The `ls` displays a listing of the all files in the `vdw` folder as

```
run.sh                          vc-relax.vdw-df3-opt1.out
vc-relax.pbe.in                 vc-relax.vdw-df3-opt2.in
vc-relax.pbe.out                vc-relax.vdw-df3-opt2.out
vc-relax.rvv10.in               vc-relax.vdw-df-C6.in
vc-relax.rvv10.out              vc-relax.vdw-df-C6.out
vc-relax.vdw-df2.in             vc-relax.vdw-df.in
vc-relax.vdw-df2.out            vc-relax.vdw-df.out
vc-relax.vdw-df3-opt1.in
```

❑ **Input file:** All input files are generated automatically by a bash script file `run.sh` as

QE-SSP/bi-gr/vdw/run.sh

```
#!/bin/bash
# Set a variable vdw for 1 GGA and 6 vdW functionals
```

```
for vdw in pbe vdw-df vdw-df2 vdw-df3-opt1 \
           vdw-df3-opt2 vdw-df-C6 rvv10; do
# Make input file for the vc-relax calculation.
cat > vc-relax.$vdw.in << EOF
&CONTROL
calculation    = 'vc-relax'
pseudo_dir     = '../pseudo/'
outdir         = '../tmp/'
prefix         = 'bi-gr'
etot_conv_thr = 1.0D-5
forc_conv_thr = 1.0D-4
/
&SYSTEM
ibrav          = 4
a              = 2.4639055825
c              = 20.0
nat            = 4
ntyp           = 1
occupations    = 'smearing'
smearing       = 'mv'
degauss        = 0.02
ecutwfc        = 60
input_dft      = '$vdw'
/
&ELECTRONS
mixing_beta    = 0.7
conv_thr       = 1.0D-9
/
&IONS
ion_dynamics   = 'bfgs'
/
&CELL
cell_dynamics = 'bfgs'
press_conv_thr= 0.05
cell_dofree    = '2Dxy'
/
ATOMIC_SPECIES
C 12.0107 C.pbe-n-rrkjus_psl.0.1.UPF
ATOMIC_POSITIONS (crystal)
C  0.000000000  0.000000000  0.412500000
C  0.333333333  0.666666666  0.412500000
C  0.000000000  0.000000000  0.587500000
C  0.666666666  0.333333333  0.587500000
K_POINTS (automatic)
8 8 1 0 0 0
EOF
# Run pw.x for vc-relax calculation.
mpirun -np 4 pw.x <vc-relax.$vdw.in> vc-relax.$vdw.
    out
```

```
51 # End of for loop.
52 done
```

☞ **Explanation of run.sh:** Six nonlocal vdW functionals (vdW-DF, vdW-DF2, vdW-DF-opt1, vdW-DF-opt2, vdW-DF-C6, and rVV10) are calculated in this tutorial. These vdW functionals are controlled by setting `input_dft = $vdw` (line 26) in the namelist `SYSTEM`, in which the values of `$vdw` (line 3) are selected as a loop of `vdw`. For the case of `input_dft = pbe`, the bilayer graphene is calculated by the GGA without the vdW correction. It is noted that, in order to apply other methods DFT-D, DFT-D3, TS, and EMD, `input_dft = $vdw` should be replaced by `vdw_corr = $vdw` with `vdw = dft-d`, `dft-d2`, `ts`, and `emd`, respectively.

☞**Note:** If the readers are using the Quantum ESPRESSO version $<$ 6.5, the readers might get an error in the output as follows:

```
%%%%%%%%%%%%%%%%%%%%%%%%%%%%%%%%%%%%%%%%%%%%%%%%%%%%%
     from read_kernel_table : error #         1
     No \"vdW_kernel_table\" file could be found
%%%%%%%%%%%%%%%%%%%%%%%%%%%%%%%%%%%%%%%%%%%%%%%%%%%%%
```

In order to solve this error, the readers need to run: <Quantum ESPRESSO-PATH>/bin/generate_vdW_kernel_table.x to generate the `vdW_kernel_table` file. For the the Quantum ESPRESSO version $>$ 6.5, the `vdW_kernel_table` is no longer needed for calculation.

❑ **Output file:** The interlayer distance d can be obtained from the optimized structures from the output files `vc-relax.*.out`. In Fig. 3.28, we show $d - d_{\text{exp}}$ for the PBE and the vdW functionals, in which $d_{\text{exp}} = 3.48$ Å is taken by the experimental value of bilayer graphene [Razado-Colambo *et al.* (2018)]. Without the vdW correction (or the PBE case), d is larger than the experimental value $d_{\text{exp}} = 3.48$ Å about 0.77 Å. With the vdW correction, d becomes close to d_{exp}, and the closest value is 0.06 Å of vdW-DF2. We note that even vdW-DF2 shows a good vdW functional for the case of bilayer graphene, there is no strict guideline for the proper use of the vdW functionals. Therefore, for any system, the readers should carefully select several vdW functionals to find the suitable one.

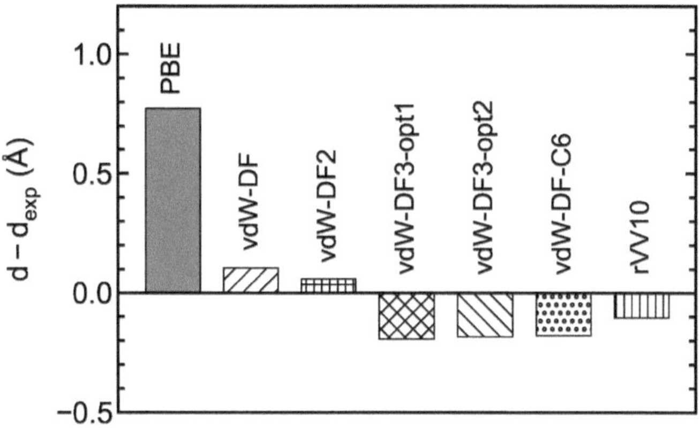

Figure 3.28 Interlayer distance d of bilayer graphene is calculated with PBE and 6 nonlocal vdW functionals. The experimental value $d_{\rm exp}$ of bilayer graphene is 3.48 Å [Razado-Colambo *et al.* (2018)].

Try It Yourself

Calculate the interlayer distance d of the 3D MoS_2 with the vdW correction, and find the suitable vdW functional for the 3D MoS_2. The experimental value of d for MoS_2 is 6.15 Å [Rasamani *et al.* (2017)].

3.5.3 External electric field

❑ **Purpose:** The bilayer graphene in Sec. 3.5.3 has a zero band gap. However, the band gap can be tuned by applying an external electric field perpendicular to the graphene layer, in which a band gap is opened at the Dirac point. In this tutorial, we show how to obtain the electrostatic potential and the band structure of the bilayer graphene under the external electric field.

❑ **Background:** In Quantum ESPRESSO, an external electric field E can be applied for a 2D system by using a saw-tooth potential $\mathcal{V}_{ext}$, as shown in Fig. 3.29. Assuming that the 2D material is set to be in the

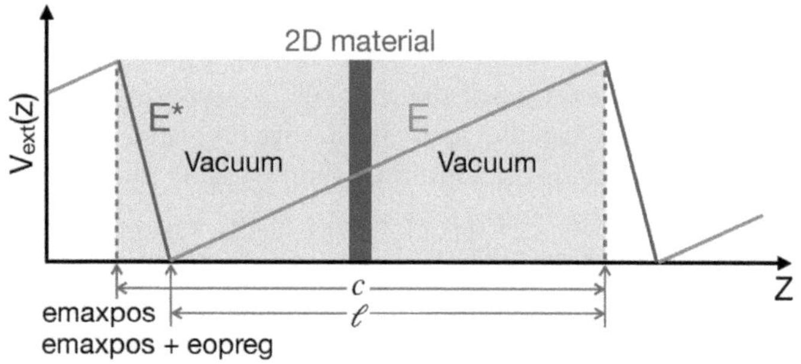

Figure 3.29 Schematics of saw-tooth potential. The dark area represents the 2D material and the light gray spaces are the vacuums in the unit cell. The solid lines with labels E and E^* denote the linear external electric potential and its counterpart, respectively. c is the size of the periodic cell along z direction, and ℓ is the length of external electric potential. `emaxpos` and `eopreg` are input parameters for Quantum ESPRESSO.

xy plane and E is applied in the z direction, $\mathcal{V}_{ext}$ is defined by

$$\mathcal{V}_{ext}(z) = -eEz. \tag{3.27}$$

If $\mathcal{V}_{ext}(z)$ changes monotonically, $\mathcal{V}_{ext}(z)$ would violate periodic boundary conditions. In order to satisfy periodic boundary conditions, we need to change the electric field in some irrelevant region (e.g., in the vacuum far from any surfaces), so that the potential restores its original value on the boundary. In Fig. 3.29, we show the schematic of the saw-tooth potential $\mathcal{V}_{ext}(z)$ in the z direction with the unit cell containing 2D material (shaded area). As shown in Fig. 3.29, the external electric field E changes abruptly in the vacuum. The saw-tooth potential also leads to an artificial field E^*, which is always opposite to E and given by

$$E^* = -E\frac{\ell}{c-\ell}, \tag{3.28}$$

where ℓ and c are the length of the external electric field E and the size of the unit cell in the z direction, respectively.

We note that the unit cell of the 2D material contains a dipole moment in the vacuum space. In particular, when the bottom and top layers of the 2D material are not equivalent to each other, there is a dipole moment between these layers due to the periodic boundary conditions. This dipole moment introduces an artificial electric field in the z direction of the unit cell. In Quantum ESPRESSO, a dipole correction can be applied to suppress the artificial dipole field by introducing an additional ramp-shaped potential $\mathcal{V}_{dip}$. Assuming that the 2D material is set to be in the xy plane, $\mathcal{V}_{dip}$ is given by [Bengtsson (1999)]:

$$\mathcal{V}_{dip}(z) = 4\pi m \left(\frac{z}{c} - \frac{1}{2} \right), \tag{3.29}$$

where m is a surface dipole moment. The dipole correction introduces a jump in electrostatic potential which should be placed within the vacuum region of the cell.

❑ **How to run:** To run this tutorial, the readers should type the following command lines:

```
1 $ cd ~/QE-SSP/bi-gr/elec-field/
2 $ mpirun -np 4 pw.x < scf.in > scf.out &
3 $ mpirun -np 4 pp.x < pp.in > pp.out &
4 $ average.x < average.in > average.out &
5 $ mpirun -np 4 pw.x < nscf.in > nscf.out &
6 $ mpirun -np 4 bands.x < bands.in > bands.out &
```

- Line 1: Go to `elec-field` directory.
- Line 2: Run SCF calculation by using `pw.x`.
- Line 3: Run `pp.x` to obtain the electrostatic potential.
- Line 4: Run `average.x` to obtain the planar-average potential along the z direction from the electrostatic potential. It is noted that this program supports for *only single processor.*
- Line 5: Run non-SCF calculation by using `pw.x`.
- Line 6: Run `bands.x` to obtain the band structure.

❑ **How to check:** When the calculations finish, a message `JOB DONE` is written at the end of each output file `scf.out`, `pp.out`, `average.out`, `nscf.out`, and `bands.out`.

❑ **Input file:** For the input file `scf.in`, the readers can open with `vi` editor as follows:

```
$ vi scf.in
```

QE-SSP/bi-gr/elec-field/scf.in

```
1 &CONTROL
2 calculation    = 'scf'
3 pseudo_dir     = '../pseudo/'
4 outdir         = '../tmp/'
5 prefix         = 'bi-gr'
6 tefield        = .true.
7 dipfield       = .true.
8 /
9 &SYSTEM
10 ibrav          = 4
11 a              = 2.4857910097
12 c              = 20.0
13 nat            = 4
14 ntyp           = 1
15 occupations    = 'smearing'
16 smearing       = 'mv'
17 degauss        = 0.02
18 ecutwfc        = 80
19 input_dft      = 'vdw-df2'
20 edir           = 3
21 emaxpos        = 0
22 eopreg         = 0.05
23 eamp           = 0.0097234527
24 /
25 &ELECTRONS
26 mixing_beta    = 0.7
27 conv_thr       = 1.0D-9
28 /
29 ATOMIC_SPECIES
30 C 12.0107 C.pbe-n-rrkjus_psl.0.1.UPF
31 ATOMIC_POSITIONS (crystal)
32 C    0.0000000000    0.0000000000    0.4115083725
33 C    0.3333333333    0.6666666667    0.4114773543
34 C    0.0000000000    0.0000000000    0.5884916275
35 C    0.6666666667    0.3333333333    0.5885226457
36 K_POINTS (automatic)
37 8 8 1 0 0 0
```

☞**Explanation of scf.in:** The external electric field and the dipole correction are performed by setting `tefield = .true.` and `dipfield = .true.` (lines 6 and 7), respectively, in the namelist `CONTROL`. For the direction of the electric field or dipole correction is parallel to the *z*-direction, we set `edir = 3` (line 20). The atomic structure (lines 32–35) of the bilayer graphene was optimized by using the van der Walls correction (vdW-DF2), as shown in Sec. 3.5.2, in which the positions of the C atoms are located at the center of the unit cell. Thus, as shown in Fig. 3.29, the position of the maximum of the saw-tooth potential will be set to 0 by `emaxpos = 0` (line 21). We set the width of the window, where the saw-tooth potential decreases, about 1 Å by `eopreg = 0.05` (line 22). We note that `emaxpos` and `eopreg` have the unit of $c = 20$ Å. The magnitude of the electric field is set to 0.5 V/Å by `eamp = 0.0097234527` (line 23), in which `eamp` has the unit of a.u. (1 V/Å= 0.0194469054 a.u.).

☞**Note:** For the external electric field, the readers should use a large value of `ecutwfc` (line 18). For example, if the readers use `ecutwfc = 60`, the readers might get an error in `scf.out` as follows:

```
%%%%%%%%%%%%%%%%%%%%%%%%%%%%%%%%%%%%%%%%%%%%%%%%%%%%
     Error in routine electrons (1):
     charge is wrong
%%%%%%%%%%%%%%%%%%%%%%%%%%%%%%%%%%%%%%%%%%%%%%%%%%%%
```

For the input file `pp.in`, the readers can open as follows:

```
$ vi pp.in
```

QE-SSP/bi-gr/elec-field/pp.in

```
1 &INPUTPP
2 outdir   = '../tmp/'
3 prefix   = 'bi-gr'
4 filplot  = 'bi-gr.dat'
5 plot_num = 11
6 /
```

☞**Explanation of pp.in:** In order to plot the electrostatic potential that considers the bare potential (local part of the ionic potential) and Hartree potential, we set `plot_num = 11` (line 5). The name of the output file of the potential is set by `filplot = 'bi-gr.dat'` (line 4).

For the input file `average.in`, the readers can open as follows:

```
$ vi average.in
```

QE-SSP/bi-gr/elec-field/average.in

```
1 1
2 bi-gr.dat
3 1.0
4 300
5 3
6 3.0
```

☞**Explanation of average.in:** The program `average.x` from Quantum ESPRESSO is used to calculate the planar- and macroscopic-average potential from the potential file `bi-gr.dat` that is generated by pp.x. The input variables of `average.in` are given as follows:

- Line 1: The number of files containing the desired quantities.
- Line 2: The name of the file.
- Line 3: The weight of the quantity read from file.
- Line 4: The number of points for the final interpolation of the planar- and macroscopic-average.
- Line 5: The direction (1, 2, and 3 correspond to *x*-, *y*-, and *z*-direction, respectively), in which the planar-average is calculated in the plane orthogonal to this direction.
- Line 6: The size of the window for macroscopic average (a.u.).

For the input file `nscf.in`, the readers can open as follows:

```
$ vi nscf.in
```

QE-SSP/bi-gr/elec-field/nscf.in

```
&CONTROL
calculation     = 'bands'
pseudo_dir      = '../pseudo/'
outdir          = '../tmp/'
prefix          = 'bi-gr'
tefield         = .true.
dipfield        = .true.
/
&SYSTEM
ibrav           = 4
a               = 2.4857910097
c               = 20.0
nat             = 4
ntyp            = 1
```

```
nbnd            = 30
occupations     = 'smearing'
smearing        = 'mv'
degauss         = 0.02
ecutwfc         = 80
input_dft       = 'vdw-df2'
edir            = 3
emaxpos         = 0
eopreg          = 0.05
eamp            = 0.0097234527
/
&ELECTRONS
mixing_beta     = 0.7
conv_thr        = 1.0D-9
/
ATOMIC_SPECIES
C 12.0107 C.pbe-n-rrkjus_psl.0.1.UPF
ATOMIC_POSITIONS (crystal)
C    0.0000000000    0.0000000000    0.4115083725
C    0.3333333333    0.6666666667    0.4114773543
C    0.0000000000    0.0000000000    0.5884916275
C    0.6666666667    0.3333333333    0.5885226457
K_POINTS (crystal_b)
4
gG  40
K   20
M   30
gG  0
```

For the input file `nscf.in`, the readers can open as follows:

```
$ vi bands.in
```

QE-SSP/bi-gr/elec-field/bands.in

```
&BANDS
 outdir  = '../tmp/'
 prefix  = 'bi-gr'
 filband = 'bi-gr.bands'
/
```

❑ **Output file:** The planar-average of electrostatic potential is written in `avg.dat`, which is output file of `average.x`. The readers can use `vi` editor to see the `avg.dat` file as following:

```
$ vi avg.dat
```

QE-SSP/bi-gr/elec-field/avg.dat

```
    0.000000000    0.550530876    0.441707911
    0.125981742    0.519365693    0.424317560
    0.251963483    0.477007077    0.405271088
    0.377945225    0.443319530    0.384610570
...
```

- Column 1: The coordinate (a.u) along *z* direction.
- Column 2: The planar-averaged potential in units of Ry.
- Column 2: The macroscopic-averaged potential in units of Ry.

❑ **Plotting data from avg.dat:** The planar-averaged electrostatic potential is plotted by running `plot-potential.ipynb`, which reads the data from the `avg.dat` file:

```
$ jupyter-lab plot-potential.ipynb
```

QE-SSP/bi-gr/elec-field/plot-potential.ipynb

```
# Import the necessary packages and modules
import matplotlib.pyplot as plt
plt.style.use('../../matplotlib/sci.mplstyle')
import numpy as np

# Load data
z, V = np.loadtxt('avg.dat', usecols=(0,1), unpack=True)
# Convert a.u. to Angstrom
au2a = 0.529177249

# Create figure object
plt.figure(figsize=(4.5,4.5))
# Plot the data, using blue color
plt.plot(z*au2a, V, c='b')
# Plot a dashed line at zero
plt.axhline(0, c='gray', ls='--')
# Set the axis limits
plt.xlim(0, 20)
plt.ylim(-3, 1)
# Add the x and y-axis labels
plt.xlabel('z ($\AA$)')
plt.ylabel('Electrostatic potential (Ry)')
```

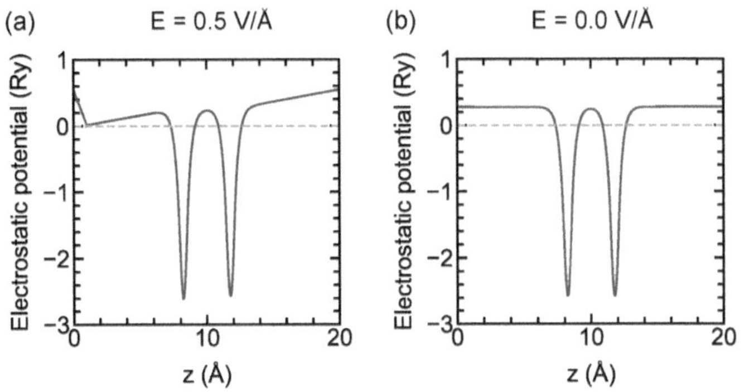

Figure 3.30 Planar-averaged electrostatic potential along the z direction of bilayer graphene with an external electric field E of (a) 0.5 V/Å and (b) 0.0 V/Å.

```
# Save the figure
plt.savefig('plot-V.pdf')
# Show the figure
plt.show()
```

By running `plot-potential.ipynb`, we obtain the planar-averaged electrostatic potential along the z direction of bilayer graphene in the presence of $E = 0.5$ V/Å, as shown in Fig. 3.30 (a). There are two minimum values of the electrostatic potential at $z =$ 8.2 Å and 11.8 Å, which correspond to the positions of each graphene layer. We can see that the saw-tooth potential decreases from $z = 0$ Å to 1 Å and increases from $z = 1$ Å to 20 Å. In contrast, for $E =$ 0.0 V/Å, the electrostatic potential does not change in the vacuum region, as shown in Fig. 3.30 (b). If the readers want to calculate the electrostatic potential without the external electric field ($E = 0.0$ V/Å), the readers can set `eamp = 0.0` (line 23) in the `scf.in` file.

❑ **Plotting data band structure:** The band structure is given by `bi-gr.bands.gnu`, which is the output file of `bands.x`. In a similar way to plot band structure of single-layer graphene in Sec. 3.2.2, the readers can run the file `plot-bands.ipynb` to plot the band structure of the bilayer graphene. In Fig. 3.31 (a) and (b), we show the band structures of the bilayer graphene for $E = 0.5$ V/Å and 0.0 V/Å, respectively. If there is no external electric field (i.e., $E = 0.0$ V/Å), the bilayer graphene has zero-gap, as shown in Fig. 3.31 (b),

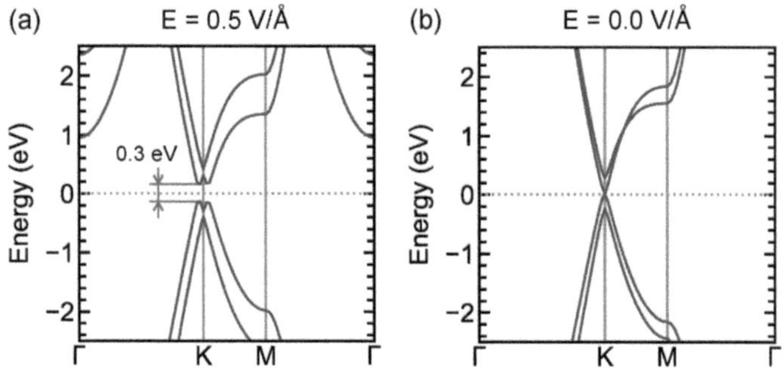

Figure 3.31 Electronic band structures of bilayer graphene with an external electric field E of (a) 0.5 V/Å and (b) 0.0 V/Å. In the presence of E, a band gap is opened at the K point for the bilayer graphene.

which is similar to the case of single-layer graphene. However, if the electric field is applied perpendicular to bilayer graphene, the two layers are no longer identical. Therefore, the two $2p_z$ orbitals around the Fermi energy change the energies since the electric field affects the p_z orbital, but not for the p_x and p_y orbitals, which are parallel to the graphene plane. The band gap thus increases by increasing the external electric field. In particular, the band gap is opened about 0.3 eV for $E = 0.5$ V/Å, as shown in Fig. 3.31 (a). It is noted that the vacuum size will affect the electric field screening, which modifies the energy band gap. Therefore, the readers should consider a sufficiently large vacuum size by changing the value of c in the input files.

Try It Yourself

Plot the planar-average of electrostatic potential with the size of the unit cell $c = 30$, 40, and 50 Å. The positions of the C atoms in the namelist `ATOMIC_POSITIONS` in `scf.in` must rescale for each c value since these postions are in crystal coordinates.

3.6 Maximally-localized Wannier functions

Maximally-localized Wannier functions (MLWFs) are helpful to compute the properties of the materials that require very dense grid of $\boldsymbol{k}$-points and to build a tight-binding model (see Sec. 5.14). In Quantum ESPRESSO, the MLWFs are employed by Wannier90 code [Mostofi *et al.* (2008), Pizzi *et al.* (2020)] (see Chapter 2 for the Wannier90 installation). In this section, we learn how to calculate the MLWFs from Wannier90 and Quantum ESPRESSO (Sec. 3.6.1) and apply the Wannier interpolation for hybrid functional (Sec. 3.6.2).

3.6.1 Wannier functions, energy dispersion, and tight-binding parameters

❑ **Purpose:** In this tutorial, we visually check the Wannier functions (WFs), and we show how to plot energy dispersion and obtain tight-binding parameters based on the MLWFs of graphene.

❑ **Background:** The WFs can be obtained from the Bloch functions by using Eq. (5.59). As discussed in Sec. 5.14, there is a freedom of choice in the Bloch function that corresponds to the shape and the spatial distribution of the WF. We can select any phases of the Bloch function, which modifies the spatial distribution of the WF.

In order to obtain the MLWFs, we minimize the spread of the WF, Ω in Eq. (5.66), by using a steepest-descent algorithm [Marzari *et al.* (2012a)]. After a ground state $\psi_{n\boldsymbol{k}}$ is obtained from the SCF calculation, the Wannier90 code will calculate the MLWFs, which require the following two matrices:

- The overlap matrix $M_{mn}^{\boldsymbol{k},\boldsymbol{b}} = \langle u_{m\boldsymbol{k}}|u_{n,\boldsymbol{k}+\boldsymbol{b}}\rangle$ between the periodic part of the Bloch functions $|u_{n,\boldsymbol{k}+\boldsymbol{b}}\rangle$ (see Eq. (5.72)).
- A starting guess the projection matrix $A_{mn}^{\boldsymbol{k}} = \langle \psi_{m\boldsymbol{k}}|g_n\rangle$ of the Bloch function $\psi_{n\boldsymbol{k}}$ onto trial localized orbitals $|g_n\rangle$ (see Eq. (5.78)).

The calculation of the MLWF of graphene consists of the following five steps:

1. First, we perform the SCF to take the Bloch functions from a ground-state calculation by using `pw.x`. In the case that we need the Bloch functions in a more dense $\boldsymbol{k}$-mesh compared with the previous SCF step, we can run non-SCF after the SCF is finished.

2. It is important to note that Wannier90 requires a full list of $\boldsymbol{k}$-points in the Brillouin zone, while `pw.x` in step (1) uses a reduced $\boldsymbol{k}$-points. Therefore, an executable `open_grid.x` in Quantum ESPRESSO is used to obtain the full grid of $\boldsymbol{k}$-points from the reduced one.
3. Next, we run Wannier90 in post-processing mode (`wannier90.x -pp`) to obtain the `.nnkp` file, which contains the relevant information from the Wannier90 input file in a format to be used in the next step.
4. Then, we run the executable `pw2wannier90.x` (of the Quantum ESPRESSO distribution), which reads the Bloch functions from the SCF (or non-SCF) and the `.nnkp` file to compute $M_{mn}^{\boldsymbol{k},\boldsymbol{b}}$ and $A_{mn}^{\boldsymbol{k}}$ that will be written in the `.mmn` and `.amn` files, respectively.
5. Finally, we run Wannier90 (`wannier90.x`) by reading the files produced by the previous step to minimize the spread to calculate the MLWFs.

❑ **How to run:** To run this tutorial, the readers should type the following command lines:

```
1 $ cd ~/QE-SSP/gr/w90/
2 $ mpirun -np 4 pw.x < scf.in > scf.out &
3 $ mpirun -np 4 open_grid.x < open_grid.in >
     open_grid.out &
4 $ wannier90.x -pp gr &
5 $ mpirun -np 4 pw2wannier90.x < pw2wan.in > pw2wan.
     out &
6 $ wannier90.x gr &
```

- Line 1: Go to `w90` directory.
- Line 2: Run `pw.x` for the SCF calculation for the step (1).
- Line 3: Run `open_grid.x` to unfold the reduced $\boldsymbol{k}$-points onto the full $\boldsymbol{k}$-points for the step (2).
- Line 4: Run Wannier90 post-processing for the step (3). It requires the second input file `gr.win`.
- Line 5: Run Wannier90 interface `pw2wannier90.x` with the input file `pw2wan.in` for the step (4).
- Line 6: Run `wannier90.x` with the input file `gr.win` for the step (5).

☞ **Note**: wannier90.x can only run on a single processor (mpirun can not be used), while pw2wannier90.x can run in parallel, in which the number of processors -np 4 must to be the same as pw.x.

❑ **How to check:** When the calculations finish, you can find a message JOB DONE at the end of scf.out, open_grid.out, and pw2wan.out files, and a message All done: wannier90 exiting at the end of the output file gr.wout. To obtain the final value of the spread of the WFs (see Sec. 5.14.2), the readers can use grep command as follows:

```
$ grep 'Final Spread' gr.wout
```

The total spread is printed in the terminal as

```
Final Spread (Ang^2)        Omega Total  =    3.773324125
```

❑ **Input file:** The input file scf.in is similar to scf.in in Sec. 3.1.1 for graphene, but we add nbnd = 16 in the namelist SYSTEM. For the input file open_grid.in, the readers can open with vi editor as follows:

```
$ vi open_grid.in
```

QE-SSP/gr/w90/open_grid.in

```
&INPUTPP
outdir   = '../tmp/'
prefix   = 'gr'
```

☞ **Note for open_grid.in:** outdir and prefix for open_grid.in must be the same as that for scf.in in the step (1).

For the input file gr.win, the readers can open as

```
$ vi gr.win
```

QE-SSP/gr/w90/gr.win

```
# CONTROL
num_bands           = 16
num_wann            = 5

dis_win_min         = -22
dis_win_max         =  11
dis_froz_min        = -22
dis_froz_max        = 0.5

num_iter            = 20

# TIGHT-BINDING
write_hr            = .true.

# PLOTTING
wannier_plot        = .true.
bands_plot          = .true.

wannier_plot_supercell = 3

begin kpoint_path
G 0.0000 0.0000 0.0000  K 0.3333 0.3333 0.0000
K 0.3333 0.3333 0.0000  M 0.5000 0.0000 0.0000
M 0.5000 0.0000 0.0000  G 0.0000 0.0000 0.0000
end kpoint_path

# SYSTEM
begin unit_cell_cart
ang
  2.463906    0.000000    0.000000
 -1.231953    2.133805    0.000000
  0.000000    0.000000   15.000000
end unit_cell_cart

begin atoms_frac
C  0.333333333  0.666666666  0.500000000
C  0.666666666  0.333333333  0.500000000
end atoms_frac

# PROJECTIONS
guiding_centres  = .true.

begin projections
C: pz
f= 0.5, 0.5, 0.5: s
f= 0.5, 0.0, 0.5: s
f= 0.0, 0.5, 0.5: s
```

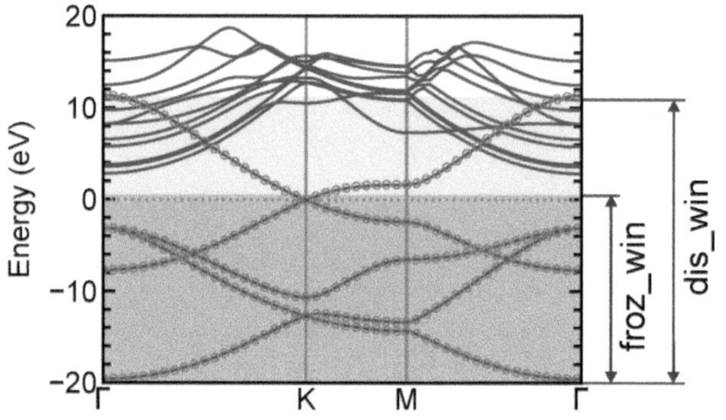

Figure 3.32 An optimized selection for the frozen `froz_win` and the disentanglement `dis_win` for the band structure of graphene. Solid lines are the energy bands that are calculated by the DFT. Solid lines with circles are "fitted energy bands" to the WFs, while only solid lines denote "not-fitted energy bands". `froz_win` should be taken as large as possible with the condition that we should not include "not-fitted bands" (only solid lines). `dis_win` should be taken as small as possible with the condition that we should include all "fitted bands" (lines with circles).

```
48 end projections
49
50 # KPOINTS
51 mp_grid   = 12 12 1
52
53 begin kpoints
54   0.00000000  0.00000000  0.00000000  0.0069444
55   0.00000000  0.08333333  0.00000000  0.0069444
56   ...
57 -0.083333333 -0.08333333  0.00000000  0.0069444
58 end kpoints
```

☞**Explanation of gr.win:** The input variables in `gr.win` are listed in Table 3.10. The `gr.win` file can be divided into 6 parts starting with a hash mark "#". It is noted that the comments in Wannier90 can begin with "#", "%", or "!". Each part is explained as follows:

1. **CONTROL**: This part contains the parameters for minimizing the spread of the WF. These parameters are essential to obtain well-localized WFs. Here, we show how to choose these parameters. First, the number of energy bands `num_bands` (line 2) should be equal to `nbnd = 16` in `scf.in`. Second, the number of the

Table 3.10 Meaning of input variables in `gr.win` file.

Line	Syntax	Meaning
2	`num_bands`	Number of energy bands, which is equal to the number of the Kohn-Sham states `nbnd = 16` in `scf.in`.
3	`num_wann`	Number of the WFs, which is equal to the number of projections.
5	`dis_win_min`	Minimum energy (in eV) of the disentanglement window.
6	`dis_win_max`	Maximum energy (in eV) of the disentanglement window.
7	`dis_froz_min`	Minimum energy (in eV) of the frozen window.
8	`dis_win_max`	Maximum energy (in eV) of the frozen window.
10	`num_iter`	Maximum number of iterations for the minimization of spread.
13	`write_hr`	If `write_hr = .true.`, the code will write the tight-binding Hamiltonian in output file `*_hr.dat`. The default is `.false.`
16	`wannier_plot`	If `wannier_plot = .true.`, the code will write out the Wannier functions in in output files `*.xsf`, which are opened by XCrySDen or VESTA codes. The default is `false`.
17	`bands_plot`	If `bands_plot = .true.`, the code will calculate the band structure, through Wannier interpolation, and write in output file `*_band.dat`. The default is `false`.
19	`wannier_plot_supercell`	Size of the supercell for plotting the WF. The default value is 2.
21–25	`kpoint_path`	Defines the path in ***k***-space along which to calculate the bandstructure.
28–32	`unit_cell_cart`	Lattice vectors in the Cartesian coordinates.

Continued

Table 3.10 – *Continued*

Line	Syntax	Meaning
33	ang	The unit of the lattice vectors (Angstrom).
35–38	atoms_frac	Atomic positions in fractional coordinates.
41	guiding_centres	If guiding_centres = .true., the projection centres are used as the guiding centres during the minimization to avoid local minima. The default is false.
43–48	projections	Defines a set of localized functions used to generate the projections matrix A^{k}_{mn}.
51	mp_grid	Dimensions of the **k**-points grid.
53–58	kpoints	The positions of each **k** point.

WFs num_wann (line 3) should be equal to the number of the projectors, which will be explained in part (5). We can see that num_wann is not equal to num_bands. In order to solve this problem, Souza *et al.* [Souza *et al.* (2001)] introduced the disentanglement method in Wannier90. This method requires the disentanglement dis_win and frozen froz_win energy windows. The optimized way to choose dis_win and froz_win is shown in Fig. 3.32. dis_win is a window to extract the target bands for the WFs. Therefore, dis_win should cover the energy range of all WFs, and dis_win should be as small as possible to reduce the computational cost. For graphene, the WFs include four valence bands and one conduction band since we have five WFs. These bands have the energy range from −20 to 12 eV (see Fig. 3.14). Therefore, after shifting with the Fermi energy (−1.6786 eV, see in scf.out), we can choose dis_win_min = -22 and dis_win_max = 11 in units of eV (lines 5 and 6), respectively. It is noted that these values do not need to be very precise. However, if dis_win is too small, the readers will get an error in gr.wout as follows:

```
Exiting......
dis_windows: Energy window contains fewer states
     than number of target WFs
```

froz_win is a window used for the disentanglement procedure. This window should not contain "not-fitted energy bands", and it should be as large as possible for optimizing the disentanglement procedure. For graphene, the lowest energy of 6-th band at 2.5 eV, as shown in Fig. 3.14. Therefore, after shifting with the Fermi energy, we can choose froz_win_min = -22 and froz_win_max = 0.5 in units of eV (lines 5 and 6), respectively. We recommend to run the energy band structure in Sec. 3.2.2 to estimate the values for the froz_win and dis_win.

Finally, we choose the maximum number of iterations num_iter = 20 for the minimization of spread. Large number of num_iter would mix the WFs such that the output WFs are not atomic like. Therefore, if you want atomic like WFs, it is better to set num_iter less than 20.

2. **TIGHT-BINDING**: By setting write_hr = .true., the tight-binding Hamiltonian in the WF basis (see Sec. 5.14.3) will be written in output file gr_hr.dat.
3. **PLOTTING**: This part contains the parameters for plotting. It is noted that the WF is not periodic in the unit cell. Therefore, it is not generally sufficient to plot it in a single unit cell. wannier_plot_supercell controls the number of units cell in which we construct the WF. For graphene, we select wannier_plot_supercell = 3 (line 19), which is sufficient to give a good picture of a WF. In order to plot the band structure with bands_plot = .true. (line 17), the path in ***k***-space need to list in the block of kpoint_path from line 22 to line 24. Each line gives the start and end points (with labels) for a section of the path. Values are in fractional coordinates with respect to the primitive reciprocal lattice vectors. The number of ***k*** points along each path is given in output file gr_band.labelinfo.dat.
4. **SYSTEM**: This part includes two blocks unit_cell_cart (lines 28–33) and atoms_frac (lines 35–38), in which unit_cell_cart are the lattice vectors (in Cartesian coordinates) and atoms_frac are the atomic positions (in

fractional coordinates). These values can be obtained from `scf.in`. The unit of lattice vectors is defined by `ang` (line 29) for Angstrom or `bohr` for Bohr.

5. **PROJECTIONS**: This part includes the parameters for selecting the initial projections, which need to generate the matrix $A^{\boldsymbol{k}}_{mn}$. For graphene, we project the Bloch functions onto the s and p_z orbitals of the C atoms. The positions of these orbitals are set in the block `projections` from line 44 to line 47. `C: pz` (line 44) means that the p_z orbitals are located at the positions of the C atoms, while s orbitals are located at the center of the C-C bonds. Since we have two C atoms and three C-C bonds in the unit cell, we have the 5 WFs. Therefore, we can set `num_wann = 5` (line 3). It is noted that the option `guiding_centres = .true.` (line 41) is used during the minimization to avoid local minima. For this reason, we recommend to use `guiding_centres = .true.` where the block `projections` is defined.
6. **KPOINTS**: The grid of $\boldsymbol{k}$-points is `mp_grid = 12 12 1` (line 51), which is same as $\boldsymbol{k}$-points in `scf.in`. The full list of $\boldsymbol{k}$-points in the block `kpoints` is copied and pasted from the output file `open_grid.out` in the step (2).

For the input file `pw2wan.in`, the readers can open by

```
$ vi pw2wan.in
```

QE-SSP/gr/w90/pw2wan.in

```
1 &INPUTPP
2 outdir     = '../tmp/'
3 prefix     = 'gr_open'
4 seedname   = 'gr'
5 write_unk  = .true.
6 /
```

☞ **Explanation of pw2wan.in:** The executable `open_grid.x` in the step (2) generates a new folder `gr_open.save` (in the `tmp` directory), to store the wavefunctions of the full $\boldsymbol{k}$-points, while `gr.save` stores the wavefunctions of the reduced $\boldsymbol{k}$-points. Therefore, we must to setting `prefix = 'gr_open'` (line 3). In order to plot the Wannier functions in real space, we need to set `write_unk = .true.` (line

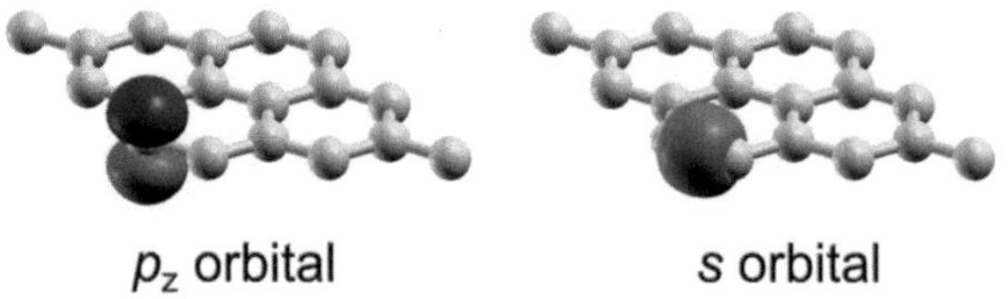

Figure 3.33 The Wannier functions of graphene includes p_z (left) and s (right) orbitals.

5). This option also produces a set of wavefunctions $u_{n\boldsymbol{k}}(\boldsymbol{r})$ on a real-space grid in the output files `UNK00001.1`, `UNK00002.1`, ...
☞**Note:** If the $\boldsymbol{k}$-points in `gr.win` and `open_grid.out` are not the same or `prefix` in `pw2wan.in` is not correct, the readers will get an error in `pw2wan.out` as follows:

```
%%%%%%%%%%%%%%%%%%%%%%%%%%%%%%%%%%%%%%%%%%%%%%%%%%%
     Error in routine pw2wannier90 (144):
     Wrong number of k-points
%%%%%%%%%%%%%%%%%%%%%%%%%%%%%%%%%%%%%%%%%%%%%%%%%%%
```

❑ **Output file:** The main output files for this tutorial are `gr_*.xsf`, `gr_band.dat`, and `gr_hr.dat`. We will visualize the WFs from `gr_*.xsf`, plot the band structure from `gr_band.dat`, and extract the tight-binding parameters from `gr_hr.dat`.
☞ **Visualizing the Wannier functions:** The `gr_*.xsf` file can open directly by using VESTA. For XCrySDen, we can open as **File → Open Structure → Open XSF (XCrySDen Structure File)** and select `gr_*.xsf` file. Then, we select **Tools → Data Grid**, click on the **OK** button, enter a value in the box Isovalue (2 is a good choice for graphene), click on the **Submit** button. In Fig. 3.33 (a) and (b), we show the p_z and s orbitals for the output files `gr_00001.xsf` and `gr_00003.xsf`, respectively.
☞ **Plotting energy dispersion:** By running step (5), the band structure based on the Wannier interpolation (see Sec. 5.14.3) is written in the output file `gr_band.dat`. In a similar way to plot band structure of graphene in Sec. 3.2.2, the readers can run the file `plot-bands.ipynb` to plot the band structure from output file `gr_band.dat`. The band structure is plotted in Fig. 3.32, in which

```
$ vi gr_hr.dat
...
     0    0    0     5    1    -0.000004     0.000014
     0    0    0     1    2    -2.910519     0.000006
     0    0    0     2    2    -1.391550    -0.000000
...
     0    0    0     5    5   -12.557218     0.000000
     0    1    0     1    1     0.224064    -0.000001
     0    1    0     2    1     0.019864     0.000001
...
```

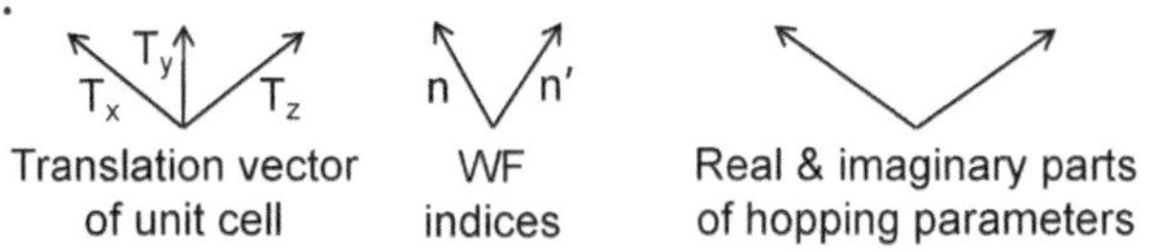

Figure 3.34 The meaning of output file `gr_hr.dat`.

the open dots and solid lines denote the band structure obtained from `wannier90.x` and `pw.x`, respectively. For five lowest energy-bands, which correspond to three s- and two p_z-orbitals, the Wannier interpolation can reproduce the band structure from SCF calculation. The advantage of the Wannier interpolation is that the band structure can be accurately calculated for a dense $\boldsymbol{k}$-mesh based on the SCF calculation for a coarse $\boldsymbol{k}$-mesh. This is helpful for the case that the SCF calculation is time-consuming for a dense $\boldsymbol{k}$-mesh, such as the hybrid functional calculation, which will be explained in Sec. 3.6.2.

☞ **Extracting tight-binding parameters:** The tight-binding parameters of graphene are given by the output file `gr_hr.dat`, as shown in Fig. 3.34. The readers can use the `vi` editor to open this file. Each line in `gr_hr.dat` includes 7 values as follows: the first three values give the translation vector $\boldsymbol{T}$ of the unit cell in x, y, and z directions, respectively, the next two values give the WF indices n and n', and the least two values are the real and imaginary parts of the hopping parameters in units of eV $t_{nn'} = \langle w_{n,0}|\mathcal{H}|w_{n',\boldsymbol{T}}\rangle$ (see Eq. 5.80). For graphene, we consider the hopping parameters up to the 1st nearest neighbor unit cell, i.e., t_{12} and t_{11} for $T = (0, 0, 0)$ and $(0, 1, 0)$, respectively. Then, we obtain $t_{12} = -2.91$ eV and $t_{11} = 0.224$ eV, as shown in Fig. 3.34.

Try It Yourself

1. Recompute band structure of graphene by changing the values of `dis_win_max` and `dis_froz_max`.
2. Calculate the MLWFs and band structure of monolayer MoS_2.

3.6.2 Wannier interpolation for hybrid functional

❑ **Purpose:** In this tutorial, we show how to obtain the band structure of the monolayer MoS_2 with the HSE (Heyd-Scuseria-Ernzerhof) hybrid functional and Wannier interpolation.

❑ **Background:** As discussed in Sec. 4.11.3, the LDA and GGA approximations often underestimate the value of the energy band gap. In order to obtain a reasonable energy gap, the HSE functional can be used in the SCF calculation in Quantum ESPRESSO. However, Quantum ESPRESSO only supports the HSE functional for a uniform $\boldsymbol{k}$-mesh. In addition, the HSE functional requires a non-local term of the Hartree-Fock exchange interaction, which is time-consuming for a dense $\boldsymbol{k}$-mesh. For this reason, the Wannier interpolation is necessary to plot the band structure along with the high-symmetry points using a non-uniform and dense $\boldsymbol{k}$-mesh for HSE functional.

❑ **How to run:** To run this tutorial, the readers should type the following command lines:

```
1 $ cd ~/QE-SSP/mos2/w90/
2 $ mpirun -np 8 pw.x < scf.in > scf.out &
3 $ mpirun -np 8 open_grid.x < open_grid.in >
      open_grid.out &
4 $ wannier90.x -pp mos2 &
5 $ mpirun -np 8 pw2wannier90.x < pw2wan.in > pw2wan.
      out &
6 $ wannier90.x mos2 &
```

The meaning of each step is given by the tutorial in Sec. 3.6.1. Since the SCF calculation with the HSE functional (line 2) is time-consuming, we run 8 processes in parallel (`mpirun -np 8`).

❑ **How to check:** When the calculations finish, the messages `JOB DONE` and `All done: wannier90 exiting` are written at the end of the output files, as shown in Sec. 3.6.1.

❑ **Input file:** For the input file `scf.in`, the readers can open as follows:

```
$ vi scf.in
```

QE-SSP/mos2/w90/scf.in

```
&CONTROL
calculation      = 'scf'
pseudo_dir       = '../pseudo/'
outdir           = '../tmp/'
prefix           = 'mos2'
/
&SYSTEM
ibrav            = 4
a                = 3.1825188839
c                = 20.0
nat              = 3
ntyp             = 2
nbnd             = 30
ecutwfc          = 60.0
input_dft        = 'hse'
exx_fraction     = 0.25
nqx1             = 2
nqx2             = 2
nqx3             = 1
/
&ELECTRONS
mixing_beta      = 0.7
conv_thr         = 1.0d-6
/
ATOMIC_SPECIES
Mo   95.94    Mo.pbe-spn-rrkjus_psl.1.0.0.UPF
S    32.065   S.pbe-nl-rrkjus_psl.1.0.0.UPF
ATOMIC_POSITIONS (crystal)
Mo    0.0000000000    0.0000000000    0.5000000000
S     0.3333333333    0.6666666667    0.4217548051
S     0.3333333333    0.6666666667    0.5782450789
K_POINTS (automatic)
6 6 1 0 0 0
```

☞**Explanation of scf.in:** The HSE functional is calculated by setting `input_dft = 'hse'` (line 15) in the namelist `SYSTEM`. `exx_fraction` (line 16) is the mixing coefficient, which is 0.25 for the HSE functional, as shown in Table 4.3. `nqx1`, `nqx2`, and `nqx3` (lines 17, 18, and 19), respectively, are dimensions of the ***q***-grid for the Hartree-Fock exchange interaction. The readers should check the convergence of the energy band by changing the ***q***-grid. The calculation of the ***q***-grid convergence will be very time-consuming compared to the ***k***-grid convergence in Sec. 3.1.3. Therefore, we recommend increasing the number of processes in parallel, which might require a workstation. Here, we select a coarse $2 \times 2 \times 1$ grid for ***q***-points, which is appropriate to run with a PC with 8 processes, such as Intel Core i7 or i9.

For the input file `open_grid.in`, the readers can open as follows:

```
$ vi open_grid.in
```

QE-SSP/mos2/w90/open_grid.in

```
&INPUTPP
outdir          = '../tmp/'
prefix          = 'mos2'
/
```

For the input file `mos2.win`, the readers can open as follows:

```
$ vi mos2.win
```

QE-SSP/mos2/w90/mos2.win

```
# CONTROL
num_bands           = 30
num_wann            = 11

num_iter            = 20

dis_win_min         = -8
dis_win_max         = 5
dis_froz_min        = -8
dis_froz_max        = 3.5
```

```
11
12 # PLOTTING
13 bands_plot        = .true.
14
15 begin kpoint_path
16 G 0.0000 0.0000 0.0000  K 0.3333 0.3333 0.0000
17 K 0.3333 0.3333 0.0000  M 0.5000 0.0000 0.0000
18 M 0.5000 0.0000 0.0000  G 0.0000 0.0000 0.0000
19 end kpoint_path
20
21 #SYSTEM
22 begin unit_cell_cart
23 ang
24    3.182518    0.000000    0.000000
25   -1.591259    2.756142    0.000000
26    0.000000    0.000000   20.000000
27 end unit_cell_cart
28
29 begin atoms_frac
30 Mo    0.0000000000    0.0000000000    0.5000000000
31 S     0.3333333333    0.6666666667    0.4217548051
32 S     0.3333333333    0.6666666667    0.5782450789
33 end atoms_frac
34
35 # PROJECTIONS
36 guiding_centres   = .true.
37
38 begin projections
39 Mo: dxy; dyz; dxz; dx2-y2; dz2
40 S: px; py; pz
41 end projections
42
43 # KPOINTS
44 mp_grid   = 6 6 1
45
46 begin kpoints
47    0.00000000  0.00000000  0.00000000   0.0277777
48    0.00000000  0.16666666  0.00000000   0.0277777
49    ...
50   -0.16666666 -0.16666666  0.00000000   0.0277777
51 end kpoints
```

☞**Explanation of mos2.win:** The detail of parameters in the `mos2.win` file is given in Table 3.10. Here, five *d* orbitals ($d_{xy}, d_{yz}, d_{xz}, d_{x^2-y^2}$, and d_{z^2}) and three *p* orbitals (p_x, p_y, and p_z) are selected as the initial projectors for the Mo and S atoms, respectively. Thus, we have 11 WFs `num_wann = 11` (line 3).

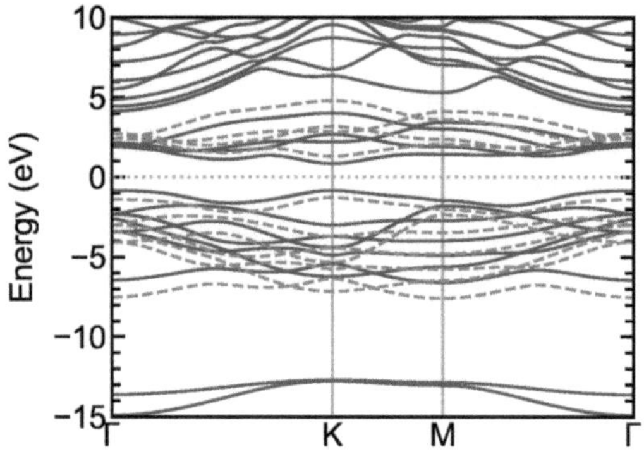

Figure 3.35 Energy band structure of monolayer MoS_2. Solid and dashed lines denote the band structure with the GGA and HSE functionals, respectively.

For the input file `pw2wan.in`, the readers can open as follows:

```
$ vi pw2wan.in
```

QE-SSP/mos2/w90/pw2wan.in

```
&INPUTPP
outdir     = '../tmp/'
prefix     = 'mos2_open'
seedname   = 'mos2'
/
```

❑ **Output file:** The band structure is written in the output file `mos2_band.dat`. The readers can run the file `plot-bands.ipynb` to plot the band structures from `mos2_band.dat` (with HSE functional) and `mos2_bands.gnu` (with only GGA functional). It is noted that the file `mos2_bands.gnu` is obtained by `bands.x` for a non-SCF calculation (see Sec. 3.2.2). In Fig. 3.35, we show the band structure of the monolayer MoS_2 with the GGA (solid lines) and HSE (dashed lines) functionals. The energy gaps are 1.688 and 2.578 eV for the GGA and HSE functionals, respectively. Compared with the experimental value (2.40 ± 0.05 eV [Huang *et al.* (2015)]), the HSE functional gives a better value of the energy gap.

Try It Yourself

1. Calculate the energy gap of the bulk Si with the LDA, GGA, and HSE functionals, and compare to the experimental value (1.12 eV).
2. Plot the Wannier function of the bulk Si with the HSE functional.

Chapter 4

Density-Functional Theory

In this chapter, we provide a density-functional theory (DFT) that becomes the basis of Quantum ESPRESSO. In order to understand how the DFT code works, we explain and apply the DFT on a simple example of the ground-state calculation of a helium atom. A simple Python code is also given for the helium atom calculation. This chapter will help the readers to understand Quantum ESPRESSO's techniques in Chapter 3, such as the self-consistent field (SCF) calculation, the ground-state total energy, the pseudopotential, etc.

4.1 "Black box" Quantum ESPRESSO

By giving input files to Quantum ESPRESSO, we can obtain output files containing the materials' information and properties. Therefore, Quantum ESPRESSO works like a black box, as shown in Fig. 4.1 (a). The readers may run a computer program successfully without understanding the black box. However, a successful run does not mean that you have run it properly. To understand what you are running, the readers need to know what is inside the black box.

Quantum ESPRESSO Course for Solid-State Physics
Nguyen Tuan Hung, Ahmad R. T. Nugraha, and Riichiro Saito

ISBN 978-981-4968-37-9 (Hardcover), 978-981-4968-63-8 (Paperback), 978-1-003-29096-4 (eBook)
www.jennystanford.com

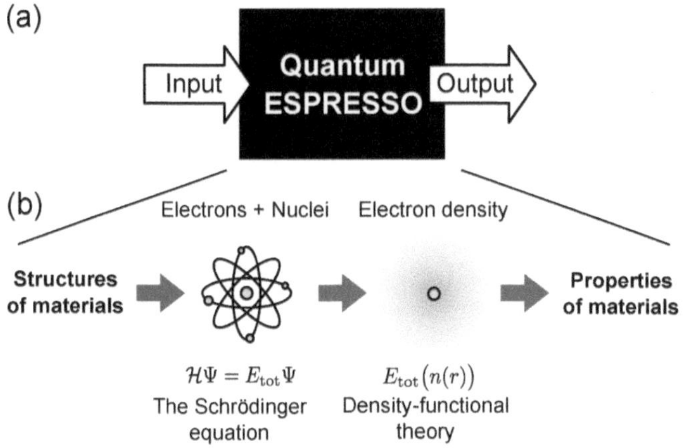

Figure 4.1 (a) Quantum ESPRESSO as a black box. (b) Density-functional theory is used in Quantum ESPRESSO to solve the Schrödinger equation of materials.

The purpose of Quantum ESPRESSO is to calculate the properties of materials at the atomic scale. Since "*materials = electrons + nuclei*", the properties of the materials are given by the complicated interactions of the electrons and nuclei. The electrons and nuclei hold together in the materials with a detailed balance between the repulsive and attractive Coulomb interactions between them. These interactions are described by an equation in quantum mechanics, the so-called **Schrödinger equation** (Eq. (4.1)) [Schrödinger (1926)]. An analytical solution of the Schrödinger equation only exists for systems of one electron (e.g., a hydrogen atom), while most atoms, molecules, and materials consist of many electrons and nuclei. For any system of more than two electrons, the Schrödinger equation can be solved only after using some approximations due to the complexity of the Coulomb interactions. Therefore, a good approximation is necessary to solve the Schrödinger equation. In Quantum ESPRESSO, the **density-functional theory (DFT)** is used as an approximation to solve the Schrödinger equation, as shown in Fig. 4.1 (b).

4.2 The Schrödinger equation

A system of electrons and nuclei is described by the Schrödinger equation as

$$\mathcal{H}\Psi = E_{\text{tot}}\Psi, \tag{4.1}$$

where $\mathcal{H}$ is the Hamiltonian of the system, E_{tot} is *the total energy of the electrons and nuclei,* and $\Psi(\boldsymbol{r}_1, ..., \boldsymbol{r}_{N_e}, \boldsymbol{R}_1, ..., \boldsymbol{R}_{N_n})$ is the wavefunction of many particles (electrons and nuclei). By using the atomic units listed in Table 4.1, a general Hamiltonian in Eq. (4.1) is given by kinetic energies $\mathcal{T}$ and potential energies $\mathcal{V}$ of the electrons and nuclei as

$$\begin{aligned} \mathcal{H} &= \mathcal{T}_n + \mathcal{V}_n + \mathcal{T}_e + \mathcal{V}_e + \mathcal{V}_{en} \\ &= \underbrace{-\sum_{I=1}^{N_n} \frac{\nabla^2_{\boldsymbol{R}_I}}{2M_I} + \frac{1}{2}\sum_{I\neq J}^{N_n} \frac{Z_I Z_J}{|\boldsymbol{R}_I - \boldsymbol{R}_J|}}_{\text{nuclei}} \\ &\quad \underbrace{-\sum_{i=1}^{N_e} \frac{\nabla^2_{\boldsymbol{r}_i}}{2} + \frac{1}{2}\sum_{i\neq j}^{N_e} \frac{1}{|\boldsymbol{r}_i - \boldsymbol{r}_j|}}_{\text{electrons}} + \underbrace{\sum_{i=1}^{N_e}\sum_{I=1}^{N_n} \frac{-Z_I}{|\boldsymbol{r}_i - \boldsymbol{R}_I|}}_{\text{mixed}}. \end{aligned} \tag{4.2}$$

$\mathcal{T}_n$:	kinetic energy of nuclei	$\mathcal{T}_e$:	kinetic energy of electrons
$\mathcal{V}_n$:	Coulomb repulsion between a pair of nuclei	$\mathcal{V}_e$:	Coulomb repulsion between a pair of electrons
$\mathcal{V}_{en}$:	Coulomb attraction between electrons and nuclei	∇:	nabla operators
$\boldsymbol{R}$:	nuclear position	$\boldsymbol{r}$:	electronic position
N_n:	number of nuclei	N_e:	number of electrons
I, J:	label for a nucleus	i, j:	label for an electron
M:	nuclear mass	Z:	nuclear charge (atomic number)

Notice that a factor of $1/2$ appears in the sums of $\mathcal{V}_n$ and $\mathcal{V}_e$ to take into account the double counting on the summation i and j $(i \neq j)$. Here, we adopt the atomic units $e = \hbar = c = m_e = 1$ (see Table 4.1) and cgs unit for the Coulomb interaction. If SI unit is adopted, $\mathcal{T}_e + \mathcal{V}_e$

Table 4.1 Atomic unit (a.u.).

Symbol	Quantity	a.u.	Value in SI
t	Time	1	2.419×10^{-17} s
$c\ (1/\alpha)$	Speed of light	1	2.998×10^{8} m/s
$\hbar$	$h/2\pi$ (angular momentum)	1	1.055×10^{-34} Js
h	Planck's constant	2π	6.626×10^{-31} Js
Ha (E_h)	Hartree (atomic energy)	1	4.360×10^{-18} J
m_e	Electron mass	1	9.110×10^{-31} kg
e	Electron charge	1	1.602×10^{-19} C
a_0	Bohr radius (atomic distance)	1	5.292×10^{-11} m
e/a_0^3	Charge density	1	1.081×10^{12} C/m^3
$eE_h/\hbar$	Current	1	6.623×10^{-3} A
$E_h/(ea_0^2)$	Electric field	1	9.717×10^{21} V/m^2
μ_B	Bohr magneton	1/2	9.274×10^{-24} J/T
$4\pi\epsilon_0$	Vacuum permittivity $\times 4\pi$	1	1.113×10^{-10} C^2/Jm
E_h/a_0	Force	1	8.238×10^{-8} N
E_h/a_0^3	Pressure	1	2.942×10^{13} Pa

Helpful conversions: 1 Ha = 27.2114 eV = 2 Ry.

is expressed by

$$-\sum_{i=1}^{N_e} \frac{\hbar^2 \nabla_{\mathbf{r}_i}^2}{2m_e} + \frac{1}{2} \sum_{i \neq j}^{N_e} \frac{e^2}{4\pi\epsilon_0 |\mathbf{r}_i - \mathbf{r}_j|}. \tag{4.3}$$

The Hamiltonian in Eq. (4.2) is considered as a coupled nuclear and electronic problem. However, the nuclei are much heavier than the electrons (e.g., $M_H/m_e = 1,836$ in hydrogen or $M_C/m_e = 21,868$ in carbon). Moreover, the average speed of the nuclei is much smaller than that of the electrons. Therefore, we can assume that the electrons will follow the nuclear motions without any delay, and we can decouple the electronic and nuclear motions to the first approximation, which is known as the **Born-Oppenheimer**

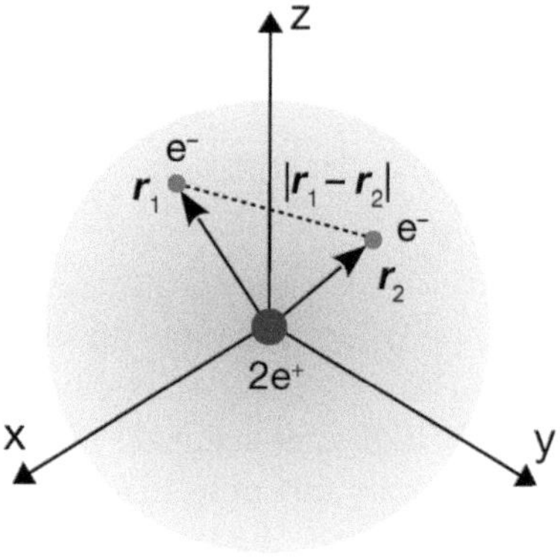

Figure 4.2 Two electrons in a helium atom, where $\boldsymbol{r}_i$ $(i = 1, 2)$ are the positions of the i-th electron, and $|\boldsymbol{r}_1 - \boldsymbol{r}_2|$ is the distance between two electrons.

approximation (BOA) [Born and Oppenheimer (1927)]. The nuclei in the material are almost immobile except for hydrogen atoms, and their positions can be determined precisely by X-ray crystallography (see Sec. 5.2). Thus, we can fix the positions of the nuclei in the real space with zero kinetic energy (i.e., $\mathcal{T}_n = 0$) for calculating the electronic energy band. Then $\mathcal{V}_n$ becomes an external potential (or simply a constant). In this case, the problem that we have to solve becomes a purely electronic one as

$$\mathcal{H}_e \psi = (\mathcal{T}_e + \mathcal{V}_e + \mathcal{V}_{en})\psi = E\psi, \tag{4.4}$$

where $E = E_{\text{tot}} - \mathcal{V}_n$ is *the total energy of the electrons* and ψ is a wavefunction of the N electrons, in which ψ is a many-body function of the positions of N electrons $\boldsymbol{r}_i$ $(i = 1, 2, ..., N)$: $\psi(\boldsymbol{r}_1, \boldsymbol{r}_2, ..., \boldsymbol{r}_N)$. Since an electron interacts with the rest of the electrons due to the Coulomb repulsion $\mathcal{V}_e$, the many-body state has to account for all the degrees of freedom of N electrons in the system (or the $3N$-dimensional configuration space).

❑ **The Schrödinger equation for a helium atom:** Let us consider the case of a helium (He) atom, which consists of one nucleus with $Z = 2$ and two electrons at $\boldsymbol{r}_1$ and $\boldsymbol{r}_2$, as shown in Fig. 4.2.

Equation (4.4) for a helium atom is given by

$$\left(-\frac{1}{2}\nabla_{r_1}^2 - \frac{1}{2}\nabla_{r_2}^2 + \frac{1}{|\boldsymbol{r}_1 - \boldsymbol{r}_2|} - \frac{2}{|\boldsymbol{r}_1|} - \frac{2}{|\boldsymbol{r}_2|}\right)\psi(\boldsymbol{r}_1, \boldsymbol{r}_2) = E\psi(\boldsymbol{r}_1, \boldsymbol{r}_2). \quad (4.5)$$

Although Eq. (4.5) looks simple, an analytical solution cannot be found because Eq. (4.5) is a six-dimensional partial differential equation (PDE). High-dimensional PDEs are hard to solve in physics, in which they occur not only in quantum mechanics but also in classical mechanics (e.g., the Hamilton-Jacobi equation[1] with a large number of degrees of freedom).

Since the Schrödinger equation is a PDE where all terms are known, one can try to solve it numerically on a grid.[2] This approach can solve Eq. (4.5) for an electron, but it fails for a system with more than two electrons. For example, if we take 3D silicon as an example, it has a unit cell volume of $a^3/4$, where $a = 5.431$ Å is the lattice constant. If we consider a grid of the unit cell with points spaced by $\Delta d \sim 0.2$ Å, this grid consists of $(a^3/4)/(\Delta d^3) \sim 5000$ points. Since two silicon atoms in the unit cell contain 28 electrons, the grid has $28 \times 3 = 84$ coordinates. Therefore, a complete specification of a wavefunction $\psi_{\text{Si}}(\boldsymbol{r}_1, \boldsymbol{r}_2, ..., \boldsymbol{r}_{28})$ requires 5000^{84} complex numbers. We need a stack of DVDs from the earth to the sun to store this wavefunction. To overcome this problem, many researchers have been developing methods of approximation since the late 1920s. The most successful method in physics is the density-functional theory (DFT). The DFT is defined by the theory of approximation in which the Coulomb interaction between electrons is expressed by a functional[3] of the electron density. With the DFT, the Schrödinger equation can be reformulated in terms of the electron density. Therefore, the $3N$-dimensional PDE is reduced to N PDE's of 3-dimension, so that we can solve Eq. (4.2) for silicon and store their wavefunction by using a PC.

[1] The Hamilton-Jacobi equation is an alternative formulation of classical mechanics, equivalent to other formulations such as the Newton second law.
[2] In a grid approach, functions are represented by their values at certain grid points and derivatives are approximated through differences in these values.
[3] A functional is a function whose variable is another function such as $V(n(\boldsymbol{r}))$.

4.3 Systems of non-interacting electrons

Before explaining the DFT, we would like to show some approximations for solving Eq. (4.5), which are regarded as the conceptual origin of the DFT. As discussed in Sec. 4.2, the problem to solve the Schrödinger equation comes from the Coulomb repulsion between electrons $\mathcal{V}_e$. If we assume that the N_e electrons in the system are non-interacting to one another, we can set $\mathcal{V}_e = 0$, and Eq. (4.4) can be rewritten as a function of $\boldsymbol{r}_i$ $(i = 1, ..., N_e)$

$$(\mathcal{T}_e + \mathcal{V}_{en})\psi = \left(-\sum_{i=1}^{N_e} \frac{\nabla_{\boldsymbol{r}_i}^2}{2} + \sum_{i=1}^{N_e}\sum_{I=1}^{N_n} \frac{-Z_I}{|\boldsymbol{r}_i - \boldsymbol{R}_I|} \right)\psi = E\psi. \quad (4.6)$$

Here we define a **single-particle Hamiltonian** of one electron as

$$h(\boldsymbol{r}) = -\frac{\nabla_{\boldsymbol{r}}^2}{2} + \sum_{I=1}^{N_n} \frac{-Z_I}{|\boldsymbol{r} - \boldsymbol{R}_I|}, \quad (4.7)$$

so that Eq. (4.6) can be rewritten as

$$h(\boldsymbol{r}_i)\psi(\boldsymbol{r}_i) = E\psi(\boldsymbol{r}_i), \text{ with } (i = 1, ..., N_e). \quad (4.8)$$

From Eq.(4.8) we can say that, for the case of the non-interacting electrons, the Hamiltonian of the system can be written as N_e independent single-particle Hamiltonian, and the wavefunction is written by $\psi(\boldsymbol{r}_i)$. Eq. (4.8) is used to obtain the energy of electrons for the following two cases: (1) **distinguishable**[4] and (2) **indistinguishable**[5] electrons. For simplicity, we consider the two electrons in a helium atom as follows:

(1) **Two distinguishable electrons**: Let us consider electron 1 in one state $\phi_a(\boldsymbol{r}_1)$ and electron 2 in another state $\phi_b(\boldsymbol{r}_2)$. Since two electrons are independent of each other, the probability of finding (electron 1 at $\boldsymbol{r}_1$) *and* (electron 2 at $\boldsymbol{r}_2$) can be given by the *product* as $|\psi(\boldsymbol{r}_1, \boldsymbol{r}_2)|^2 = |\phi_a(\boldsymbol{r}_1)|^2 |\phi_b(\boldsymbol{r}_2)|^2$. In this case, the wavefunction can

[4] If the electrons are distinguishable, the physical properties of the system are changed by switching the positions of two electrons.

[5] If the electrons are indistinguishable, switching the positions of two electrons makes no physical change.

be expressed by the product of one-body wavefunction, $\psi(\boldsymbol{r}_1, \boldsymbol{r}_2) = \phi_a(\boldsymbol{r}_1)\phi_b(\boldsymbol{r}_2)$, which is known as **one-body approximation**. In this case, Eq. (4.8) can be written as

$$[h(\boldsymbol{r}_1) + h(\boldsymbol{r}_2)]\,\phi_a(\boldsymbol{r}_1)\phi_b(\boldsymbol{r}_2) = E\phi_a(\boldsymbol{r}_1)\phi_b(\boldsymbol{r}_2). \tag{4.9}$$

Since $h(\boldsymbol{r}_i)$ $(i = 1, 2)$ acts only on $\phi_\alpha(\boldsymbol{r}_i)$ $(\alpha = a, b)$ as

$$h(\boldsymbol{r}_i)\phi_\alpha(\boldsymbol{r}_i) = \epsilon_\alpha\phi_\alpha(\boldsymbol{r}_i), \tag{4.10}$$

where ϵ_α is the energy of one electron, the left-hand side of Eq. (4.9) is rewritten as

$$\begin{aligned} &[h(\boldsymbol{r}_1) + h(\boldsymbol{r}_2)]\,\phi_a(\boldsymbol{r}_1)\phi_b(\boldsymbol{r}_2) \\ &= [h(\boldsymbol{r}_1)\phi_a(\boldsymbol{r}_1)]\phi_b(\boldsymbol{r}_2) + \phi_a(\boldsymbol{r}_1)[h(\boldsymbol{r}_2)\phi_b(\boldsymbol{r}_2)] \\ &= \epsilon_a\phi_a(\boldsymbol{r}_1)\phi_b(\boldsymbol{r}_2) + \phi_a(\boldsymbol{r}_1)\epsilon_b\phi_b(\boldsymbol{r}_2) \\ &= (\epsilon_a + \epsilon_b)\phi_a(\boldsymbol{r}_1)\phi_b(\boldsymbol{r}_2). \end{aligned} \tag{4.11}$$

By substituting Eq. (4.11) into Eq. (4.9), we obtain

$$E = \epsilon_a + \epsilon_b. \tag{4.12}$$

Thus, for the case of the non-interacting system with the distinguishable electrons, the total energy of the electrons is the sum of the energy of each electron.

(2) **Two indistinguishable electrons**: In quantum mechanics, two electrons are indistinguishable (i.e., $\phi_a(\boldsymbol{r}_1)\phi_b(\boldsymbol{r}_2)$ is equivalent to $\phi_a(\boldsymbol{r}_2)\phi_b(\boldsymbol{r}_1)$). Thus, one possible shape of the wavefunction $\psi(\boldsymbol{r}_1, \boldsymbol{r}_2)$ is a **linear combination** of two states as [Kittel (1976)]

$$\psi(\boldsymbol{r}_1, \boldsymbol{r}_2) = \frac{1}{\sqrt{2}}[\phi_a(\boldsymbol{r}_1)\phi_b(\boldsymbol{r}_2) - \phi_b(\boldsymbol{r}_1)\phi_a(\boldsymbol{r}_2)], \tag{4.13}$$

where the prefactor $\frac{1}{\sqrt{2}}$ is for normalizing the wave function, $\int |\psi(\boldsymbol{r}_1, \boldsymbol{r}_2)|^2 d\boldsymbol{r}_1\boldsymbol{r}_2 = 1$. Eq. (4.13) shows that the wavefunction is anti-symmetric for exchanging $\boldsymbol{r}_1$ and $\boldsymbol{r}_2$ as $\psi(\boldsymbol{r}_1, \boldsymbol{r}_2) = -\psi(\boldsymbol{r}_2, \boldsymbol{r}_1)$. When we put $\boldsymbol{r}_1 = \boldsymbol{r}_2$ for $\psi(\boldsymbol{r}_1, \boldsymbol{r}_2)$, we get $\psi = 0$. It means that two electrons cannot occupy the same position, in accordance with the

Pauli exclusion principle [Pauli (1925)]. Eq. (4.13) can also be written by using a matrix determinant as

$$\psi(\boldsymbol{r}_1, \boldsymbol{r}_2) = \frac{1}{\sqrt{2}} \begin{vmatrix} \phi_a(\boldsymbol{r}_1) & \phi_b(\boldsymbol{r}_1) \\ \phi_a(\boldsymbol{r}_2) & \phi_b(\boldsymbol{r}_2) \end{vmatrix}, \tag{4.14}$$

which is known as the **Slater determinant** [Slater (1929)].

By substituting Eq. (4.13) into Eq. (4.8), we get the same energy as the case of two distinguishable electrons (i.e., $E = \epsilon_a + \epsilon_b$). The probability of finding an electron at position $\boldsymbol{r}$, i.e., the **electron density**, is given by

$$n(\boldsymbol{r}) = P_1(\boldsymbol{r}) + P_2(\boldsymbol{r}), \tag{4.15}$$

where $P_1(\boldsymbol{r}) = \int |\psi(\boldsymbol{r}, \boldsymbol{r}_2)|^2 \mathrm{d}\boldsymbol{r}_2$ and $P_2(\boldsymbol{r}) = \int |\psi(\boldsymbol{r}_1, \boldsymbol{r})|^2 \mathrm{d}\boldsymbol{r}_1$ are probabilities of finding first and second electrons at $\boldsymbol{r}$, respectively, while other electrons can exist anywhere except for $\boldsymbol{r}$. Since two electrons are indistinguishable (i.e., $P_1(\boldsymbol{r}) = P_2(\boldsymbol{r})$), Eq. (4.15) can be rewritten as

$$n(\boldsymbol{r}) = 2 \int |\psi(\boldsymbol{r}, \boldsymbol{r}_2)|^2 \mathrm{d}\boldsymbol{r}_2. \tag{4.16}$$

By substituting Eq. (4.13) into Eq. (4.16) and using the normalization and orthogonality conditions, respectively, as

$$\int \phi_a^*(\boldsymbol{r})\phi_a(\boldsymbol{r})\mathrm{d}\boldsymbol{r} = \int \phi_b^*(\boldsymbol{r})\phi_b(\boldsymbol{r})\mathrm{d}\boldsymbol{r} = 1 \tag{4.17}$$

and

$$\int \phi_b^*(\boldsymbol{r})\phi_a(\boldsymbol{r})\mathrm{d}\boldsymbol{r} = \int \phi_a^*(\boldsymbol{r})\phi_b(\boldsymbol{r})\mathrm{d}\boldsymbol{r} = 0, \tag{4.18}$$

the electron density is expressed by

$$n(\boldsymbol{r}) = |\phi_a(\boldsymbol{r})|^2 + |\phi_b(\boldsymbol{r})|^2. \tag{4.19}$$

Therefore, the electron density of the non-interacting electrons is simply equal to the sum of the squares of the occupied states.

❑ **Electron density and energy of a helium atom without**

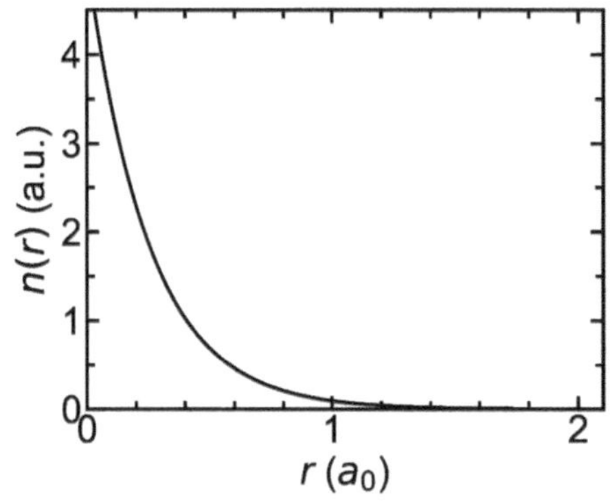

Figure 4.3 Electron density in atomic unit of a helium atom as a function of r in the approximation of non-interacting electrons.

Coulomb repulsion: Let us recall that the wavefunction of s-orbital and the energy of a hydrogen-like atom with a nucleus $+Ze$ are given by

$$\phi(\boldsymbol{r}) = \frac{Z^{3/2}}{\sqrt{\pi}} \exp(-Z|\boldsymbol{r}|) \text{ and } \epsilon = -\frac{Z^2}{2}, \tag{4.20}$$

respectively. We note that the hydrogen atom is a special case that contains only one electron and Eq. (4.20) gives an analytical solution for $Z = 1$. By substituting Eq. (4.20) into Eqs. (4.19) and (4.12), the electron density and energy of the helium atom are given by

$$n(\boldsymbol{r}) = \frac{2Z^3}{\pi} \exp(-2Z|\boldsymbol{r}|) \text{ and } E = -Z^2, \tag{4.21}$$

respectively, when we neglect the interaction between two electrons. Since the atomic number of a helium atom is $Z = 2$, the electron density and total energy are

$$n(\boldsymbol{r}) = \frac{16}{\pi} \exp(-4|\boldsymbol{r}|) \tag{4.22}$$

and $E = -4$ Ha, respectively. In Fig. 4.3, we plot $n(\boldsymbol{r})$ of Eq. (4.22) as a function of the distance $r = |\boldsymbol{r}|$ from the nucleus, respectively.

To compare the calculated energy E with the experimental values, we remove two electrons from the helium atom as

$$\text{He} \xrightarrow{I_1} \text{He}^+ + e^- \xrightarrow{I_2} \text{He}^{++} + 2e^-, \tag{4.23}$$

where I_1 and I_2 are the first and second **ionization energies**, respectively. I_1 is the minimum energy required to remove the first electron from the helium atom and is experimentally given as 0.9 Ha (24.59 eV) [Sucher (1958)]. I_2 can be calculated exactly since He^+ is a hydrogen-like ion, thus $I_2 = -\epsilon(\text{He}^+) = 2$ Ha. Therefore, the energy of the helium atom is given by $-(I_1 + I_2) = -2.9$ Ha, which is larger than the calculated energy (-4 Ha). In other words, the non-interacting electrons approximation gives a large deviation of the total energy. The reason for the deviation is that we do not consider the energy of the Coulomb repulsion between the two electrons.

4.4 Hartree potential

In Sec. 4.3, we show that the non-interacting electrons without the Coulomb repulsion give a large deviation of the energy. Here we would like to keep the expression of $n(\boldsymbol{r})$ as the for the non-interacting electrons (Eq. (4.19)), and we will take the Coulomb repulsion into account in the energy E. When we consider the Coulomb interaction, Eq. (4.4) can be rewritten as

$$\left[h(\boldsymbol{r}_1) + h(\boldsymbol{r}_2) + \frac{1}{|\boldsymbol{r}_1 - \boldsymbol{r}_2|}\right] \phi_a(\boldsymbol{r}_1)\phi_b(\boldsymbol{r}_2) = E\phi_a(\boldsymbol{r}_1)\phi_b(\boldsymbol{r}_2). \tag{4.24}$$

If we multiply $\phi_b^*(\boldsymbol{r}_2)$ to Eq. (4.24) and integrate both sides of Eq. (4.24) on $\boldsymbol{r}_2$ with using the normalization condition in Eq. (4.17), we can obtain the **single-particle Schrödinger equation** for the first electron as

$$\left[h(\boldsymbol{r}_1) + \int \frac{|\phi_b(\boldsymbol{r}_2)|^2}{|\boldsymbol{r}_1 - \boldsymbol{r}_2|} \mathrm{d}\boldsymbol{r}_2\right] \phi_a(\boldsymbol{r}_1) = \epsilon_a^H \phi_a(\boldsymbol{r}_1), \tag{4.25}$$

where ϵ_a^H is expressed by

$$\epsilon_a^H = E - \int \phi_b^*(\boldsymbol{r}_2)h(\boldsymbol{r}_2)\phi_b(\boldsymbol{r}_2)\mathrm{d}\boldsymbol{r}_2 = E - \epsilon_b, \tag{4.26}$$

where ϵ_b is the energy of the second electron in the non-interacting Hamiltonian $h(\boldsymbol{r}_2)$ as defined by Eq. (4.10). The single-particle Schrödinger equation for the second electron also can be written by multiplying $\int \phi_a^*(\boldsymbol{r}_1)\mathrm{d}\boldsymbol{r}_1$ as

$$\left[h(\boldsymbol{r}_2) + \int \frac{|\phi_a(\boldsymbol{r}_1)|^2}{|\boldsymbol{r}_1 - \boldsymbol{r}_2|}\mathrm{d}\boldsymbol{r}_1\right] \phi_b(\boldsymbol{r}_2) = \epsilon_b^H \phi_b(\boldsymbol{r}_2), \tag{4.27}$$

with

$$\epsilon_b^H = E - \epsilon_a, \tag{4.28}$$

where ϵ_a is the energy of the first electron.

The integration in Eq. (4.25) can be rewritten in term of electron density, $n(\boldsymbol{r}_2) = |\phi_a(\boldsymbol{r}_2)|^2 + |\phi_b(\boldsymbol{r}_2)|^2$, as

$$\int \frac{|\phi_b(\boldsymbol{r}_2)|^2}{|\boldsymbol{r}_1 - \boldsymbol{r}_2|}\mathrm{d}\boldsymbol{r}_2 = \underbrace{\int \frac{n(\boldsymbol{r}_2)}{|\boldsymbol{r}_1 - \boldsymbol{r}_2|}\mathrm{d}\boldsymbol{r}_2}_{\mathcal{V}_H(\boldsymbol{r}_1)} - \underbrace{\int \frac{|\phi_a(\boldsymbol{r}_2)|^2}{|\boldsymbol{r}_1 - \boldsymbol{r}_2|}\mathrm{d}\boldsymbol{r}_2}_{\mathcal{V}_H^{\mathrm{sic}}(\boldsymbol{r}_1)}. \tag{4.29}$$

The first term $\mathcal{V}_H$ is called the **Hartree potential** [Hartree (1928)], which describes the Coulomb repulsion potential as a function of $\boldsymbol{r}_1$. The second term $\mathcal{V}_H^{\mathrm{sic}}(\boldsymbol{r}_1)$ is called the *self-interaction correction* of the Hartree potential, which takes into account that the electron at the state ϕ_a shall not interact with itself, but with the remaining electrons. When the system has many electrons ($\sim 10^{23}$), $\mathcal{V}_H^{\mathrm{sic}}$ can be neglected since $\mathcal{V}_H^{\mathrm{sic}}$ might contribute negligibly small to the total energy [Parr and Yang (1989)].

From Eq. (4.25), we can write the energy as a functional of the wavefunctions as

$$\begin{aligned}\epsilon_a^H &= \int \phi_a^*(\boldsymbol{r}_1)\left[h(\boldsymbol{r}_1) + \int \frac{|\phi_b(\boldsymbol{r}_2)|^2}{|\boldsymbol{r}_1 - \boldsymbol{r}_2|}\mathrm{d}\boldsymbol{r}_2\right]\phi_a(\boldsymbol{r}_1)\mathrm{d}\boldsymbol{r}_1 \\ &= \epsilon_a + \iint \frac{|\phi_a(\boldsymbol{r}_1)|^2|\phi_b(\boldsymbol{r}_2)|^2}{|\boldsymbol{r}_1 - \boldsymbol{r}_2|}\mathrm{d}\boldsymbol{r}_1\mathrm{d}\boldsymbol{r}_2.\end{aligned} \tag{4.30}$$

By substituting Eq. (4.26) into Eq. (4.30), the total energy E is given by

$$E = \epsilon_a + \epsilon_b + \iint \frac{|\phi_a(\boldsymbol{r}_1)|^2 |\phi_b(\boldsymbol{r}_2)|^2}{|\boldsymbol{r}_1 - \boldsymbol{r}_2|} \mathrm{d}\boldsymbol{r}_1 \mathrm{d}\boldsymbol{r}_2. \tag{4.31}$$

Now the total energy E is not equal neither to $\epsilon_a + \epsilon_b$ nor to $\epsilon_a^H + \epsilon_b^H$. This is because the positive Coulomb interaction between the two electrons is included twice in $\epsilon_a^H + \epsilon_b^H$ (Eqs. (4.25) and (4.27)). Therefore, only one term from the inter-electron interaction is added to $\epsilon_a + \epsilon_b$ in the expression for E as shown in Eq. (4.31).

4.5 Self-consistent field

A difficulty in solving the single-particle Schrödinger equation is that the Hartree potential $\mathcal{V}_H$ in Eq. (4.29) requires the value of the state ϕ, which is not known before we solve the single-particle Schrödinger equation in Eq. (4.25). This is called a *self-consistent problem*.

The Hartree potential can be expressed in a differential form of the Maxwell equation as $\nabla \cdot \boldsymbol{E} = 4\pi n(\boldsymbol{r})$, where $n(\boldsymbol{r})$ is the electron density at $\boldsymbol{r}$ and $\boldsymbol{E} = -\nabla \mathcal{V}_H(\boldsymbol{r})$, where $\mathcal{V}_H(\boldsymbol{r})$ is generally called the electrostatic potential. Combining the two equations, we get the **Possion equation** [Jackson (1999)]:

$$\nabla^2 \mathcal{V}_H(\boldsymbol{r}) = -4\pi n(\boldsymbol{r}). \tag{4.32}$$

The solution of the Poisson equation is given as follows:

$$\mathcal{V}_H(\boldsymbol{r}) = \int \frac{n(\boldsymbol{r}')}{|\boldsymbol{r} - \boldsymbol{r}'|} \mathrm{d}\boldsymbol{r}'. \tag{4.33}$$

Now let us introduce a **self-consistent field** (SCF) method that can be applied to solve the single-particle Schrödinger equation by using the Poisson equation. As shown in Fig. 4.4, first, guessing an initial electron density $n(\boldsymbol{r})$ for the solution, we put $n(\boldsymbol{r})$ into Eq. (4.33) to get the Hartree potential $\mathcal{V}_H(\boldsymbol{r})$. Then we put the $\mathcal{V}_H$ into the single-particle Schrödinger equation to get eigenstate $\phi(\boldsymbol{r})$ to obtain the new $n(\boldsymbol{r}) = \sum |\phi(\boldsymbol{r})|^2$, where the sum runs over the occupied states. We repeat the procedure until $n(\boldsymbol{r})$ does not change

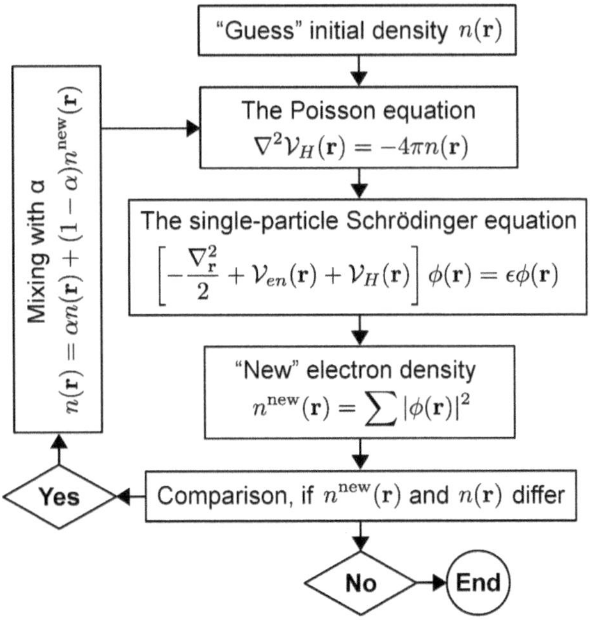

Figure 4.4 Flow-chart of the self-consistent field method for the solution of the single-particle Schrödinger equation with the Hartree potential $\mathcal{V}_H$.

appreciably, that is, the convergence of the solution. This procedure is called the SCF method. Here, we construct the next guess of $n(\boldsymbol{r})$ by mixing $n^{\text{new}}(\boldsymbol{r})$ with the previous $n(\boldsymbol{r})$. If the mixing parameter α $(0 < \alpha < 1)$ is large, we might find oscillating behavior of $n(\boldsymbol{r})$. We note that the self-interaction term $\mathcal{V}_H^{\text{sic}}$ in Eq. (4.29) is neglected in this procedure.

Let us consider the example of silicon in Sec. 4.2. By using the SCF method, the 3N-dimensional configuration space of the Schrödinger equation is reduced to an N-single-particle Schrödinger equations of three dimensions. In this case, the characteristic size of the arrays needed for describing the wavefunctions becomes 28×5000^3 for silicon with 28 electrons in a unit cell, instead of 5000^{84} in Sec. 4.2. These arrays correspond approximately to a few GB of computer storage.

❑ **Hartree potential for helium:** In order to solve the Poisson equation for a helium atom, it is more convenient to use spherical coordinates (r, θ, φ) rather than Cartesian coordinates $\boldsymbol{r} = (x, y, z)$. These are related to each other by $r = |\boldsymbol{r}|$, $x = r\cos\theta\sin\varphi$, $y = r\sin\theta\cos\varphi$, and $z = r\cos\theta$. Referring back to Eq. (4.33), the Laplacian ∇^2 is expressed in spherical coordinates as follows:

$$\begin{aligned}\nabla^2 =& \frac{1}{r^2}\frac{\partial}{\partial r}\left(r^2\frac{\partial}{\partial r}\right) \\ &+ \frac{1}{r^2\sin\theta}\frac{\partial}{\partial\theta}\left(\sin\theta\frac{\partial}{\partial\theta}\right) + \frac{1}{r^2\sin^2\theta}\frac{\partial^2}{\partial\varphi^2}.\end{aligned} \tag{4.34}$$

When we assume that the Hartree potential $\mathcal{V}_H$ has a spherical symmetry, the derivatives of $\mathcal{V}_H$ with respect to θ and φ become zero. Then the Poisson equation is given by a differential equation on r as follows:

$$\frac{1}{r^2}\frac{\partial}{\partial r}\left(r^2\frac{\partial\mathcal{V}_H}{\partial r}\right) = -4\pi n(r). \tag{4.35}$$

Integrating Eq. (4.35) on r we obtain

$$\mathcal{V}_H(r) = -4\pi\int_0^r \frac{1}{r'^2}\int_0^{r'} r''^2 n(r'')\mathrm{d}r'\mathrm{d}r'' - \frac{C}{r}. \tag{4.36}$$

It is noted that C/r in Eq. (4.36) is a solution of Eq. (4.35) for any value of C. Then by substituting the electron density of the helium atom in Eq. (4.21) into Eq. (4.36) and using

$$\int_0^x \frac{1}{x'^2}\int_0^{x'} x''^2 \exp(x'')\mathrm{d}x'\mathrm{d}x'' = (1 - \frac{2}{x})\exp(x), \tag{4.37}$$

Eq. (4.36) becomes

$$\mathcal{V}_H(r) = -2\left(Z + \frac{1}{r}\right)\exp(-2Zr) - \frac{C}{r}. \tag{4.38}$$

At the limit of $r \to \infty$, from Eq. (4.33), we have $\mathcal{V}_H(r) \xrightarrow{r\to\infty} \frac{1}{r}\int n(r')\mathrm{d}r' = \frac{Z}{r}$. Applying this boundary condition for Eq. (4.38), we obtain $C = -Z$, and Eq. (4.38) can be rewritten for the helium atom

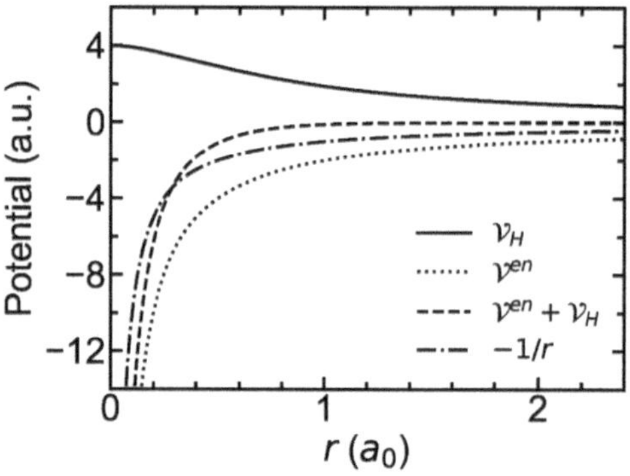

Figure 4.5 The Hartree potential $\mathcal{V}_H(r)$ (solid line), the Coulomb attraction between electrons and nuclei $\mathcal{V}_{en}$ (dotted line), and the total potential $\mathcal{V}_{en} + \mathcal{V}_H(r)$ (dashed line) of helium atom are plotted in atomic unit as functions of r.

as

$$\mathcal{V}_H(r) = -2\left(Z + \frac{1}{r}\right)\exp(-2Zr) + \frac{Z}{r}. \tag{4.39}$$

In Fig. 4.5, we show $\mathcal{V}_H(r)$ with $Z = 2$ (solid line), the Coulomb attractive interaction between electrons and nuclei $\mathcal{V}_{en}(r) = -2/r$ (dotted line), and the total potential $\mathcal{V}_{en}(r) + \mathcal{V}_H(r)$ (dashed line) of the helium atom as functions of r. When r increases, the total potential decays exponentially. This trend is incorrect since the total potential of the helium atom decays as $-1/r$ at a large distance [Umrigar and Gonze (1994)]. This is because we neglect some interactions, which will be introduced in the next section.

4.6 Exchange potential

In Sec. 4.5, we show that the Schrödinger equation can be solved by using the Hartree approximation and the SCF method. In the Hartree approximation, the Coulomb repulsion between two electrons in the system is approximated by the Hartree potential, in which each individual electron moves independently of each other, only feeling the averaged electrostatic potential by the other electrons. However, in 1930, Slater pointed out that the Hartree method did not take into account the Pauli exclusion principle for many electrons [Slater (1930b)]. That is, if two electrons have parallel spins to each other, they are not allowed to occupy the same states at the same time. Since the two electrons can not be close to each other by the Pauli principle, the Coulomb repulsion can not be very large. In order to take account of the Pauli exclusion principle, we introduce an **exchange potential**, which is an attractive Coulomb interaction between two electrons with parallel spins. We note that the Hartree potential is a repulsive Coulomb interaction between an electron and electron density of the other electrons ($\mathcal{V}_H > 0$, see Fig. 4.5). Exchange potential is a correction term to $\mathcal{V}_H$, which is over-estimated compared with the real case.

By using the Slater determinant in Eq. (4.14), Eq. (4.8) can be rewritten as

$$\begin{aligned}\left[h(\boldsymbol{r}_1) + h(\boldsymbol{r}_2) + \frac{1}{|\boldsymbol{r}_1 - \boldsymbol{r}_2|}\right][\phi_a(\boldsymbol{r}_1)\phi_b(\boldsymbol{r}_2) - \phi_b(\boldsymbol{r}_1)\phi_a(\boldsymbol{r}_2)] \\ = E[\phi_a(\boldsymbol{r}_1)\phi_b(\boldsymbol{r}_2) - \phi_b(\boldsymbol{r}_1)\phi_a(\boldsymbol{r}_2)].\end{aligned} \tag{4.40}$$

By multiplying $\phi_b^*(\boldsymbol{r}_2)$ in Eq. (4.40) and integrating over $\boldsymbol{r}_2$ with the normalization and orthogonality conditions (Eqs. (4.17) and (4.18)), we obtain:

$$\begin{aligned}h(\boldsymbol{r}_1)\phi_a(\boldsymbol{r}_1) + \phi_a(\boldsymbol{r}_1)\int \frac{\phi_b^*(\boldsymbol{r}_2)\phi_b(\boldsymbol{r}_2)}{|\boldsymbol{r}_1 - \boldsymbol{r}_2|}\mathrm{d}\boldsymbol{r}_2 \\ - \int \frac{\phi_b^*(\boldsymbol{r}_2)\phi_b(\boldsymbol{r}_1)}{|\boldsymbol{r}_1 - \boldsymbol{r}_2|}\phi_a(\boldsymbol{r}_2)\mathrm{d}\boldsymbol{r}_2 = \epsilon_a^{HF}\phi_a(\boldsymbol{r}_1),\end{aligned} \tag{4.41}$$

where ϵ_a^{HF} is defined by

$$\epsilon_a^{HF} = E - \int \phi_b^*(\boldsymbol{r}_2)h(\boldsymbol{r}_2)\phi_b(\boldsymbol{r}_2)\mathrm{d}\boldsymbol{r}_2 = E - \epsilon_b. \tag{4.42}$$

By substituting Eq. (4.29) into Eq. (4.41), we obtain the single-particle Schrödinger equation for $\phi_a(\boldsymbol{r}_1)$ as

$$[h(\boldsymbol{r}_1) + \mathcal{V}_H(\boldsymbol{r}_1)]\,\phi_a(\boldsymbol{r}_1) - \int \frac{\gamma(\boldsymbol{r}_2, \boldsymbol{r}_1)}{|\boldsymbol{r}_1 - \boldsymbol{r}_2|}\phi_a(\boldsymbol{r}_2)\mathrm{d}\boldsymbol{r}_2 = \epsilon_a^{HF}\phi_a(\boldsymbol{r}_1), \quad (4.43)$$

where $\gamma(\boldsymbol{r}_2, \boldsymbol{r}_1)$ is defined as

$$\gamma(\boldsymbol{r}_2, \boldsymbol{r}_1) = \phi_a^*(\boldsymbol{r}_2)\phi_a(\boldsymbol{r}_1) + \phi_b^*(\boldsymbol{r}_2)\phi_b(\boldsymbol{r}_1). \quad (4.44)$$

The single-particle Schrödinger equation also can be written for $\phi_a(\boldsymbol{r}_2)$ as

$$[h(\boldsymbol{r}_2) + \mathcal{V}_H(\boldsymbol{r}_2)]\,\phi_b(\boldsymbol{r}_2) - \int \frac{\gamma(\boldsymbol{r}_1, \boldsymbol{r}_2)}{|\boldsymbol{r}_1 - \boldsymbol{r}_2|}\phi_b(\boldsymbol{r}_1)\mathrm{d}\boldsymbol{r}_1 = \epsilon_b^{HF}\phi_b(\boldsymbol{r}_2), \quad (4.45)$$

in which ϵ_b^{HF} is symmetrically given by

$$\epsilon_b^{HF} = E - \epsilon_a. \quad (4.46)$$

Eq. (4.43) or (4.45) is known as the **Hartree-Fock equation** [Fock (1930)], in which the exchange potential $\mathcal{V}_x$ in Eq. (4.43) is defined by

$$\mathcal{V}_x(\boldsymbol{r}_1, \boldsymbol{r}_2) = -\frac{\gamma(\boldsymbol{r}_2, \boldsymbol{r}_1)}{|\boldsymbol{r}_1 - \boldsymbol{r}_2|}. \quad (4.47)$$

$\mathcal{V}_x$ exists only for two electrons at $\boldsymbol{r}_1$ and $\boldsymbol{r}_2$, which have the same quantum state ϕ_a or ϕ_b including spin, i.e., the exchange potential is the effective Coulomb interaction between the electrons with parallel spins in the system. The origin of $\mathcal{V}_x$ is that two-electrons with the same spin can not exist at the same position ($\boldsymbol{r}_1 = \boldsymbol{r}_2$), in which $\mathcal{V}_H$ should be small by the restriction of the Pauli principle. By considering both $\mathcal{V}_H$ and $\mathcal{V}_x$ in Eq. (4.39), two electrons are interacted not only by their electronic charge but also by their spins. Therefore, the *quantum* behavior of electrons is included by $\mathcal{V}_x$ term in the Hartree-Fock equation. It is important to note that $\mathcal{V}_x$ is a *non-local* potential on $\boldsymbol{r}_1$ since it depends on an integration over the additional variable $\boldsymbol{r}_2$. Therefore, the practical solution of the Hartree-Fock equation becomes more complicated and time-consuming.

Let us discuss the energy that is obtained by the Hartree-Fock equation. From Eq. (4.43), we can write the energy by integrating the wavefunctions as

$$\begin{aligned}\epsilon_a^{HF} =& \int \phi_a^*(\boldsymbol{r}_1)[h(\boldsymbol{r}_1) + \mathcal{V}_H(\boldsymbol{r}_1)]\phi_a(\boldsymbol{r}_1)\mathrm{d}\boldsymbol{r}_1 \\ &+ \iint \phi_a^*(\boldsymbol{r}_1)\frac{\gamma(\boldsymbol{r}_2,\boldsymbol{r}_1)}{|\boldsymbol{r}_1 - \boldsymbol{r}_2|}\phi_a(\boldsymbol{r}_2)\mathrm{d}\boldsymbol{r}_1\boldsymbol{r}_2 \\ =& \epsilon_a + J - K,\end{aligned} \tag{4.48}$$

where J and K are called the **Coulomb and exchange integrals**, respectively, which are given by

$$J = \iint \frac{|\phi_a(\boldsymbol{r}_1)|^2|\phi_b(\boldsymbol{r}_2)|^2}{|\boldsymbol{r}_1 - \boldsymbol{r}_2|}\mathrm{d}\boldsymbol{r}_1\mathrm{d}\boldsymbol{r}_2 \tag{4.49}$$

and

$$K = \iint \frac{\phi_a^*(\boldsymbol{r}_1)\phi_b(\boldsymbol{r}_1)\phi_b^*(\boldsymbol{r}_2)\phi_a(\boldsymbol{r}_2)}{|\boldsymbol{r}_1 - \boldsymbol{r}_2|}\mathrm{d}\boldsymbol{r}_1\mathrm{d}\boldsymbol{r}_2. \tag{4.50}$$

J represents the Coulomb repulsion between the electron at the state $\phi_a(\boldsymbol{r}_1)$ and the other electron at the state $\phi_b(\boldsymbol{r}_2)$. K has a physical interpretation of the overestimate of the Coulomb interaction for two electrons in the case of the parallel spins. It means that with the parallel spins, the two electrons can not be close to each other by the Pauli exclusion principle. It is noted that both values of J and K are the positive values and $J \geq K$, as shown below [Roothaan (1951)]:

By substituting Eq. (4.46) into Eq. (4.48), the total electron energy E in the Hartree-Fock approximation is expressed by

$$E = \epsilon_a + \epsilon_b + J - K. \tag{4.51}$$

On the other hand, from Eq. (4.40), the total electron energy can be obtained, too, as

$$\begin{aligned} E &= \iint \psi^*(\boldsymbol{r}_1,\boldsymbol{r}_2)[h(\boldsymbol{r}_1)+h(\boldsymbol{r}_2)]\psi(\boldsymbol{r}_1,\boldsymbol{r}_2)\mathrm{d}\boldsymbol{r}_1\mathrm{d}\boldsymbol{r}_2 \\ &\quad + \iint \psi^*(\boldsymbol{r}_1,\boldsymbol{r}_2)\frac{1}{|\boldsymbol{r}_1-\boldsymbol{r}_2|}\psi(\boldsymbol{r}_1,\boldsymbol{r}_2)\mathrm{d}\boldsymbol{r}_1\boldsymbol{r}_2 \\ &= \epsilon_a+\epsilon_b+\iint \frac{|\psi(\boldsymbol{r}_2,\boldsymbol{r}_1)|^2}{|\boldsymbol{r}_1-\boldsymbol{r}_2|}\mathrm{d}\boldsymbol{r}_1\boldsymbol{r}_2. \end{aligned} \tag{4.52}$$

By subtracting Eq. (4.52) from Eq. (4.51), we obtain:

$$J-K=\iint \frac{|\psi(\boldsymbol{r}_1,\boldsymbol{r}_2)|^2}{|\boldsymbol{r}_1-\boldsymbol{r}_2|}\mathrm{d}\boldsymbol{r}_1\mathrm{d}\boldsymbol{r}_2 \geq 0. \tag{4.53}$$

Therefore, $J \geq K$ and the total electron energy E is always larger than the sum of the one-electron energies $(\epsilon_a+\epsilon_b)$ because of the net repulsion Coulomb interaction between two electrons.

❑ **Total energy of helium atom:** In the case of the helium atom, two electrons at the 1s state ϕ must have different spin directions, up $\uparrow$ and down $\downarrow$ because of the Pauli principle. Let us consider the spin states of two electrons, respectively, as

$$\phi_\uparrow(\boldsymbol{r})=\phi(\boldsymbol{r})\begin{pmatrix}1\\0\end{pmatrix} \text{ and } \phi_\downarrow(\boldsymbol{r})=\phi(\boldsymbol{r})\begin{pmatrix}0\\1\end{pmatrix}, \tag{4.54}$$

where $\binom{1}{0}$ and $\binom{0}{1}$ correspond to up- and down-spin wavefunctions, respectively. From Eq. (4.54), we define the product of two spin-functions as follows:

$$\phi_\uparrow^*(\boldsymbol{r})\phi_\uparrow(\boldsymbol{r})=\phi^2(\boldsymbol{r})\,(1\ 0)\begin{pmatrix}1\\0\end{pmatrix}=\phi^2(\boldsymbol{r}), \tag{4.55}$$

and

$$\phi_\uparrow^*(\boldsymbol{r})\phi_\downarrow(\boldsymbol{r})=\phi^2(\boldsymbol{r})\,(1\ 0)\begin{pmatrix}0\\1\end{pmatrix}=0. \tag{4.56}$$

By substituting Eqs. (4.55) and (4.56) into Eqs. (4.49) and (4.50), respectively, we obtain:

$$J = \iint \frac{\phi_\uparrow^*(\mathbf{r})\phi_\uparrow(\mathbf{r})\phi_\downarrow^*(\mathbf{r}')\phi_\downarrow(\mathbf{r}')}{|\mathbf{r}-\mathbf{r}'|^2} \mathrm{d}\mathbf{r}\mathrm{d}\mathbf{r}' = \iint \frac{|\phi(\mathbf{r})|^2\,|\phi(\mathbf{r}')|^2}{|\mathbf{r}-\mathbf{r}'|^2} \mathrm{d}\mathbf{r}\mathrm{d}\mathbf{r}', \tag{4.57}$$

and

$$K = \iint \frac{\phi_\uparrow^*(\mathbf{r})\phi_\downarrow(\mathbf{r})\phi_\downarrow^*(\mathbf{r}')\phi_\uparrow(\mathbf{r}')}{|\mathbf{r}-\mathbf{r}'|^2} \mathrm{d}\mathbf{r}\mathrm{d}\mathbf{r}' = 0. \tag{4.58}$$

$K = 0$ is consistent with fact that the exchange potential is zero since there are no parallel spins. Since the electronic density is expressed as $n(\mathbf{r}') = |\phi_\uparrow(\mathbf{r}')|^2 + |\phi_\downarrow(\mathbf{r}')|^2 = 2|\phi(\mathbf{r}')|^2$, the Coulomb integral in Eq. (4.57) can be rewritten as

$$J = \frac{1}{2}\iint \frac{|\phi(\mathbf{r})|^2 n(\mathbf{r}')}{|\mathbf{r}-\mathbf{r}'|^2} \mathrm{d}\mathbf{r}\mathrm{d}\mathbf{r}' = \frac{1}{2}\int |\phi(\mathbf{r})|^2 \mathcal{V}_H(\mathbf{r})\mathrm{d}\mathbf{r}. \tag{4.59}$$

By substituting J in Eq. (4.59) and $K = 0$ in Eq. (4.58) into Eq. (4.51), the total electron energy E is expressed as

$$E = 2\epsilon + \frac{1}{2}\int |\phi(\mathbf{r})|^2 \mathcal{V}_H(\mathbf{r})\mathrm{d}\mathbf{r}. \tag{4.60}$$

In order to apply Eq. (4.60) for the helium atom, we define an *effective* atomic number Z, i.e., $\phi(\mathbf{r}, Z)$, which gives minimized E when subject to the Hartree potential $\mathcal{V}_H(\mathbf{r})$. Eq. (4.60) can be obtained by the following steps:

- Step 1: Using a beginning value $Z = 2$, from Eq. (4.39), the Hartree potential $\mathcal{V}_H(\mathbf{r})$ is given by

$$\mathcal{V}_H(r) = -2\left(2 + \frac{1}{r}\right)\exp(-4r) + \frac{2}{r}. \tag{4.61}$$

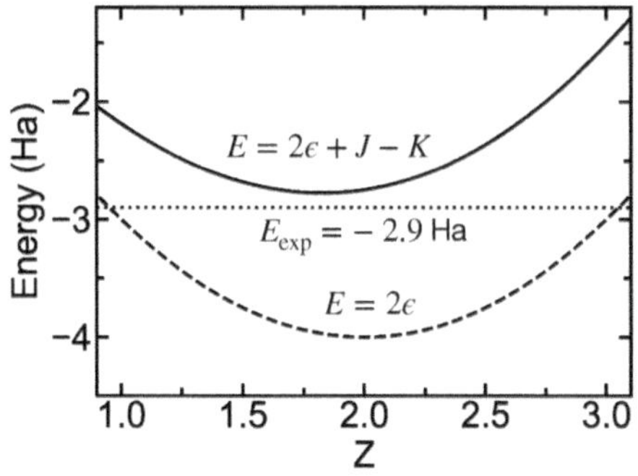

Figure 4.6 Total electron energy of helium atom as a function of effective atomic number Z. $E = 2\epsilon$ and $E = 2\epsilon + J - K$ are corresponding to the cases of non-interacting electrons and Hartree-Fock approximation, respectively. Experimental value E_{exp} of the helium atom is -2.9 Ha [Sucher (1958)].

- Step 2: By substituting Eq. (4.61) and $\phi(\mathbf{r}, Z) = \frac{Z^{3/2}}{\sqrt{\pi}} \exp(-Z|\mathbf{r}|)$ in Eq. (4.20) into Eq. (4.60), we obtain:

$$
\begin{aligned}
E &= \underbrace{Z^2 - 4Z}_{2\epsilon} + 2\pi \int r^2 |\phi(\mathbf{r}, Z)|^2 \mathcal{V}_H(r) \mathrm{d}r \\
&= Z^2 - 3Z + \frac{Z^3(Z+4)}{(Z+2)^3}.
\end{aligned}
\tag{4.62}
$$

- Step 3: According to the variational principle,[6] we minimize E with respect to the variational parameter Z, i.e., $\mathrm{d}E/\mathrm{d}Z = 0$ for E in Eq. (4.62). We can obtain the effective atomic number $Z_{\text{eff}} = 1.826$ and the value of E at $Z = Z_{\text{eff}}$ is about -2.777 Ha.

In Fig. 4.6, we show the energies for the two cases of the non-interacting electrons, $E = 2\epsilon$, and the Hartree-Fock approximation, $E = 2\epsilon + J - K$, as functions of the effective atomic number Z. For $E = 2\epsilon$, the electrons feel a potential corresponding to the nucleus with a charge $Z = 2$, but it makes the error for E as discussed in Sec. 4.3. For $E = 2\epsilon + J - K$, the electrons feel a potential with a reduced charge $Z = 1.826$. Eq. (4.60) can also be solved by the SCF method. This

[6] The variational principle states that the energy of any approximate wavefunction is higher than or equal to the energy of ground-state wavefunction. The lower the energy is the better the approximation to the ground state.

could be done by recalculating the Hartree potential with new state $\phi(\mathbf{r}, Z_{\text{eff}} = 1.826)$ in Step 1, then we recalculate Steps 2 and 3 until the value of Z_{eff} does not change significantly. Finally, we can obtain the $E_{\text{eff}} = -2.85$ Ha, which is not too far from the exact value from the experiment about -2.9 Ha. Slater derived a formula to determine Z_{eff} as $Z_{\text{eff}} = Z - S$, where S is the shielding constant, which is known as **Slater's rule** [Slater (1930a)]. For the 1s state in the helium atom, $S = 0.3$ and $Z_{\text{eff}} = 1.7$, which is close to the result of the SCF method ($Z_{\text{eff}} = 1.826$).

4.7 Correlation potential

In Sec. 4.6, we show that due to the Pauli exclusion principle, the electrons with parallel spins have an exchange term of the Coulomb interaction $\mathcal{V}_x$. The two electrons with the parallel spin move to avoid each other because the electrons can not overlap. Avoiding picture should exist for the two electrons with anti-parallel spins because these electrons also move with keeping apart to lower the Coulomb repulsion. This behavior of the electrons with anti-parallel spins is missing in the Hartree approximation as well as in the Hartree-Fock approximation. A complete picture of the motion of an electron in a system is shown in Fig. 4.7. In general, the correlation interaction, $\mathcal{V}_c$, is defined by the remaining term that is missing in the Hartree-Fock method.[7] Thus, an **effective potential** in the single-particle Schrödinger equation is defined by

$$\mathcal{V}_{\text{eff}} = \mathcal{V}_{en} + \mathcal{V}_H + \mathcal{V}_{xc}, \tag{4.63}$$

where $\mathcal{V}_{xc} \equiv \mathcal{V}_x + \mathcal{V}_c$ is called the **exchange-correlation potential.**

Since $\mathcal{V}_H$ is obtained by the total electron density $n(\mathbf{r})$, it suggests that if $\mathcal{V}_{xc}$ can be expressed by a functional of $n(\mathbf{r})$, the single-particle Schrödinger equation can be written as

$$\left\{-\frac{\nabla_r^2}{2} + \mathcal{V}_{\text{eff}}[n(\mathbf{r})]\right\}\phi(\mathbf{r}) = \epsilon\phi(\mathbf{r}). \tag{4.64}$$

[7] The correlation interaction is defined in the field of the density-functional theory.

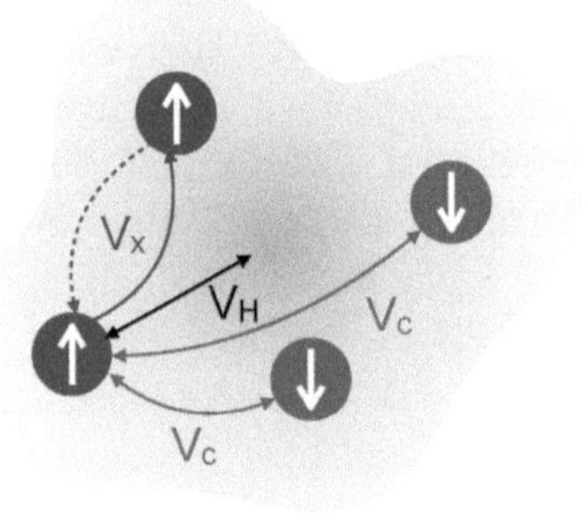

Figure 4.7 The motion of an electron with up-spin in a system is affected by three interactions, including the Hartree $\mathcal{V}_H$, exchange $\mathcal{V}_x$, and correlation $\mathcal{V}_c$ potentials. $\mathcal{V}_H$ is the Coulomb interaction between one electron and the total charge density, which is generated by all electrons in the system. $\mathcal{V}_x$ arises due to the Pauli exclusion principle, in which the electrons with parallel spins can not exist at the same position at the same time. $\mathcal{V}_c$ arises due to the fact that the electrons with anti-parallel spins also keep apart to lower their mutual Coulomb repulsion.

When we know the value of $\mathcal{V}_{xc}[n(\boldsymbol{r})]$, Eq. (4.64) can be solved by using the SCF method as discussed in Sec. 4.5. However, we do not know the exact shape of $\mathcal{V}_{xc}[n(\boldsymbol{r})]$. Many people proposed the many functionals for $\mathcal{V}_{xc}[n(\boldsymbol{r})]$ over the past few decades. The efforts to develop the accurate approximations for $\mathcal{V}_{xc}[n(\boldsymbol{r})]$ led to a theory, the so-called density-functional theory (DFT). We will discuss these approximations in the next section.

4.8 Early DFT for free-electron gas

Let us consider how to obtain $\mathcal{V}_{xc}[n(\boldsymbol{r})]$ for a free-electron gas, where the contribution of ions is treated as a uniform-positive-charge background, which is called the Jellium model [Brack (1993)], as shown in Fig. 4.8. Since the potential is uniform, the electronic states are expressed by plane waves as (see Sec. 5.3)

$$\phi_{\boldsymbol{k},\sigma}(\boldsymbol{r}) = \frac{1}{\sqrt{V}} \exp(i\boldsymbol{k}\boldsymbol{r})\chi_{\sigma}, \tag{4.65}$$

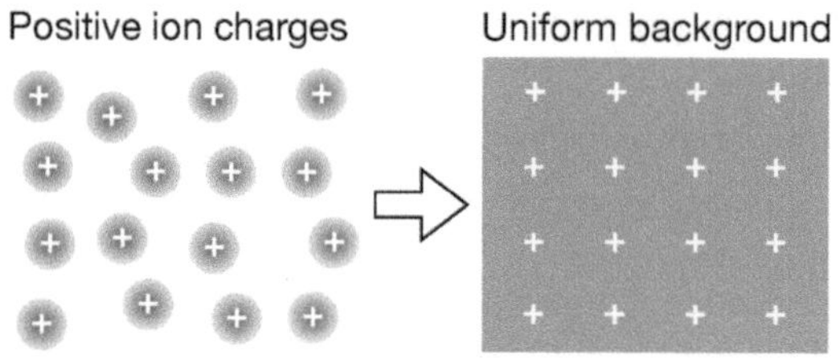

Figure 4.8 A free-electron gas, where the contribution of nuclear ions is treated as a uniform-positive-charge background.

where V is the volume of the system, $\boldsymbol{k}$ is the wavevector, and χ_σ is the spin function, in which $\binom{1}{0}$ for spin-up and $\binom{0}{1}$ for spin-down. Since the unit cell can not be defined by the uniform background, the value of $\boldsymbol{k}$ is taken not from the Brillouin zone but all the $\boldsymbol{k}$ space. Since $\phi_{\boldsymbol{k},\sigma}(\boldsymbol{r})$ is a one-electron wavefunction, $\phi_{\boldsymbol{k},\sigma}(\boldsymbol{r})$ can be a solution of the Hartree-Fock equation in Eq. (4.43). Moreover, in the free-electron gas, the electron density of the ground state is also a uniform-negative-charge background. Therefore, the electron density cancels the positive charge background, i.e., $\mathcal{V}_{en} + \mathcal{V}_H = 0$ and only the exchange term $\mathcal{V}_x$ survives in Eq. (4.43). Thus, the single-particle Schrödinger equation of a free-electron gas can be written as

$$-\frac{\nabla_{\boldsymbol{r}}}{2}\phi_{\boldsymbol{k},\sigma}(\boldsymbol{r}) + \underbrace{\int \mathcal{V}_x(\boldsymbol{r}',\boldsymbol{r})\phi_{\boldsymbol{k},\sigma}(\boldsymbol{r}')\mathrm{d}\boldsymbol{r}'}_{I_x} = \epsilon_{\boldsymbol{k}}\phi_{\boldsymbol{k},\sigma}(\boldsymbol{r}), \tag{4.66}$$

where the integral I_x is given by

$$\begin{aligned} I_x &= -\sum_{\sigma'}\sum_{\boldsymbol{k}'}\int \frac{\phi^*_{\boldsymbol{k}',\sigma'}(\boldsymbol{r}')\phi_{\boldsymbol{k}',\sigma'}(\boldsymbol{r})}{|\boldsymbol{r}-\boldsymbol{r}'|}\phi_{\boldsymbol{k},\sigma}(\boldsymbol{r}')\mathrm{d}\boldsymbol{r}' \\ &= -\sum_{\sigma'}\sum_{\boldsymbol{k}'}\phi_{\boldsymbol{k}',\sigma'}(\boldsymbol{r})\int \frac{\phi^*_{\boldsymbol{k}',\sigma'}(\boldsymbol{r}')\phi_{\boldsymbol{k},\sigma}(\boldsymbol{r}')}{|\boldsymbol{r}-\boldsymbol{r}'|}\mathrm{d}\boldsymbol{r}'. \end{aligned} \tag{4.67}$$

By substituting the electronic states in Eq. (4.65) into Eq. (4.67), we obtain:

$$\begin{aligned} I_x &= -\sum_{\boldsymbol{k}'} \frac{1}{V^{3/2}} \exp(i\boldsymbol{k}'\boldsymbol{r})\chi_\sigma \int \frac{\exp\left(-i(\boldsymbol{k}'-\boldsymbol{k})\boldsymbol{r}'\right)}{|\boldsymbol{r}-\boldsymbol{r}'|} \mathrm{d}\boldsymbol{r}' \\ &= \left[-\sum_{\boldsymbol{k}'} \frac{1}{V} \int \frac{\exp\left(-i(\boldsymbol{k}'-\boldsymbol{k})(\boldsymbol{r}'-\boldsymbol{r})\right)}{|\boldsymbol{r}-\boldsymbol{r}'|} \mathrm{d}\boldsymbol{r}'\right] \phi_{\boldsymbol{k},\sigma}(\boldsymbol{r}) \\ &= \left[-\sum_{\boldsymbol{k}'} f(\boldsymbol{k}'-\boldsymbol{k})\right] \phi_{\boldsymbol{k},\sigma}(\boldsymbol{r}), \end{aligned} \tag{4.68}$$

where the function f is defined as

$$f(\boldsymbol{q}) = \frac{1}{V} \int \frac{\exp(-i\boldsymbol{q}\boldsymbol{x})}{|\boldsymbol{x}|} \mathrm{d}\boldsymbol{x} = \frac{1}{V}\frac{4\pi}{\boldsymbol{q}^2}, \tag{4.69}$$

with $\boldsymbol{q} = \boldsymbol{k}' - \boldsymbol{k}$ and $\boldsymbol{x} = \boldsymbol{r}' - \boldsymbol{r}$. By substituting Eq. (4.69) into Eq. (4.68), we obtain:

$$\begin{aligned} I_x &= \left[-\frac{1}{V}\sum_{\boldsymbol{k}'} \frac{4\pi}{|\boldsymbol{k}'-\boldsymbol{k}|^2}\right] \phi_{\boldsymbol{k},\sigma}(\boldsymbol{r}) \\ &= \left[-\frac{1}{(2\pi)^3} \int_0^{k_F} \frac{4\pi}{|\boldsymbol{k}'-\boldsymbol{k}|^2} \mathrm{d}\boldsymbol{k}'\right] \phi_{\boldsymbol{k},\sigma}(\boldsymbol{r}) \\ &= -\frac{k_F}{\pi} F\left(\frac{k}{k_F}\right) \phi_{\boldsymbol{k},\sigma}(\boldsymbol{r}), \end{aligned} \tag{4.70}$$

where $k = |\boldsymbol{k}|$, k_F is the Fermi wavevector, and F is called the **Lindhard function**:

$$F(x) = 1 + \frac{1-x^2}{2x} \ln\left|\frac{1+x}{1-x}\right|, \text{ with } x = \frac{k}{k_F}. \tag{4.71}$$

The function $F(x)$ is shown in Fig. 4.9. $F(x)$ monotonically decreases from 2 at $x = 0$ (or $k = 0$) to 1 at $x = 1$ (or $k = k_F$).

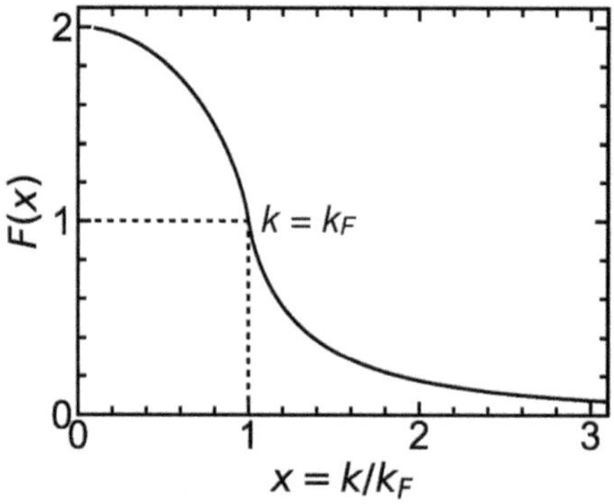

Figure 4.9 The Lindhard function $F(x)$ is plotted as a function of the dimensionless $x = k/k_F$.

By substituting Eq. (4.70) into Eq. (4.66), Eq. (4.66) can be rewritten as

$$\left[-\frac{\nabla_{\boldsymbol{r}}}{2} - \frac{k_F}{\pi}F\left(\frac{k}{k_F}\right)\right]\phi_{\boldsymbol{k},\sigma}(\boldsymbol{r}) = \epsilon_{\boldsymbol{k}}\phi_{\boldsymbol{k},\sigma}(\boldsymbol{r}). \tag{4.72}$$

Then we can obtain the energy as a function of wavevector $\boldsymbol{k}$ as

$$\epsilon_{\boldsymbol{k},\sigma} = \frac{k^2}{2} - \frac{k_F}{\pi}F\left(\frac{k}{k_F}\right). \tag{4.73}$$

By taking the sum of all occupied states, we obtain the total energy as

$$\begin{aligned}
E &= \sum_{\sigma}\sum_{|\boldsymbol{k}|<k_F}\frac{k^2}{2} - \frac{1}{2}\sum_{\sigma}\sum_{|\boldsymbol{k}|<k_F}\frac{k_F}{\pi}F\left(\frac{k}{k_F}\right) \\
&= \frac{2V}{(2\pi)^3}\int_0^{k_F}\frac{k^2}{2}\mathrm{d}\boldsymbol{k} - \frac{k_F}{\pi}\frac{V}{(2\pi)^3}\int_0^{k_F}F\left(\frac{k}{k_F}\right)\mathrm{d}\boldsymbol{k} \\
&= \frac{V}{4\pi^3}4\pi\int_0^{k_F}\frac{k^4}{2}\mathrm{d}k - \frac{k_F V}{8\pi^4}4\pi\int_0^{k_F}k^2F\left(\frac{k}{k_F}\right)\mathrm{d}k \\
&= \frac{V}{10\pi^2}k_F^5 - \frac{V}{2\pi^3}k_F^4\underbrace{\int_0^1 x^2F(x)\mathrm{d}x}_{=1/2} \\
&= \frac{V}{10\pi^2}k_F^5 - \frac{V}{4\pi^3}k_F^4,
\end{aligned} \tag{4.74}$$

with the factor of 1/2 in the second term of the right-hand side in the first line of Eq. (4.74) is needed to compensate for double counting of the exchange interaction.

For the free-electron gas, the number of electrons N_e is equal to the number of occupied states in the systems:

$$N_e = \sum_{\sigma}\sum_{\boldsymbol{k}} = 2V\int_0^{k_F}\frac{\mathrm{d}\boldsymbol{k}}{(2\pi)^3} = \frac{V}{\pi^2}\int_0^{k_F}k^2\mathrm{d}k = V\frac{k_F^3}{3\pi^2}. \quad (4.75)$$

Therefore, the electron density n is expressed by

$$n = \frac{N_e}{V} = \frac{k_F^3}{3\pi^2}. \quad (4.76)$$

By substituting Eqs. (4.75) and (4.76) into Eq. (4.74), the total energy per electron is expressed by

$$\frac{E}{N_e} = C_1 n^{2/3} + C_2 n^{1/3}, \quad (4.77)$$

where C_1 and C_2 are constants given by

$$C_1 = \frac{3}{10}(3\pi^2)^{2/3} = 2.871 \text{ and } C_2 = -\frac{3}{4}\left(\frac{3}{\pi}\right)^{1/3} = -0.738. \quad (4.78)$$

The first term ($C_1 n^{2/3}$) in Eq. (4.77) represents the kinetic energy, and the second term ($C_2 n^{1/3}$) represents the effective electron-electron interaction due to exchange potential. We can see that the exchange potential reduces the total energy of the system from the kinetic energy.

For a system with the free electron gas, n is a constant. However, for a system with the non-uniform-electron charge, n becomes a function of $\boldsymbol{r}$. Slater [Slater (1951)] introduced the exchange potential as a function of $n(\boldsymbol{r})$ as

$$\mathcal{V}_x[n(\boldsymbol{r})] = 2C_2 n^{1/3}(\boldsymbol{r}) = -\frac{3}{2}\left(\frac{3}{\pi}\right)^{1/3} n^{1/3}(\boldsymbol{r}), \quad (4.79)$$

where an extra factor of 2 is introduced to account in Eq. (4.79) because the $\mathcal{V}_x$ term in the single-particle equation (Eq. (4.64)) is twice as large as the corresponding term in the total energy per particle (Eq. (4.77)).

For an approximation to include the correlation effects, Slater and Johnson [Slater and Johnson (1972)] introduced a *"fudge factor"* α in Eq. (4.79) as

$$\mathcal{V}_{xc}[n(\mathbf{r})] = 2\alpha C_2 n^{1/3}(\mathbf{r}), \tag{4.80}$$

where α usually taken in the range $2/3 < \alpha < 1$. For example, $\alpha =$ 0.772 and 0.978 [Schwarz (1972)] for He and H atoms, respectively. Eq. (4.80) is called the $\boldsymbol{X\alpha}$ **potential**.

4.9 Thomas-Fermi-Dirac theory

In Sec. 4.8, we consider a free-electron gas with a uniform-positive background. However, the ions are not represented by a uniform-positive background in the real materials. The Thomas-Fermi-Dirac theory gives a first approximation to calculate the total energy of non-uniform systems. It is important to note that both papers [Thomas (1927), Fermi (1928)] published by Thomas and Fermi in 1927 and 1928, respectively, before the development of the Hartree-Fock theory in 1930. Although Thomas and Fermi neglected exchange and correction potentials, the approximation was extended by Dirac [Dirac (1930)] in 1930 based on the Hartree-Fock theory.

Let us consider an atom with N_e electrons, and thus the atomic number is $Z = N_e$. We assume that the presence of the non-uniform-positive charge does not change the Hartree-Fock results significantly (in Sec. 4.8) for kinetic and exchange energies of the uniform background. In this case, the total energy per atom in Eq. (4.77) is modified by adding two terms as follows:

$$\frac{E}{N_e} = C_1 n^{2/3} + C_2 n^{1/3} - \frac{Z}{|\mathbf{r}|} + \frac{1}{2}\int \frac{n(\mathbf{r}')}{|\mathbf{r}-\mathbf{r}'|}\mathrm{d}\mathbf{r}', \tag{4.81}$$

where $Z/|\mathbf{r}|$ is the external potential that each electron feels the presence of the ions and the last term is the Coulomb repulsion between an electron at position $\mathbf{r}$ and all other electrons at $\mathbf{r}'$, which is expressed by the density $n(\mathbf{r}')$. The factor of 1/2 of the last term is given by avoiding the double-counting of the Coulomb interaction. For the non-uniform systems, N_e is defined by

$$N_e = \int n(\boldsymbol{r})\mathrm{d}\boldsymbol{r}. \tag{4.82}$$

By substituting Eq. (4.82) into Eq. (4.81), the total energy of the system is expressed as

$$\begin{aligned} E[n(\boldsymbol{r})] = C_1 \int n^{5/3}(\boldsymbol{r})\mathrm{d}\boldsymbol{r} + C_2 \int n^{4/3}(\boldsymbol{r})\mathrm{d}\boldsymbol{r} \\ - Z\int \frac{n(\boldsymbol{r})}{|\boldsymbol{r}|}\mathrm{d}\boldsymbol{r} + \frac{1}{2}\iint \frac{n(\boldsymbol{r})n(\boldsymbol{r}')}{|\boldsymbol{r}-\boldsymbol{r}'|}\mathrm{d}\boldsymbol{r}'\mathrm{d}\boldsymbol{r}, \end{aligned} \tag{4.83}$$

where the first and second terms of the right-hand side are the kinetic and exchange energies of the free-electron gas, respectively, the third term is the Coulomb attraction between the ions and the electron density, and the last term describes Coulomb repulsion between the electrons. $n(\boldsymbol{r})$ for the ground-state and the total energy of the system can be obtained by minimizing the energy functional $E[n(\boldsymbol{r})]$. Although the Thomas-Fermi-Dirac approximation is not accurate good-enough compared with the present electronic structure calculation, the approach shows a prototype expression of the DFT, in which $n(\boldsymbol{r})$ is a crucial physical quantity to calculate the ground-state properties of a many-electron system.

4.10 DFT: Hohenberg-Kohn-Sham

The DFT is established by two papers by Hohenberg and Kohn in 1964 [Hohenberg and Kohn (1964)] and Kohn and Sham [Kohn and Sham (1965)] in 1965. The original formulation discussed in the first paper of Hohenberg and Kohn, which is known as **the Hohenberg-Kohn theorem**. The second paper of Kohn and Sham developed a method to apply the Hohenberg-Kohn theorem, which is referred to as **the Kohn-Sham equation**.

4.10.1 Hohenberg-Kohn theorem

> **Theorem:** There is a one-to-one correspondence between an external potential $\mathcal{V}_{en}(\boldsymbol{r})$ and an electron density $n(\boldsymbol{r})$ [Hohenberg and Kohn (1964)].

Proof: To prove the theorem, we suppose that two different external potentials, $\mathcal{V}_{en}(\boldsymbol{r})$ and $\mathcal{V}'_{en}(\boldsymbol{r})$, give the same electron density $n(\boldsymbol{r})$. We will show that this situation is not possible. Let us consider the two Hamiltonians $\mathcal{H}$ and $\mathcal{H}'$, which contain $\mathcal{V}_{en}$ and $\mathcal{V}'_{en}$, respectively, as follows:

$$\mathcal{H} = \mathcal{F} + \mathcal{V}_{en} \text{ and } \mathcal{H}' = \mathcal{F} + \mathcal{V}'_{en}, \tag{4.84}$$

where $\mathcal{F}$ includes all the terms in the Hamiltonian except for the external potential. That means that $\mathcal{F}$ contains the kinetic energy and electron-electron interaction terms. Therefore, $\mathcal{F}$ has the same shape for all N_e-electrons systems with any external potentials.

The total energies of the ground states of the Hamiltonians is given by

$$E = \langle\psi|\mathcal{H}|\psi\rangle \text{ and } E' = \langle\psi'|\mathcal{H}'|\psi'\rangle, \tag{4.85}$$

where ψ and ψ' are the wavefunctions of the ground states of $\mathcal{H}$ and $\mathcal{H}'$, respectively. Here we can say that $E \neq E'$ and $\psi \neq \psi'$ because the potentials are not the same. According to the variational principle, we have:

$$\begin{aligned} E < \langle\psi'|\mathcal{H}|\psi'\rangle &= \langle\psi'|\mathcal{H}' - \mathcal{V}'_{en} + \mathcal{V}_{en}|\psi'\rangle \\ &= E' + \langle\psi'|\mathcal{V}_{en} - \mathcal{V}'_{en}|\psi'\rangle, \end{aligned} \tag{4.86}$$

and

$$\begin{aligned} E' < \langle\psi|\mathcal{H}'|\psi\rangle &= \langle\psi|\mathcal{H} - \mathcal{V}_{en} + \mathcal{V}'_{en}|\psi\rangle \\ &= E - \langle\psi|\mathcal{V}_{en} - \mathcal{V}'_{en}|\psi\rangle. \end{aligned} \tag{4.87}$$

By adding Eqs. (4.86) and (4.87), we obtain:

$$(E + E') < (E + E') + \langle \psi' | \mathcal{V}_{en} - \mathcal{V}'_{en} | \psi' \rangle - \langle \psi | \mathcal{V}_{en} - \mathcal{V}'_{en} | \psi \rangle. \quad (4.88)$$

If the both $\mathcal{V}_{en}$ and $\mathcal{V}'_{en}$ gave the same electron density $n(\boldsymbol{r})$, the two terms on the right-hand side of Eq. (4.88) would be expressed by

$$\langle \psi' | \mathcal{V}_{en} - \mathcal{V}'_{en} | \psi' \rangle = \int [\mathcal{V}_{en}(\mathbf{r}) - \mathcal{V}'_{en}(\mathbf{r})]\, n(\mathbf{r}) d\mathbf{r} \quad (4.89)$$

and

$$\langle \psi | \mathcal{V}_{en} - \mathcal{V}'_{en} | \psi \rangle = \int [\mathcal{V}_{en}(\mathbf{r}) - \mathcal{V}'_{en}(\mathbf{r})]\, n(\mathbf{r}) d\mathbf{r}. \quad (4.90)$$

Eqs. (4.88), (4.89), and (4.90) lead to the contradictory relation $E + E' < E + E'$. Therefore, our assumption that $n(\boldsymbol{r})$ is the same with the different external potentials is not correct. This proves the Hohenberg-Kohn theorem.

Corollary 1: *The electron density $n(\boldsymbol{r})$ uniquely specifies the external potential $\mathcal{V}_{en}(\boldsymbol{r})$ and hence the Hamiltonian $\mathcal{H}$.* Because the ground-state wavefunction ψ is obtained by solving the Schrödinger equation for $\mathcal{H}$, ψ must be a unique functional of $n(\boldsymbol{r})$. Therefore, we obtain the following equation:

$$\langle \psi | \mathcal{H} - \mathcal{V}_{en} | \psi \rangle = \langle \psi | \mathcal{F} | \psi \rangle = \mathcal{F}[n(\mathbf{r})], \quad (4.91)$$

which tells us that $\mathcal{F}$ must be a functional of $n(\boldsymbol{r})$.

We can also conclude that the total energy E is a functional of $n(\boldsymbol{r})$, and E is given by

$$E[n(\mathbf{r})] = \langle \psi | \mathcal{H} | \psi \rangle = \mathcal{F}[n(\mathbf{r})] + \int \mathcal{V}_{en}(\mathbf{r}) n(\mathbf{r}) d\mathbf{r}. \quad (4.92)$$

According to the variational principle, we can deduce that for a given $\mathcal{V}_{en}(\boldsymbol{r})$, Eq. (4.92) gives the global minimum value for the correct electron density $n(\boldsymbol{r})$. This is because that for any other density $n'(\boldsymbol{r})$, we get a larger $E[n'(\boldsymbol{r})]$:

$$\begin{aligned} E[n'(\boldsymbol{r})] &= \mathcal{F}(n'(\boldsymbol{r})) + \int \mathcal{V}_{en}(\boldsymbol{r})n'(\boldsymbol{r})\mathrm{d}\boldsymbol{r} \\ &= \langle\psi'|\mathcal{H}|\psi'\rangle > \langle\psi|\mathcal{H}|\psi\rangle = E[n(\boldsymbol{r})]. \end{aligned} \tag{4.93}$$

Corollary 2: *If the functional $\mathcal{F}[n(\boldsymbol{r})]$ was known, then by minimizing the total energy in Eq. (4.92), with respect to variations in the electron density $n(\boldsymbol{r})$, the ground state of the electron density and the total energy are obtained.* Therefore, it is important to develop adequate approximations for the functional $\mathcal{F}[n(\boldsymbol{r})]$ that will be discussed in Sec. 4.10.2. It is noted that the functional $\mathcal{F}[n(\boldsymbol{r})]$ determines only non-degenerated ground state, and thus the Hohenberg-Kohn theorem does not provide any guidance concerning excited states.

4.10.2 Kohn-Sham equation

In Sec. 4.10.1, we discussed the Hohenberg-Kohn theorem, which guarantees to give the ground-state energy. However, the theorem does not provide any analytical solution of the ground state since the functional $\mathcal{F}[n(\boldsymbol{r})]$ of Eq. (4.92) is not explicitly given. Kohn and Sham [Kohn and Sham (1965)] provided an explicit form for $\mathcal{F}[n(\boldsymbol{r})]$ and constructing the Schrödinger-like equation based on $\mathcal{F}[n(\boldsymbol{r})]$ with the SCF method (see Sec. 4.5), which allows the implementation to computer codes.

Let us write the functional $\mathcal{F}[n(\boldsymbol{r})]$ for the single-particle states $\phi_i(\boldsymbol{r})$ as

$$\mathcal{F}[n(\boldsymbol{r})] = \mathcal{K}_e[n(\boldsymbol{r})] + \frac{1}{2}\iint \frac{n(\boldsymbol{r})n(\boldsymbol{r}')}{|\boldsymbol{r}-\boldsymbol{r}'|}\mathrm{d}\boldsymbol{r}\mathrm{d}\boldsymbol{r}' + \mathcal{E}_{xc}[n(\boldsymbol{r})], \tag{4.94}$$

where electron density $n(\boldsymbol{r})$ is given by Eq. (4.19) as

$$n(\boldsymbol{r}) = \sum_i |\phi_i(\boldsymbol{r})|^2, \tag{4.95}$$

$\mathcal{K}_e[n(\boldsymbol{r})]$ represents the kinetic energy of single-particle states:

$$\mathcal{K}_e[n(\boldsymbol{r})] = \sum_i \left\langle \phi_i \left| -\frac{\nabla_{\boldsymbol{r}}^2}{2} \right| \phi_i \right\rangle, \tag{4.96}$$

the second term of the right-hand side represents the electrostatic Coulomb repulsion between electrons, with a factor of 1/2 due to the double counting, and $\mathcal{E}_{xc}[n(\boldsymbol{r})]$ is the **exchange-correlation functional**, which contains all other contributions to the many-body energy of electrons. It is noted that $\mathcal{K}_e[n(\boldsymbol{r})]$ is not the exact kinetic energy of the many-body system of electrons; thus $\mathcal{E}_{xc}[n(\boldsymbol{r})]$ is needed to reproduce the correct functional $\mathcal{K}_e[n(\boldsymbol{r})]$.

When we apply the Hohenberg-Kohn theorem by substituting Eq. (4.94) into Eq. (4.92), the total energy is given by

$$\begin{aligned} E[n(\boldsymbol{r})] = \mathcal{K}_e[n(\boldsymbol{r})] + \frac{1}{2}\iint \frac{n(\boldsymbol{r})n(\boldsymbol{r}')}{|\boldsymbol{r}-\boldsymbol{r}'|}\mathrm{d}\boldsymbol{r}\mathrm{d}\boldsymbol{r}' \\ + \mathcal{E}_{xc}[n(\boldsymbol{r})] + \int \mathcal{V}_{en}(\boldsymbol{r})n(\boldsymbol{r})\mathrm{d}\boldsymbol{r}, \end{aligned} \tag{4.97}$$

that we need to minimize with respect to $n(\boldsymbol{r})$ in order to obtain the ground-state energy. We consider a variation in the electron density with the constraint that the total number of electrons does not change:

$$\int \delta n(\boldsymbol{r})\mathrm{d}\boldsymbol{r} = 0. \tag{4.98}$$

By using the Lagrange multiplier, ϵ_i, with the constraint in Eq. (4.98), we obtain the **Kohn-Sham equation** as

$$\left\{-\frac{\nabla_{\boldsymbol{r}}^2}{2} + \mathcal{V}_{\mathrm{eff}}[n(\boldsymbol{r})]\right\}\phi_i(\boldsymbol{r}) = \epsilon_i\phi_i(\boldsymbol{r}), \tag{4.99}$$

where the effective potential $\mathcal{V}_{\mathrm{eff}}[n(\boldsymbol{r})]$ is given by

$$\mathcal{V}_{\mathrm{eff}}[n(\boldsymbol{r})] = \mathcal{V}_{en}(\boldsymbol{r}) + \int \frac{n(\boldsymbol{r}')}{|\boldsymbol{r}-\boldsymbol{r}'|}\mathrm{d}\boldsymbol{r}' + \frac{\delta\mathcal{E}_{xc}[n(\boldsymbol{r})]}{\delta n(\boldsymbol{r})}, \tag{4.100}$$

where the first term $\mathcal{V}_{en}(\boldsymbol{r})$ is the external potential due to the ions, the second term is the Hartree potential $\mathcal{V}_H$ (see Eq. (4.33)) and the last term is the variational functional derivative of the exchange-correlation interaction $\mathcal{E}_{xc}[n(\boldsymbol{r})]$. The last term in Eq. (4.100) is defined as the **exchange-correlation potential**:

$$\mathcal{V}_{xc}[n(\boldsymbol{r})] = \frac{\delta \mathcal{E}_{xc}[n(\boldsymbol{r})]}{\delta n(\boldsymbol{r})}. \tag{4.101}$$

By assuming that one knows $\mathcal{E}_{xc}[n(\boldsymbol{r})]$ or at least an adequate approximation, the Kohn-Sham equation in Eq. (4.99) can be solved by the SCF method (see Sec. 4.5) by the following steps:

1. Make an initial guess of $n^{\text{in}}(\boldsymbol{r})$.
2. Calculate $\mathcal{V}_{xc}[n^{\text{in}}(\boldsymbol{r})]$ in Eq. (4.101) and therewith $\mathcal{V}_{\text{eff}}[n^{\text{in}}(\boldsymbol{r})]$ in Eq. (4.100).
3. Solve Eq. (4.99) to get $\phi_i(\boldsymbol{r})$.
4. Calculate the new electron density $n^{\text{new}}(\boldsymbol{r})$ via Eq. (4.95).
5. If $n^{\text{new}}(\boldsymbol{r})$ is not equal to $n^{\text{in}}(\boldsymbol{r})$, we go back to step (1).
6. If $n^{\text{new}}(\boldsymbol{r})$ is equal to $n^{\text{in}}(\boldsymbol{r})$, we calculate the total energy by using Eq. (4.97).

The SCF calculation is an essential step in Quantum ESPRESSO. The run-time tutorial for the SCF calculation is shown in Sec. 3.1.1.

4.10.3 Relationship between Kohn-Sham energy and total energy

Multiplying the Kohn-Sham equation in Eq. (4.99) by $\phi_i^*(r)$ from the left and summing over all occupied states, we obtain **the Kohn-Sham energy** as

$$\sum_i \epsilon_i = \mathcal{K}_e[n(\boldsymbol{r})] + \int \mathcal{V}_{\text{eff}}(\boldsymbol{r}) n(\boldsymbol{r}) d\boldsymbol{r}. \tag{4.102}$$

By substituting $\mathcal{V}_{\text{eff}}(\boldsymbol{r})$ from Eq. (4.100) into Eq. (4.102), then subtracting the total energy $E[n(\boldsymbol{r})]$ in Eq. (4.97), we find:

$$\begin{aligned}E[n(\boldsymbol{r})] &= \sum_i \epsilon_i - \frac{1}{2}\iint \frac{n(\boldsymbol{r})n(\boldsymbol{r}')}{|\boldsymbol{r}-\boldsymbol{r}'|}\mathrm{d}\boldsymbol{r}\mathrm{d}\boldsymbol{r}' - \Delta\mathcal{E}_{\text{xc}}[n(\boldsymbol{r})] \\ &= \sum_i \epsilon_i - \frac{1}{2}\int \mathcal{V}_H(\boldsymbol{r})n(\boldsymbol{r})\mathrm{d}\boldsymbol{r} - \Delta\mathcal{E}_{\text{xc}}[n(\boldsymbol{r})],\end{aligned} \tag{4.103}$$

where $\Delta\mathcal{E}_{\text{xc}}[n(\boldsymbol{r})]$ is the difference between exchange and correlation potentials as

$$\Delta\mathcal{E}_{\text{xc}}[n(\boldsymbol{r})] = \int \mathcal{V}_{\text{xc}}(\boldsymbol{r})n(\boldsymbol{r})\mathrm{d}\boldsymbol{r} - \mathcal{E}_{\text{xc}}[n(\boldsymbol{r})]. \tag{4.104}$$

4.11 Exchange-correlation functional

In Sec. 4.10, we show that how the ground-state properties are calculated by the Kohn-Sham equation. Unfortunately, the exact form of the exchange-correlation functional $\mathcal{E}_{xc}[n(\boldsymbol{r})]$ is not known, and we should take an approximation from many proposed approximations. The most common approximations are the **local-density approximation** (LDA) and the **generalized gradient approximation** (GGA), which will be discussed in Sec. 4.11.1 and Sec. 4.11.2, respectively. We also discuss the **hybrid functionals** in Sec. 4.11.3, which is beyond the LDA and GGA.

4.11.1 Local-density approximation

The LDA is an approximation of $\mathcal{E}_{xc}[n(\boldsymbol{r})]$ based on the uniform-electron system as discussed in Sec. 4.8. As shown in Eq. (4.77), the contribution of exchange interaction to the total energy for the uniform-electron charge is:

$$E_x(n) = N_e C_2 n^{1/3}, \tag{4.105}$$

with $C_2 = -0.738$ (see Eq. (4.78)).

Now, let us consider Eq. (4.105) for the case of the non-uniform-electron charge, in which n is a function of $\boldsymbol{r}$. If $n(\boldsymbol{r})$ is a slowly varying function, the exchange functional $\mathcal{E}_x[n(\boldsymbol{r})]$ is approximated as

$$\mathcal{E}_x^{\text{LDA}}[n(\boldsymbol{r})] \approx N_e C_2 n^{1/3}(\boldsymbol{r}). \tag{4.106}$$

By substituting the number of electrons N_e in Eq. (4.82) into Eq. (4.83), we obtain:

$$\mathcal{E}_x^{\text{LDA}}[n(\boldsymbol{r})] = \int \epsilon_x(\boldsymbol{r}) n(\boldsymbol{r}) \mathrm{d}\boldsymbol{r}, \tag{4.107}$$

where $\epsilon_x(\boldsymbol{r})$ is the exchange energy, which is defined by

$$\epsilon_x(\boldsymbol{r}) = C_2 n^{1/3}(\boldsymbol{r}). \tag{4.108}$$

Eq. (4.108) allows to calculate the exchange potential $\mathcal{V}_x[n(\boldsymbol{r})]$ as

$$\mathcal{V}_x[n(\boldsymbol{r})] = \frac{\delta \mathcal{E}_x^{\text{LDA}}[n(\boldsymbol{r})]}{\delta n(\boldsymbol{r})} = \frac{\partial[\epsilon_x(\boldsymbol{r}) n(\boldsymbol{r})]}{\partial n(\boldsymbol{r})} = \frac{4}{3} C_2 n^{1/3}(\boldsymbol{r}). \tag{4.109}$$

Note that the ratio of the exchange potential in Eq. (4.109) to the Slater exchange potential in Eq. (4.79) is 2/3.

Since Eq. (4.107) is based on the free-electron gas (see Sec. 4.8), the correlation interaction is needed to capture accurately the many-body system. Therefore, the energy of correlation interaction $\epsilon_c(\boldsymbol{r})$ is added to Eq. (4.107) as

$$\mathcal{E}_{xc}^{\text{LDA}}[n(\boldsymbol{r})] = \int [\epsilon_x(\boldsymbol{r}) + \epsilon_c(\boldsymbol{r})] n(\boldsymbol{r}) \mathrm{d}\boldsymbol{r}. \tag{4.110}$$

When we consider two limit cases of high-density limit ($n \to \infty$) and low-density limit ($n \to 0$), $\epsilon_c(\boldsymbol{r})$ is proposed in analytic forms by many groups as follows:

For the high-density limit:

$$\epsilon_c(r_s) = \sum_{i=0}^{\infty} [a_i \ln r_s + b_i] r_s^i = a_0 \ln r_s + b_0 + \ldots \ (r_s \ll 1), \tag{4.111}$$

where r_s is the radius of a sphere containing a single electron in the atomic unit, which is defined by

$$\frac{4\pi r_s^3}{3} = \frac{1}{n(\mathbf{r})}, \text{ or } r_s = \left[\frac{3}{4\pi n(\mathbf{r})}\right]^{1/3}, \tag{4.112}$$

and $a_0 = 0.0311$ and $b_0 = -0.048$ are given by Gell-Mann and Brueckner [Gell-Mann and Brueckner (1957)] with neglecting contributions to the coefficients of higher order terms ($i \geq 1$).

For the low-density limit:

$$\epsilon_c(r_s) = \sum_{i=0}^{\infty} \frac{c_i}{r_s^{i/2+1}} = \frac{c_0}{r_s} + \frac{c_1}{r_s^{3/2}} + \frac{c_2}{r_s^2} + ... \ (r_s \gg 1), \tag{4.113}$$

where $c_0 = -1.792$, $c_1 = 2.65$, and $c_2 = -0.73$ are given by Carr *et al.* [Carr Jr *et al.* (1961)] for a body-centered-cubic lattice system, in which the higher order terms $i \geq 2$ are neglected.

The more accurate numerical calculation of $\epsilon_c(r_s)$ is given by Ceperley and Alder [Ceperley and Alder (1980)]. They calculated the total energy for the uniform-electron system for different values of r_s by using the quantum Monte Carlo method.[8] Then the correlation energy was obtained by subtracting the corresponding kinetic and exchange energies from the total energy. Based on fitting functions to the numerical results of Ceperley and Alder, several forms of $\epsilon_c(r_s)$ are proposed by Vosko, Wilk, and Nusair (VWN) [Vosko *et al.* (1980)], Perdew and Zunger (PZ) [Perdew and Zunger (1981)], and Perdew and Wang (PW) [Perdew and Wang (1992)], which are listed in Table 4.2. Although the expressions of $\epsilon_c(r_s)$ do not depend on the spin, the parameters of $\epsilon_c(r_s)$ depend on the relative spin polarization, which is defined by

$$\zeta = \frac{n_\uparrow(\mathbf{r}) - n_\downarrow(\mathbf{r})}{n(\mathbf{r})}, \tag{4.114}$$

[8] The quantum Monte Carlo method is a set of computational methods to provide a reliable solution (or an accurate approximation) of the quantum many-body problem, in which the anti-commutation relation of two electrons are taken into account.

Table 4.2 Correlation energy $\epsilon_c(r_s)$ in various models (VWN = Vosko-Wilk-Nusair, PZ = Perdew-Zunger, and PW = Perdew-Wang). r_s is in atomic unit ($a_0 = 1$) and $\epsilon_c(r_s)$ is in units of Ry. $\zeta = [n_\uparrow(\mathbf{r}) - n_\downarrow(\mathbf{r})]/n(\mathbf{r})$ is the relative spin polarization. Parameters are set for $\zeta = 0$ and $\zeta = 1$ in units of a.u.

Model	Correlation energy $\epsilon_c(r_s)$	for $\zeta = 0$	for $\zeta = 1$
VWN	$A\left\{ \ln \frac{x^2}{X(x)} + \frac{2b}{Q} \arctan \frac{Q}{2x+b} - \frac{bx_0}{X(x_0)} \left[\ln \frac{(x-x_0)^2}{X(x)} + \frac{2(b+2x_0)}{Q} \arctan \frac{Q}{2x+b} \right] \right\}$, where $x = \sqrt{r_s}$, $Q = \sqrt{4c - b^2}$, and $X(x) = x^2 + bx + c$	$A = 0.031091$ $b = 3.72744$ $c = 12.9352$ $x_0 = -0.10498$	0.015545 7.06042 18.0578 −0.32500
PZ	$A \ln(r_s) + B + Cr_s \ln(r_s) + Dr_s$, where $r_s < 1$ and $\gamma/(1 + \beta_1\sqrt{r_s} + \beta_2 r_s)$ with $r_s \geq 1$	$A = 0.0311$ $B = -0.048$ $C = 0.0020$ $D = -0.0116$ $\gamma = -0.1423$ $\beta_1 = 1.0529$ $\beta_2 = 0.3334$	0.0311 −0.048 0.0007 −0.0048 −0.0843 1.3981 0.2611
PW	$-2A(1 + \alpha_1 r_s) \ln \left[1 + \frac{1}{2A\left(\beta_1 r_s^{1/2} + \beta_2 r_s + \beta_3 r_s^{3/2} + \beta_4 r_s^{p+1}\right)} \right]$	$A = 0.031091$ $\alpha_1 = 0.21370$ $\beta_1 = 7.5957$ $\beta_2 = 3.5876$ $\beta_3 = 1.6382$ $\beta_4 = 0.49294$ $p = 1.0$	0.015545 0.20548 14.1189 6.1977 3.3662 0.6251) 1.0

where $n_\uparrow(\mathbf{r})$ and $n_\downarrow(\mathbf{r})$ are the electron densities of up-spin and down-spin states. $\zeta = 0$ and 1 are the spin-unpolarized and spin-polarized systems, respectively.

In Fig. 4.10, we show $-\epsilon_c$ by VWN, PZ, and PW parameters as a function of r_s for the uniform electron system with both the spin-unpolarized and spin-polarized systems by using the parameters in Table 4.2. All functions fit very well with the numerical results (the circles and diamonds symbols for the spin-unpolarized and spin-polarized cases, respectively) of Ceperley and Alder [Ceperley and Alder (1980)]. For $r_s < 5$, which is relevant for condensed metals, the correlation energy becomes more important. On the other hand, the correlation energy becomes a small

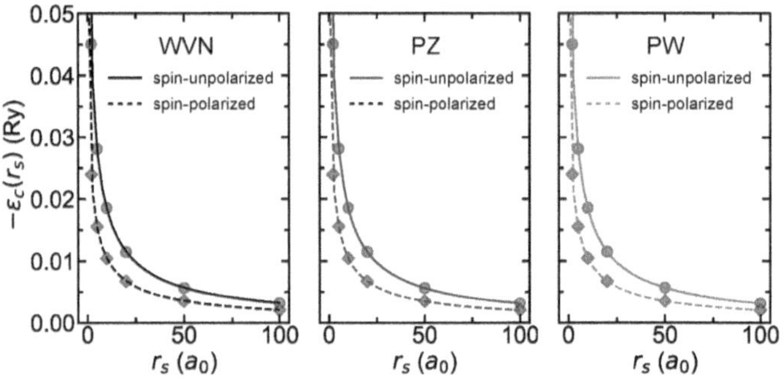

Figure 4.10 The Vosko-Wilk-Nusair (VWN), Perdew-Zunger (PZ), and Perdew-Wang (PW) correlation energies $-\epsilon_c(r_s)$ are plotted as functions of r_s for both spin-unpolarized (solid lines) and spin-polarized (dashed lines) systems. The expressions and parameters of $\epsilon_c(r_s)$ are given in Table 4.2. The circle and diamond symbols are the accurate numerical results (by Ceperley and Alder) of the uniform electron system with the spin-unpolarized and spin-polarized systems, respectively.

contribution for $r_s > 10$, which are relevant for semiconductors or insulators. Although all functions give the same results in the case of the uniform electron system, for a specific material, they can give somewhat different results of the total energy. Among these functions, the PZ function is often used in Quantum ESPRESSO.

4.11.2 Generalized gradient approximation

As discussed in Sec. 4.11.1, the LDA is an approximation in the case that $n(\boldsymbol{r})$ is a slowly varying function. However, $n(\boldsymbol{r})$ often changes rapidly in the real material. Moreover, $n(\boldsymbol{r})$ is generally spin-dependent. There are many attempts to improve the accuracy of the LDA for the real systems where $n(\boldsymbol{r})$ varies rapidly. The most successful one is the **generalized gradient approximation** (GGA). In the GGA, for the exchange term, an **enhancement factor** F_x is added in Eq. (4.107) by Perdew, Burke, and Ernzerhof (PBE) [Perdew *et al.* (1996a)] as

$$\mathcal{E}_x^{\mathrm{GGA}}[n(\boldsymbol{r})] = \int \epsilon_x(\boldsymbol{r})n(\boldsymbol{r})F_x(s)\mathrm{d}\boldsymbol{r} = C_2 \int n^{4/3}(\boldsymbol{r})F_x(s)\mathrm{d}\boldsymbol{r}, \quad (4.115)$$

where s is the dimensionless gradient of $n(\mathbf{r})$, which is defined by

$$s = \frac{|\nabla n(\mathbf{r})|}{2k_F n(\mathbf{r})}, \text{ with } k_F = \left[3\pi^2 n(\mathbf{r})\right]^{1/3}, \tag{4.116}$$

and the enhancement factor $F_x(s)$ is expressed by

$$F_x(s) = 1 + \kappa - \frac{\kappa}{1 + \mu s^2/\kappa}, \tag{4.117}$$

where $\kappa = 0.804$ and $\mu = 0.21951$ are constants. Note that the range of s for the real systems is $0 \le s \le 3$. Eqs. (4.115) and (4.117) show that

$$\mathcal{E}_x^{\text{GGA}}[n(\mathbf{r})] \ge \mathcal{E}_x^{\text{LDA}}[n(\mathbf{r})] \text{ with } s \ge 0. \tag{4.118}$$

Therefore, $\mathcal{E}_x^{\text{GGA}}[n(\mathbf{r})]$ satisfies the Lieb-Oxford lower bound[9] [Lieb and Oxford (1981)] as shown in Eq. (4.118).

For the correlation term, $\mathcal{E}_c^{\text{GGA}}[n(\mathbf{r})]$ is expressed by $\epsilon_c(\mathbf{r})$ of the uniform electron system (see Table 4.2) plus an additional term $H[n(\mathbf{r}), \zeta]$, which depends on both the gradient $\nabla n(\mathbf{r})$ and the spin polarization ζ. The $\mathcal{E}_c^{\text{GGA}}[n(\mathbf{r})]$ functional is given by PBE as

$$\mathcal{E}_c^{\text{GGA}}[n(\mathbf{r})] = \int \{\epsilon_c(\mathbf{r}) + H[n(\mathbf{r}), \zeta]\}\, n(\mathbf{r}) d\mathbf{r}, \tag{4.119}$$

where $H[n(\mathbf{r}), \zeta]$ is given by

$$H[n(\mathbf{r}), \zeta] = \gamma\phi^3 \ln\left[1 + \frac{\beta}{\gamma} t^2 \left(\frac{1 + At^2}{1 + At^2 + A^2t^4}\right)\right]. \tag{4.120}$$

Here the function A in Eq. (4.120) is given by

$$A = \frac{\beta}{\gamma}\left\{\exp\left[-\frac{\epsilon_c(\mathbf{r})}{\gamma\phi^3}\right] - 1\right\}^{-1}, \tag{4.121}$$

where $\beta = 0.066725$ and $\gamma = 0.031091$ are non-empirical constants, and $\phi(\zeta) = \left[(1+\zeta)^{2/3} + (1-\zeta)^{2/3}\right]/2$ is the spin-scaling factor.

[9] The repulsive Coulomb energy (exchange plus correlation energy) has a lower bound of the form $C_2 \int n^{4/3}(\mathbf{r}) d\mathbf{r}$, where $n(\mathbf{r})$ is the single-particle electron density.

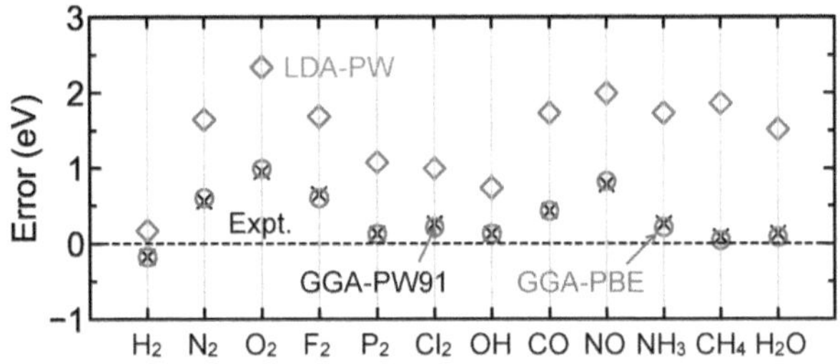

Figure 4.11 Errors of atomization energies of several molecules in units of eV. The dashed line at zero error corresponds to the experimental values. The square, cross, and circle symbols are the LDA-PW functional, the GGA-PW91 functional, and the GGA-PBE functional, respectively. All numerical data are taken from Ref. [Perdew *et al.* (1996a)].

The function t in Eq. (4.120) is given by $t = |\nabla n(\mathbf{r})|/2\phi k_s \nabla n(\mathbf{r})$, which is another dimensionless gradient of $n(\mathbf{r})$, where $k_s = \sqrt{4k_F/\pi}$ is the Thomas-Fermi screening wave number. The $F(s)$ and $H[n(\mathbf{r}), \zeta]$ functions in Eqs. (4.115) and (4.119), respectively, are also developed by Perdew and Wang (PW91) [Perdew *et al.* (1992)] with the slightly different forms and parameters. Although both PBE and PW91 are available in Quantum ESPRESSO, the PBE functional is often used for many materials (see Sec. 3.1.6).

In Fig. 4.11, we show the error of atomization energies for small molecules, in which the experimental values are set to zero error. The atomization energies are calculated by the LDA-PW, GGA-PW91, and GGA-PBE. Both GGA-PW91 and GGA-PBE give almost the same results, and they are closer to the experimental values than LDA-PW. Although LDA-PW overestimates the atomization energies of molecules and solids, LDA-PW remains a popular approximation for realistic solid-state calculations since it has a simple form of the functional compared with GGA.

4.11.3 Hybrid functionals

Hybrid functional is a functional that combines the GGA functional with a fraction of the non-local Hartree-Fock exchange interaction. The hybrid functionals are developed for solving the famous "band-gap problem" [Sham and Schlüter (1983)], in which *the LDA and*

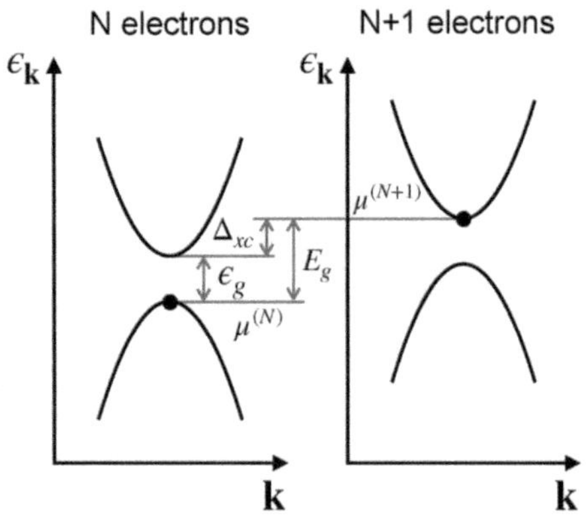

Figure 4.12 Illustration of the contribution of Δ_{xc}, the discontinuity in $\mathcal{V}_{xc}[n(\mathbf{r})]$. The energies ϵ of the Kohn-Sham equation are shown in the form of a band structure for the (N)- and $(N+1)$- electron systems. The two differ in a uniform increase of the eigenvalues by Δ_{xc}. The quasiparticle band-gap E_g is the difference between the two eigenvalues as $E_g = \epsilon_{N+1}^{(N+1)} - \epsilon_N^{(N)}$, which leads to $E_g = \epsilon_g + \Delta_{xc}$, where the Kohn-Sham gap is given by $\epsilon_g = \epsilon_{N+1}^{(N)} - \epsilon_N^{(N)}$.

GGA always underestimate the band gap. First, let us explain the reason why both the LDA and GGA underestimate the band gaps for semiconductors and insulators even though they are the basis of good approximations for the ground-state properties.

The origin of the band-gap problem is the discontinuity of chemical potential in the exchange-correlation potential $\mathcal{V}_{xc}$ [Sham and Schlüter (1985)]. This is because the density is given by the summation up to the chemical potential μ as $n(\mathbf{r}) = \sum_i \Theta(\mu - \epsilon_i)|\phi_i|^2$, where Θ is the Heaviside step function.[10] Therefore, a change $\delta n(\mathbf{r})$ leads to a discontinuous jump of μ from the top of valence band to the bottom of conduction band when we change from (N) to $(N+1)$-electron systems. Since $\mathcal{V}_{xc}[n(\mathbf{r})] = \delta\mathcal{E}_{xc}[n(\mathbf{r})]/\delta n(\mathbf{r})$ in Eq. (4.101) is given by derivative on $\delta n(\mathbf{r})$, we also expect a discontinuity Δ_{xc} (positive value) in $\mathcal{V}_{xc}[n(\mathbf{r})]$. Thus, the quasiparticle band-gap E_g can be divided into

[10] The Heaviside step function $\Theta(x)$ is a discontinuous function, whose value is zero for negative arguments $x < 0$ and one for positive arguments $x > 0$.

two components as follows [Perdew and Levy (1983)]

$$E_g = \epsilon_g + \Delta_{xc}, \tag{4.122}$$

where ϵ_g is the gap obtained from the Kohn-Sham equation for the ground state. In Fig. 4.12, we show the contribution of Δ_{xc} to $E_g = \epsilon_{N+1}^{(N+1)} - \epsilon_N^{(N)}$, where $\epsilon_J^{(N)}$ denotes J-th energy for (N)-electron system. As shown in Secs. 4.11.1 and 4.11.2, both the LDA and GGA do not consider the discontinuity in $\mathcal{V}_{xc}[n(\boldsymbol{r})]$ since they are the *smooth* and *local* functions of $n(\boldsymbol{r})$. Therefore, the LDA and GGA would give zero Δ_{xc}, which is the reason why they always underestimate the value of E_g even they give a good approximation to ϵ_g.

Several hybrid functionals have been proposed to obtain non-zero value of Δ_{xc}. The idea of the hybrid functional is based on a *non-local* functional to the exchange-correlation functional $\mathcal{E}_{xc}[n(\boldsymbol{r})]$. As discussed in Sec. 3.6, the exchange potential of the Hartree-Fock equation is a *non-local* potential (see Eq. (4.47)). The exchange energy can be obtained from the Hartree-Fock exchange potential as

$$\mathcal{E}_x^{\mathrm{HF}} = -\frac{1}{2}\sum_{i,j}\iint \frac{\phi_i^*(\boldsymbol{r}_1)\phi_j^*(\boldsymbol{r}_2)\phi_j(\boldsymbol{r}_1)\phi_i(\boldsymbol{r}_2)}{|\boldsymbol{r}_1 - \boldsymbol{r}_2|}\mathrm{d}\boldsymbol{r}_1\mathrm{d}\boldsymbol{r}_2, \tag{4.123}$$

where a factor of 1/2 accounts for double counting of the exchange interactions. Here, the exchange potential as a function of $\boldsymbol{r}_1$ is given by as a function of $\boldsymbol{r}_2$, which is the origin of non-local potential. By using $\mathcal{E}_x^{\mathrm{HF}}$ in Eq. (4.123), the expressions of the hybrid functionals are given in Table 4.3, which are usually constructed as a linear combination of a *non-local* term $\mathcal{E}_x^{\mathrm{HF}}$ and the *local* terms from the LDA or GGA. The most popular hybrid functional is B3LYP (B = Becke, 3 = three coefficients, LYP = Lee-Yang-Parr) [Becke (1993), Stephens *et al.* (1994)]. For B3LYP, $\mathcal{E}_x^{\mathrm{GGA}}$ and $\mathcal{E}_c^{\mathrm{GGA}}$ are the exchange functional of Becke (B88) [Becke (1988)] and the correlation functional of Lee, Yang, and Parr [Lee *et al.* (1988)], respectively, and the coefficients a_o, a_x, and a_c are empirically fitted to atomic and molecular data. The next hybrid functional is PBE0 [Perdew *et al.* (1996b)] (PBE = Perdew-Burke-Ernzerhof, 0 = no empirical coefficient), in which the coefficient $a = 1/4$ is determined by fourth-order perturbation theory [Adamo and Barone (1999)].

Table 4.3 Hybrid functionals of B3LYP (Becke, 3-parameter, Lee-Yang-Parr), PBE0 (Perdew-Burke-Ernzerhof), and HSE (Heyd-Scuseria-Ernzerhof).

Model	Exchange-correlation functional $\mathcal{E}_{xc}$	Mixing coefficients
B3LYP	$\mathcal{E}_{xc}^{\mathrm{VWN}} + a_0\left(\mathcal{E}_x^{\mathrm{HF}} - \mathcal{E}_x^{\mathrm{VWN}}\right) + a_x\mathcal{E}_x^{\mathrm{GGA}} + a_c\left(\mathcal{E}_c^{\mathrm{GGA}} - \mathcal{E}_c^{\mathrm{VWN}}\right)$	$a_0 = 0.20$, $a_x = 0.72$, $a_c = 0.81$
PBE0	$a\mathcal{E}_x^{\mathrm{HF}} + (1-a)\mathcal{E}_x^{\mathrm{PBE}} + \mathcal{E}_c^{\mathrm{PBE}}$	$a = 1/4$
HSE	$a\mathcal{E}_x^{\mathrm{HF,SR}}(\eta) + (1-a)E_x^{\mathrm{PBE,SR}}(\eta) + \mathcal{E}_x^{\mathrm{PBE,LR}}(\eta) + \mathcal{E}_c^{\mathrm{PBE}}$	$a = 1/4, \eta = 0.106$

Although B3LYP and PBE0 are very successful for quantitative calculations in molecules, they might not be suitable for solids. This is because the Hartree-Fock approximation is problematic for delocalized electrons such as metals. As discussed in Sec. 4.8, the Hartree-Fock equation leads to the Lindhard function $F(k/k_F)$ in the energy $\epsilon_{\boldsymbol{k}}$ of a uniform-electron system (see Eq. (4.73)). Therefore, the velocity $\mathrm{d}\epsilon_{\boldsymbol{k}}/\mathrm{d}\boldsymbol{k}$ diverges at the Fermi surface, which contradicts the experiment. The singularity of the Lindhard function at the Fermi surface, which was pointed out by Bardeen [Bardeen (1936)], is a consequence of long-range Coulomb interaction. However, the divergence of $\mathrm{d}\epsilon_{\boldsymbol{k}}/\mathrm{d}\boldsymbol{k}$ can be avoided by either if there is a finite gap (i.e., in an insulator) or if the Coulomb interaction is screened to be effectively short range, which is explained below.

The Coulomb interaction can be divided into short-range (SR) and long-range (LR) parts by using the Ewald summation [Ewald (1921)] as

$$\frac{1}{r} = \underbrace{\frac{1-\mathrm{erf}(\eta r)}{r}}_{\mathrm{SR}} + \underbrace{\frac{\mathrm{erf}(\eta r)}{r}}_{\mathrm{LR}}, \tag{4.124}$$

where erf is the error function[11] and η is an adjustable parameter. Applying Eq. (4.124), the HSE (Heyd-Scuseria-

[11] The error function is a function of a variable x, which is defined as $\mathrm{erf}(x) = \frac{2}{\sqrt{\pi}}\int_0^x \exp(-t^2)\mathrm{d}t$.

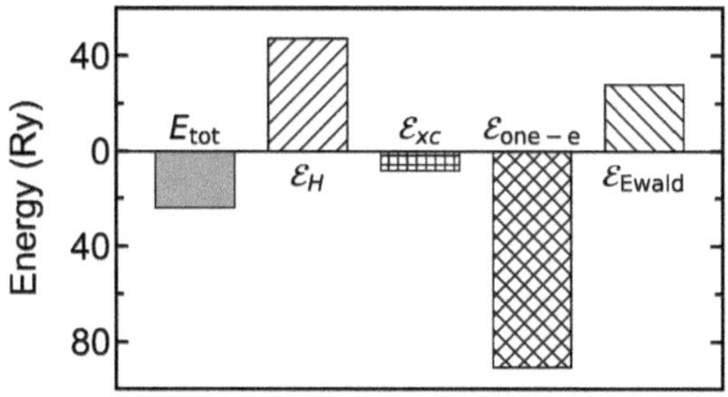

Figure 4.13 Contributions of the Hartree energy $\mathcal{E}_H$, the exchange-correlation energy $\mathcal{E}_{xc}$, the one-electron energy $\mathcal{E}_{\text{one-e}}$, and the Ewald energy $\mathcal{E}_{\text{Ewald}}$ to total energy E_{tot} of graphene.

Ernzerhof) [Heyd *et al.* (2003)] hybrid functional (the expression is given in Table. 4.3) is obtained from the PBE0 functional. The form of HSE has the advantage that it can be applied to metals, and it reduces to the PBE functional for $\eta \to \infty$ and the PBE0 hybrid functional for $\eta \to 0$. All B3LYP, PBE0, and HSE options are available in Quantum ESPRESSO. Although the hybrid functionals give a better value, they are significantly time-consuming compared with the LDA and GGA.

4.12 Total energy calculation

In the discussion up to now, we only consider the total energy of the electrons by solving the Kohn-Sham equation (see Sec. 4.10.2). However, in Quantum ESPRESSO, the total energy of a solid contains both the electron and ion contributions as

$$E_{\text{tot}} = \underbrace{\mathcal{E}_H + \mathcal{E}_{xc} + \mathcal{E}_{\text{one-e}}}_{\text{electrons}} + \underbrace{\mathcal{E}_{\text{Ewald}}}_{\text{ions}}, \tag{4.125}$$

where $\mathcal{E}_H$ is the Hartree energy, $\mathcal{E}_{xc}$ is the exchange-correlation energy, $\mathcal{E}_{\text{one-e}}$ is the kinetic energy of the electrons plus the pseudopotential energy, and $\mathcal{E}_{\text{Ewald}}$ is the Coulomb repulsion between pairs of ions. These values of the energies can be found on the output file of the Quantum ESPRESSO calculation, as shown in Fig. 4.13 for graphene (see output file in Sec. 3.1.1). In this section, we will explain each energy by using the **plane wave expansion** (Sec. 5.3). With the plane waves, the convergence of physical properties is controlled by the **cut-off energy**, which can be tested (see Sec. 3.1.2).

4.12.1 Hartree contribution

First, let us discuss the Hartree energy $\mathcal{E}_H$, which is expressed as

$$\mathcal{E}_H[n(\boldsymbol{r})] = \frac{1}{2}\int \mathcal{V}_H(\boldsymbol{r})n(\boldsymbol{r})\mathrm{d}\boldsymbol{r}, \tag{4.126}$$

where a factor of 1/2 takes into account the double counting and $\mathcal{V}_H(\boldsymbol{r})$ is the Hartree potential, which is given in Eq. (4.33).

For a periodic solid, the total electron density and the potentials can be expressed in the terms of the plane waves, $e^{i\boldsymbol{G}\boldsymbol{r}}$, with $\boldsymbol{G}$ is the reciprocal lattice vectors (see Sec. 5.3) as

$$n(\boldsymbol{r}) = \sum_{\boldsymbol{G}} e^{i\boldsymbol{G}\boldsymbol{r}} n(\boldsymbol{G}), \text{ and } \mathcal{V}_H(\boldsymbol{r}) = \sum_{\boldsymbol{G}} e^{i\boldsymbol{G}\boldsymbol{r}} \mathcal{V}_H(\boldsymbol{G}), \tag{4.127}$$

where $n(\boldsymbol{G})$ and $\mathcal{V}_H(\boldsymbol{G})$ are, respectively, given by

$$n(\boldsymbol{G}) = \int n(\boldsymbol{r})e^{-i\boldsymbol{G}\boldsymbol{r}}\mathrm{d}\boldsymbol{r}, \text{ and } \mathcal{V}_H(\boldsymbol{G}) = \int \mathcal{V}_H(\boldsymbol{r})e^{-i\boldsymbol{G}\boldsymbol{r}}\mathrm{d}\boldsymbol{r}. \tag{4.128}$$

For the neutral case, the total negative charge of the electrons is canceled by the total positive charge of the ions. Therefore, the average potential is zero. Since $\boldsymbol{G} = 0$ in Eq. (4.127) corresponds to the average potential over all the space, we can omit $\boldsymbol{G} = 0$ in the summation due to charge neutrality.[12] Then, by substituting

[12] There might be a contribution from the $\boldsymbol{G} = 0$ term if the systems have an intrinsic electric dipole moment.

Eq. (4.127) into Eq. (4.126), we obtain

$$\begin{aligned}\mathcal{E}_H[n(\boldsymbol{r})] &= \frac{1}{2}\sum_{\boldsymbol{G}\boldsymbol{G}'}\int e^{i(\boldsymbol{G}+\boldsymbol{G}')\boldsymbol{r}}\mathcal{V}_H(\boldsymbol{G})n(\boldsymbol{G}')\mathrm{d}\boldsymbol{r} \\ &= \frac{\Omega}{2}\delta_{\boldsymbol{G}+\boldsymbol{G}',0}\mathcal{V}_H(\boldsymbol{G})n(\boldsymbol{G}') \\ &= \frac{\Omega}{2}\sum_{\boldsymbol{G}\neq 0}\mathcal{V}_H(\boldsymbol{G})n(-\boldsymbol{G}) \\ &= \frac{\Omega}{2}\sum_{\boldsymbol{G}\neq 0}\mathcal{V}_H(\boldsymbol{G})n(\boldsymbol{G}),\end{aligned} \tag{4.129}$$

where Ω is the volume of the unit cell. Here we assume that $n(+\boldsymbol{G}) = n(-\boldsymbol{G})$. On the other hand, by substituting Eq. (4.127) into the Poisson equation (see Sec. 4.5), we have

$$\nabla^2\mathcal{V}_H(\boldsymbol{r}) = -4\pi n(\boldsymbol{r}) \iff \mathcal{V}_H(\boldsymbol{G}) = 4\pi\frac{n(\boldsymbol{G})}{|\boldsymbol{G}|^2}. \tag{4.130}$$

By substituting Eq. (4.130) into Eq. (4.129), we obtain

$$\mathcal{E}_H = 2\pi\Omega\sum_{\boldsymbol{G}\neq 0}\frac{n^2(\boldsymbol{G})}{|\boldsymbol{G}|^2}. \tag{4.131}$$

4.12.2 Exchange-correlation contribution

From Eq. (4.110), the exchange-correlation energy is defined by

$$\mathcal{E}_{xc}[n(\boldsymbol{r})] = \int \epsilon_{xc}(\boldsymbol{r})n(\boldsymbol{r})\mathrm{d}\boldsymbol{r}, \tag{4.132}$$

where $\epsilon_{xc}(\boldsymbol{r}) = \epsilon_x(\boldsymbol{r}) + \epsilon_c(\boldsymbol{r})$ can be expressed in the terms of the plane waves as

$$\epsilon_{xc}(\boldsymbol{r}) = \sum_{\boldsymbol{G}\neq 0} e^{i\boldsymbol{G}\boldsymbol{r}}\epsilon_{xc}(\boldsymbol{G}). \tag{4.133}$$

By substituting Eqs. (4.133) and (4.127) into Eq. (4.132), we obtain

$$\mathcal{E}_{xc}[n(\boldsymbol{r})] = \Omega \sum_{\boldsymbol{G} \neq 0} \epsilon_{xc}(\boldsymbol{G}) n(\boldsymbol{G}). \tag{4.134}$$

4.12.3 One-electron contribution and pseudopotential

In Quantum ESPRESSO, $\mathcal{E}_{\text{one-e}}$ defined in Eq. (4.125) is the sum of the kinetic energy $\mathcal{K}_e$ and the external energy $\mathcal{E}_{\text{ext}}$ of the electrons in the potential of the atom cores:

$$\mathcal{E}_{\text{one-e}}[n(\boldsymbol{r})] = \mathcal{K}_e[n(\boldsymbol{r})] + \mathcal{E}_{\text{ext}}[n(\boldsymbol{r})]. \tag{4.135}$$

The expression of $\mathcal{K}_e[n(\boldsymbol{r})]$ is given by Eq. (4.96), which can be rewritten in the terms of the plane waves as

$$\mathcal{K}_e[n(\boldsymbol{r})] = \frac{\Omega}{2} \sum_i \sum_{\boldsymbol{k}} \sum_{\boldsymbol{G}} |\boldsymbol{k} + \boldsymbol{G}|^2 |C_{i,\boldsymbol{k}}(\boldsymbol{G})|^2, \tag{4.136}$$

where $C_{i,\boldsymbol{k}}(\boldsymbol{G})$ are the coefficients of the wavefunctions ϕ_i.

The expression of $\mathcal{E}_{\text{ext}}[n(\boldsymbol{r})]$ is given by

$$\mathcal{E}_{\text{ext}}[n(\boldsymbol{r})] = \int \mathcal{V}_{en}(\boldsymbol{r}) n(\boldsymbol{r}) \mathrm{d}\boldsymbol{r}. \tag{4.137}$$

Since the external potential term $\mathcal{V}_{en}(\boldsymbol{r})$ in Eq. (4.137) is costly computation with the plane waves, $\mathcal{V}_{en}(\boldsymbol{r})$ is replaced by a **pseudopotential** $\mathcal{V}_{\text{ps}}(\boldsymbol{r})$, which is related to replacing the effects of the core electrons with an effective potential.

A local form of the pseudopotential of a single ion, $v_{\text{ps}}(\boldsymbol{r})$, is proposed by Ashcroft-Heine-Abarenkov [Ashcroft (1966), Heine and Abarenkov (1964)] as

$$v_{\text{ps}}(\boldsymbol{r}) = \begin{cases} v_0 & r < r_{\text{cut}} \\ -\frac{Z}{r} & r > r_{\text{cut}} \end{cases}, \tag{4.138}$$

where Z is the charge of the ionic core and r_{cut} is the effective radius, which is determined by the radius of core electrons, as shown in Fig. 4.14. The constants r_{cut} and v_0 are chosen such that the energy levels

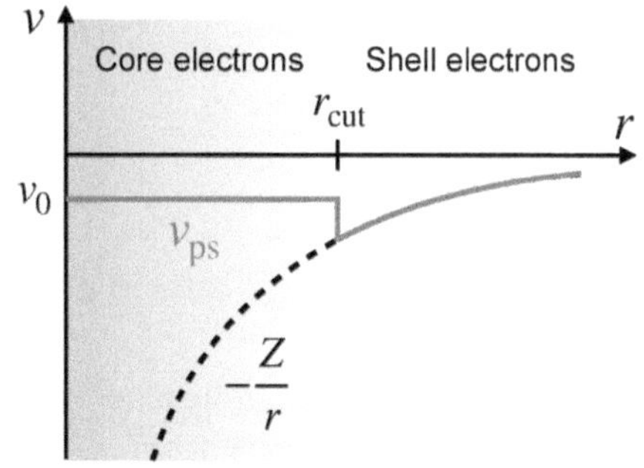

Figure 4.14 The Ashcroft-Heine-Abarenkov pseudopotential for a single ion.

of the shell electrons are reproduced for the single-atom calculations. For example, let us consider 1s- and 2s-electrons of the C atom as core electrons. Then, r_{cut} and v_0 are adjusted by solving the one-particle equation to reproduce the observed ionization energy of the 2p-electron.

A global form of the pseudopotential can be obtained by taking into account the contribution of the individual atoms of a unit cell as

$$\mathcal{V}_{\mathrm{ps}}(\boldsymbol{r}) = \sum_I \sum_{\boldsymbol{T}} v_{\mathrm{ps}}^I(|\boldsymbol{r} - \boldsymbol{T} - \boldsymbol{R}_I|), \tag{4.139}$$

where $\boldsymbol{T}$ are the lattice translation vectors, $\boldsymbol{R}_I$ denotes the relative positions of the I-th atom in the unit cell, and v_{ps}^I is the pseudopotential of the I-th atom. Eq. (4.139) can be expressed by the Fourier transform as

$$\begin{aligned}\mathcal{V}_{\mathrm{ps}}(\boldsymbol{G}) &= \frac{1}{V}\int \sum_I \sum_{\boldsymbol{T}} v_{\mathrm{ps}}^I(|\boldsymbol{r} - \boldsymbol{T} - \boldsymbol{R}_I|)e^{-i\boldsymbol{G}\boldsymbol{r}}\mathrm{d}\boldsymbol{r} \\ &= \frac{N_{\mathrm{cell}}}{V}\sum_I e^{-i\boldsymbol{G}\boldsymbol{R}_I}\int v_{\mathrm{ps}}^I(|\boldsymbol{r}|)e^{-i\boldsymbol{G}\boldsymbol{r}}\mathrm{d}\boldsymbol{r} \\ &= \sum_I e^{-i\boldsymbol{G}\boldsymbol{R}_I}\frac{1}{\Omega}\int v_{\mathrm{ps}}^I(|\boldsymbol{r}|)e^{-i\boldsymbol{G}\boldsymbol{r}}\mathrm{d}\boldsymbol{r} \\ &= \sum_I S_I(\boldsymbol{G})F_I(\boldsymbol{G}),\end{aligned} \tag{4.140}$$

in which $S_I(\boldsymbol{G})$ and $F_I(\boldsymbol{G})$ are defined by

$$S_I(\boldsymbol{G}) = e^{-i\boldsymbol{G}\boldsymbol{R}_I}, \text{ and } F_I(\boldsymbol{G}) = \frac{1}{\Omega}\int v_{\text{ps}}^I(|\boldsymbol{r}|)e^{-i\boldsymbol{G}\boldsymbol{r}}\mathrm{d}\boldsymbol{r}, \qquad (4.141)$$

where N_{cell} is the number of unit cells in the crystal, $V = N_{\text{cell}}\Omega$ is the crystal volume, and $\sum_I S_I(\boldsymbol{G})$ and $F_I(\boldsymbol{G})$ are called the **structure factor**[13] and the **atomic form factor**[14] of the I-th atom, respectively.

For a given pseudopotential such as Eq. (4.138), the external energy can be expressed as

$$\mathcal{E}_{\text{ext}}[n(\boldsymbol{r})] = \int \mathcal{V}_{\text{ps}}(\boldsymbol{r})n(\boldsymbol{r})\mathrm{d}\boldsymbol{r} = \Omega\sum_{\boldsymbol{G}} \mathcal{V}_{\text{ps}}(\boldsymbol{G})n(\boldsymbol{G}). \qquad (4.142)$$

Then by substituting Eq. (4.140) into Eq. (4.142), we obtain

$$\mathcal{E}_{\text{ext}}[n(\boldsymbol{r})] = \Omega\sum_{\boldsymbol{G}}\sum_{I} S_I(\boldsymbol{G})F_I(\boldsymbol{G})n(\boldsymbol{G}). \qquad (4.143)$$

For Quantum ESPRESSO, the $\mathcal{V}_{\text{ps}}$ is often used with norm-conserving or ultra-soft (non-local) pseudopotentials (see Sec. 5.4). In the case of non-local pseudopotential, however, the expression of the external energy is rather complicated and needs special treatment for an efficient implementation [Pickett (1989)].

4.12.4 The Ewald contribution

Finally, we will discuss the last term $\mathcal{E}_{\text{Ewald}}$ in Eq. (4.125). In Quantum ESPRESSO, $\mathcal{E}_{\text{Ewald}}$ denotes the ion-ion Coulomb energy, which is given by

$$\mathcal{E}_{\text{Ewald}} = \frac{1}{2}\sum_{I\neq J}^{N_n} \frac{Z_I Z_J}{|\boldsymbol{R}_I - \boldsymbol{R}_J|}. \qquad (4.144)$$

---

[13] The structure factor contains all the information about the lattice structure. The structure factor gives the extinction rule for X-ray diffraction (see Sec. 5.2).

[14] The atomic form factor contains the information of each atom, which can be calculated or obtained by fitting experimental data. The form factor is important for obtaining the intensity of X-ray scattering.

This sum converges slowly since the potentials of the Coulomb interaction is a long-range function of $|\boldsymbol{R}_I - \boldsymbol{R}_J|$. Thus, in order to converge Eq. (4.144) rapidly, we can apply the **Ewald summation** [Ewald (1921)], which splits the potential into short-range (SR) and long-range (LR) parts. As shown in Eq. (4.124), the Ewald summation of the ionic potential can be expressed as

$$\sum_J^{N_n} \frac{Z_J}{|\boldsymbol{R}-\boldsymbol{R}_J|} = \underbrace{\sum_J^{N_n} Z_I \frac{1-\mathrm{erf}(\eta|\boldsymbol{R}-\boldsymbol{R}_J|)}{|\boldsymbol{R}-\boldsymbol{R}_J|}}_{\mathrm{SR}} + \underbrace{\sum_J^{N_n} Z_J \frac{\mathrm{erf}(\eta|\boldsymbol{R}-\boldsymbol{R}_J|)}{|\boldsymbol{R}-\boldsymbol{R}_J|}}_{\mathrm{LR}}, \tag{4.145}$$

in which, the LR part can be expressed in reciprocal space as [Marder (2010)]

$$\sum_J^{N_n} Z_J \frac{\mathrm{erf}(\eta|\boldsymbol{R}-\boldsymbol{R}_J|)}{|\boldsymbol{R}-\boldsymbol{R}_J|} = \frac{4\pi}{V} \sum_J^{N_n} \sum_{\boldsymbol{G}\neq 0} Z_J \frac{e^{-|\boldsymbol{G}|^2/(4\eta)^2} e^{i\boldsymbol{G}(\boldsymbol{R}-\boldsymbol{R}_J)}}{|\boldsymbol{G}|^2}. \tag{4.146}$$

Then the both SR and LR parts can converge quickly with increasing $|\boldsymbol{G}|$ and $|\boldsymbol{R}|$ for a given value of η. Therefore, the ion-ion Coulomb energy can be obtained by a few terms in the summations over $|\boldsymbol{G}|$ and J.

When we insert Eqs. (4.145) and (4.146) into Eq. (4.144), we must subtract separately the term for $\boldsymbol{R} = \boldsymbol{R}_J$ in Eq. (4.146), to avoid the divergence. If we adopt the following equation:

$$\lim_{\boldsymbol{R}\to\boldsymbol{R}_J} \left[\frac{\mathrm{erf}(\eta|\boldsymbol{R}-\boldsymbol{R}_J|)}{|\boldsymbol{R}-\boldsymbol{R}_J|} \right] = 2\frac{\eta}{\sqrt{\pi}}, \tag{4.147}$$

Eq. (4.144) can be rewritten as

$$\begin{aligned}\mathcal{E}_{\text{Ewald}} =& \frac{1}{2}\sum_{I\neq J}^{N_n} Z_I Z_J \frac{1-\text{erf}(\eta|\boldsymbol{R}_I-\boldsymbol{R}_J|)}{|\boldsymbol{R}_I-\boldsymbol{R}_J|} \\ &+ \frac{2\pi}{V}\sum_{I,J}^{N_n}\sum_{\boldsymbol{G}\neq 0} Z_I Z_J \frac{e^{-|\boldsymbol{G}|^2/(4\eta)^2} e^{i\boldsymbol{G}(\boldsymbol{R}_I-\boldsymbol{R}_J)}}{|\boldsymbol{G}|^2} \\ &- \sum_{J}^{N_n} Z_J^2 \frac{\eta}{\sqrt{\pi}}.\end{aligned} \tag{4.148}$$

Thus, we can avoid the divergence, too.

4.13 Ionic forces

The ionic forces are used to study the dynamics of ions such as optimizing ionic positions (see Sec. 3.1.4). Using the total energy in Sec. 4.12, we can calculate the forces that act to the ion I, which is given by taking the derivative of the total energy with respect to individual ionic position $\boldsymbol{R}_I$ as

$$F_I = -\frac{\partial E_{\text{tot}}}{\partial \boldsymbol{R}_I} = -\frac{\partial \mathcal{E}_{\text{Ewald}}}{\partial \boldsymbol{R}_I} - \frac{\partial \mathcal{E}_{\text{ext}}}{\partial \boldsymbol{R}_I}. \tag{4.149}$$

There are two contributions to the ionic forces, one from the ion-ion interaction energy $\mathcal{E}_{\text{Ewald}}$ in Eq. (4.148), and one from the ion-electron interaction energy $\mathcal{E}_{\text{ext}}$ in Eq. (4.142). For the ion-ion interaction energy, the force of ion I is given by Eq. (4.148) as follows:

$$F_I^{\text{ion}} = -\frac{\partial \mathcal{E}_{\text{Ewald}}}{\partial \boldsymbol{R}_I} = \sum_{J\neq I}^{N_n} \frac{Z_I Z_J}{|\boldsymbol{R}_I-\boldsymbol{R}_J|^3}(\boldsymbol{R}_I-\boldsymbol{R}_J), \tag{4.150}$$

which can be solved by a method analogous to the Ewald summation that is used in Sec. 4.12.4.

As for the ion-electron interaction energy, the force is calculated by adopting the **Hellmann-Feynman theorem**[15] [Feynman (1939), Hellmann (1937)] as

$$F_I^{\text{ion-e}} = -\frac{\partial \mathcal{E}_{\text{ext}}}{\partial \boldsymbol{R}_I} = -\sum_i \left\langle \phi_i \left| \frac{\partial \mathcal{V}_{\text{ps}}}{\partial \boldsymbol{R}_I} \right| \phi_i \right\rangle. \tag{4.151}$$

For the I-th pseudopotential $\mathcal{V}_{\text{ps}}^I(\boldsymbol{r}) = \sum_{\boldsymbol{T}} v_{\text{ps}}^I(|\boldsymbol{r} - \boldsymbol{T} - \boldsymbol{R}_I|)$ (see Eq. (4.139)), Eq. (4.151) is rewritten in term of $n(\boldsymbol{r})$ as

$$F_I^{\text{ion-e}} = -\int \frac{\partial \mathcal{V}_{\text{ps}}(\boldsymbol{r})}{\partial \boldsymbol{R}_I} n(\boldsymbol{r}) \mathrm{d}\boldsymbol{r}. \tag{4.152}$$

Eq. (4.152) tells us that $F_I^{\text{ion-e}}$ does not depend on any derivative of $n(\boldsymbol{r})$. The Hellmann-Feynman force is straightforwardly calculated by the calculated $n(\boldsymbol{r})$ once the self-consistent electronic density is obtained.

4.14 A simple DFT-LDA program for an atom

This section shows a step-by-step calculation of the ground-state energy of a helium atom for understanding the DFT for an atom. The DFT calculation of an atom is also important since Quantum ESPRESSO does not support the package for an atom. We use Python language (version 3) to write the simplest DFT code for the LDA (see Sec. 4.11.1). The present tutorial contains three main calculations: (1) to solve the radial Schrödinger equation by using the Numerov algorithm, (2) to solve the radial Possion equation by using the Verlet algorithm, and (3) incorporating to calculate the ground-state energy of the helium atom by using the LDA with PZ parameters. This tutorial requires the basic Python programming and Jupyter notebook, which are given in Chapter 6.

---

[15] The Hellmann-Feynman theorem states that the forces are given by the expectation value of the derivative of the external potential.

4.14.1 Radial Schrödinger equation

❑ **Purpose:** The purpose of the first step is to calculate the electron density of the helium atom by solving a **radial Schrödinger equation**, which is the Schrödinger equation in spherical coordinates.
❑ **Background:** In spherical coordinates, the wavefunction $\phi(\mathbf{r}) = \phi(r,\theta,\varphi)$ can be written as the product of a radial function $R(r)$ and an angular function $Y(\theta,\varphi)$:

$$\phi(r,\theta,\varphi) = R(r)Y(\theta,\varphi). \tag{4.153}$$

Then we obtain the radial Schrödinger equation for $R(r)$ as follows [Kittel (1976)]

$$\left[-\frac{1}{2r^2}\frac{\mathrm{d}}{\mathrm{d}r}\left(r^2\frac{\mathrm{d}}{\mathrm{d}r}\right) + \frac{\ell(\ell+1)}{2r^2} + \mathcal{V}(r)\right]R(r) = \epsilon R(r), \tag{4.154}$$

where ℓ is azimuthal quantum number $\ell = 0, 1, 2, \ldots$. Let us define a *reduced radial wavefunction* $u(r)$ as

$$u(r) = rR(r). \tag{4.155}$$

By substituting Eq. (4.155) into Eq. (4.154), we obtain the *reduced radial equation* for $1s$ orbital by substituting $\ell = 0$ as

$$\left[-\frac{1}{2}\frac{\mathrm{d}^2}{\mathrm{d}r^2} + \mathcal{V}(r)\right]u(r) = \epsilon u(r). \tag{4.156}$$

Eq. (4.156) can be rewritten in the form of a second-order differential equation as

$$\frac{\mathrm{d}^2 u}{\mathrm{d}r^2} = -k(r)u(r), \tag{4.157}$$

where $k(r)$ is defined by

$$k(r) = 2\left[\epsilon - \mathcal{V}(r)\right]. \tag{4.158}$$

A method to solve Eq. (4.157) is the **Numerov algorithm**.[16] In the Numerov algorithm, the reduced radial wavefunction $u(r)$ is expressed as

$$u_{n+1}(r) = \frac{2c_0 u_n(r) - c_1 u_{n-1}(r)}{c_2}, \tag{4.159}$$

where c_0, c_1, and c_2 are defined by

$$\begin{cases} c_0 = 1 - \frac{5}{12}h^2 k_n^2(r) \\ c_1 = 1 + \frac{1}{12}h^2 k_{n-1}^2(r) \\ c_2 = 1 + \frac{1}{12}h^2 k_{n+1}^2(r) \end{cases}, \tag{4.160}$$

where $h = r_{n+1} - r_n$. By given two initial values, $u_0(r)$ and $u_1(r)$, Eq. (4.159) can be used to determine $u_n(r)$ for $n = 2, 3, 4, \ldots$ with an error in the order of h^6.

We normalize $u(r)$ as

$$\int u^2(r)\mathrm{d}r = 1. \tag{4.161}$$

Then, the electron density $n(r)$ of a single electron for the 1s state $(Y(\theta, \varphi) = 1/\sqrt{4\pi})$ is given by

$$n(r) = \frac{R^2(r)}{4\pi} = \frac{u^2(r)}{4\pi r^2}. \tag{4.162}$$

Note that, for the helium atom, the total electron density is $2n(r)$ since we have two electrons.

❑ **How to run:** To run this tutorial, the readers will do the following command lines:

```
$ cd ~/SSP-QE/dft-he
$ jupyter-lab check-schrodinger.ipynb
```

- Line 1: Go to the `dft-he` directory that includes the input files.
- Line 2: Run `check-schrodinger.ipynb` by JupyterLab.

[16] The Numerov algorithm is a numerical method to solve ordinary differential equations of second order in which the first-order term does not appear.

❑ **Input file:** The input files include a subroutine file (`schrodinger.py`) and a main file (`check-schrodinger.ipynb`). The subroutine contains the Numerov algorithm in Eq. (4.159) and the radial Schrödinger equation in Eq. (4.157), and the main file calculates the ground-states energy and wavefunction for the helium atom.

SSP-QE/dft-he/schrodinger.py

```
# Import the numpy modules
import numpy as np
# The Numerov algorithm for equation:
# u"(r) = -k(r)u(r) with k(r)=2(eps-V)
# Input: k(r), two initial values u0(r) and u1(r)
# Output: u(r)
def numerov(k, u0, u1, dr):
    u = np.zeros_like(k)
    u[0] = u0
    u[1] = u1
    for i in range(2, len(k)):
        dr_sqr = dr**2
        c0 = (1 + 1/12.*dr_sqr*k[i-2])
        c1 = 2*(1 - 5/12.*dr_sqr*k[i-1])
        c2 = (1 + 1/12.*dr_sqr*k[i])
        u[i] = (c1*u[i-1] - c0*u[i-2])/c2
    return u
# Solve the reduced radial Schrodinger equation:
# u"(r) = -2(eps-V)u(r)
# Inputs: r and V(r)
# Outputs: eigen energy eps and u(r)
def solve_schrodinger(r, V, eps_min=-4, eps_max=0,
    maxiter=100, stoptol = 0.0001):
    dr = r[1] - r[0]
    for i in range(maxiter):
        eps = (eps_min + eps_max)/2.
        k = 2*(eps - V)
        # starting from r*exp(-r)
        u0 = r[-1]*np.exp(-r[-1])
        u1 = r[-2]*np.exp(-r[-2])
        # call the Numerov algorithm
        u = numerov(k[::-1], u0, u1, dr)
        u = u[::-1]
        num_nodes = np.sum(u[1:]*u[:-1] < 0)
        if num_nodes == 0 and np.abs(u[0]) <=
            stoptol:
            return (eps, u)
        if num_nodes == 0 and u[0] > 0:
            eps_min = eps
```

```
        else:
            assert num_nodes > 0, 'expect #nodes>0 since u[0]<0 while u[infty]>0'
            eps_max = eps
    raise Exception('Not converged after %d iterations.' %(maxiter))
```

SSP-QE/dft-he/check-schrodinger.ipynb

```
# Import the necessary packages and modules
# sci.mplstyle is customized Matplotlib style
import matplotlib.pyplot as plt
plt.style.use('../matplotlib/sci.mplstyle')
import numpy as np
from schrodinger import solve_schrodinger

# Set r from 0 to 15 bohr with 50000 steps
r, dr = np.linspace(0, 15, 50001, retstep=True)
r = r[1:]  # Skip r = 0
# Set kinetic energy of helium
Z = 2
V_en = -Z/r

# Solve the reduced radial equation:
# u"(r) = -2(eps-V_en)u(r)
eps, u = solve_schrodinger(r, V_en)
# Normalize u(r)
u /= np.linalg.norm(u)*np.sqrt(dr)

# Total electron density of helium
n = 2*(u**2/4/np.pi/r**2)
# Compare with exact density of helium
n_exact = (2*Z**3/np.pi)*np.exp(-2*Z*r)

# Plot the comparison
plt.figure()
plt.plot(r, n, label='numerov')
plt.plot(r, n_exact, ':', lw=4, label='exact')
plt.xlabel('$r$ ($a_0$)')
plt.ylabel('$n(r)$ (a.u.)')
plt.title('Electron density of helium')
plt.legend(loc='best')
plt.xlim(0, 5)
plt.show()
```

❑ **Output data:** In main work area of the JupyterLab interface, the readers can see the plot as shown in Fig. 4.15. We can see

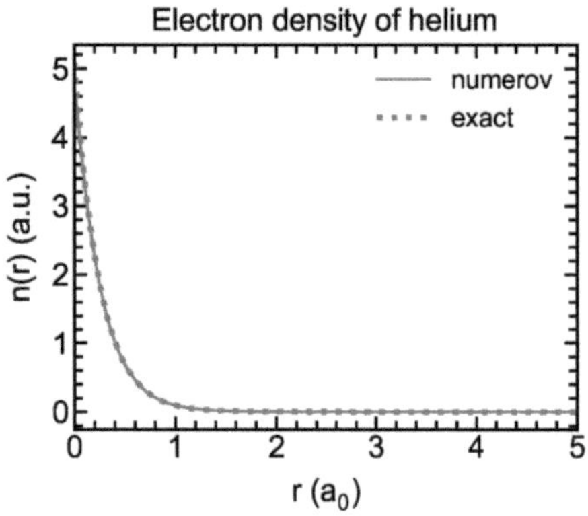

Figure 4.15 Electron density of helium. Solid and dashed lines denote the electron densities, which are obtained by the Numerov algorithm and analytical methods, respectively.

that the electron density from the numerical method (the Numerov algorithm) reproduces the exact electron density of the helium atom given by Eq. (4.21).

4.14.2 The Poisson equation

❑ **Purpose:** Next, let us calculate the Hartree potential from the reduced radial wavefunction $u(r)$.

❑ **Background:** Let us define a Hartree potential as a function of r, $U_H(r) = r\mathcal{V}_H(r)$, then the Poisson equation in Eq. (4.32) reduces to the following differential equation:

$$\nabla^2 U_H(r) = -4\pi rn(r). \tag{4.163}$$

Eq. (4.163) is an ordinary second-order differential equation which can be solved by using the **Verlet algorithm.**[17] By substituting Eq. (4.162) into Eq. (4.163), we have

$$\nabla^2 U_H(r) = -\frac{u^2(r)}{r}. \tag{4.164}$$

[17] The Verlet algorithm is a simple method for integrating second order differential equations of the form $x''(t) = f[x(t), t]$.

The Verlet algorithm for Eq. (4.164) is expressed as following:

$$U_H(r+\delta r) = 2U_H(r) - U_H(r-\delta r) + \delta r^2 \nabla^2 U_H(r). \qquad (4.165)$$

The boundary conditions are also applied to solve Eq. (4.164). For the hydrogen, the conditions are $U_H(0) = 0$ and $U_H(r_{\text{max}}) = 1$. Note that, for the helium case, the Hartree potential is $2U_H(r)/r$ since the electron density is $2n(r)$.

❑ **How to run:** To run this tutorial, the readers will do the following command lines:

```
$ cd ~/SSP-QE/dft-he
$ jupyter-lab check-poisson.ipynb
```

- Line 1: Go to the `dft-he` directory that includes the input files.
- Line 2: Run `check-poisson.ipynb` by JupyterLab.

❑ **Input file:** The input files include two subroutine files (`schrodinger.py` and `poisson.py`) and a main file (`check-poisson.ipynb`). The subroutine `poisson.py` contains the Verlet algorithm in Eq. (4.165) and the Poisson equation in Eq. (4.164), and the main file calculates the Hartree potential for the helium atom.

SSP-QE/dft-he/poisson.py

```
# Import the numpy modules
import numpy as np
# The Verlet algorithm
def verlet(f, U0, U1, dr):
    dr_sqr = dr**2
    U_H = np.zeros_like(f)
    U_H[0] = U0
    U_H[1] = U1
    for i in range(2, len(f)):
        U_H[i] = 2*U_H[i-1] - U_H[i-2] + f[i-1]*
            dr_sqr
    return U_H
# Solve the Poisson equation
# Inputs: r and u(r)
# Outputs: U_H(r)
def solve_poisson(r, u):
    # start the Verlet algorithm
    dr = r[1]-r[0]
```

```
    f = -u**2/r
    U0, U1 = r[0], r[1]
    U_H = verlet(f, U0, U1, dr)
    # fix the boundary condition
    U_H = U_H - (U_H[-1]-1)/r[-1]*r
    return U_H
```

SSP-QE/dft-he/check-poisson.ipynb

```
# Import the necessary packages and modules
# sci.mplstyle is customized Matplotlib style
import matplotlib.pyplot as plt
plt.style.use('../matplotlib/sci.mplstyle')
import numpy as np
from schrodinger import solve_schrodinger
from poisson import solve_poisson

# Set r from 0 to 15 bohr with 50000 steps
r, dr = np.linspace(0, 15, 50001, retstep=True)
r = r[1:]  # Skip r = 0
# Set kinetic energy of helium
Z = 2
V_en = -Z/r

# Solve the radial Schrodinger equation
eps, u = solve_schrodinger(r, V_en)
# Normalize the radial wave function u(r)
u /= np.linalg.norm(u)*np.sqrt(dr)

# Solve the Poisson equation
U_H = solve_poisson(r, u)
# Convert U_H(r) to the Hartree potential V_H(r)
# The factor of 2 since helium has two electrons
V_H = 2*U_H/r
# Compare with the exact Hartree potential
V_exact = -2*(Z + 1/r)*np.exp(-2*Z*r) + Z/r

# Plot the comparison
plt.plot(r, V_H, label='verlet')
plt.plot(r, V_exact, ':', lw=4, label='exact')
plt.xlabel('r (bohr)')
plt.ylabel('$\mathcal{V}_H(r)$ (a.u.)')
plt.title('Hartree potential of helium')
plt.legend(loc='best')
plt.xlim(0, 5)
plt.show()
```

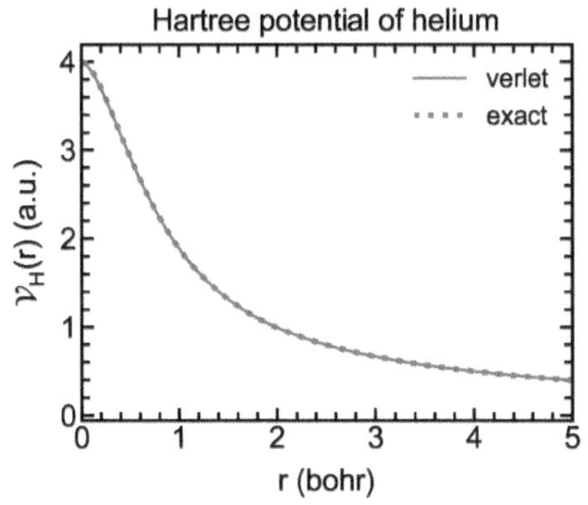

Figure 4.16 The Hartree potential of helium. Solid and dashed lines denote the Hartree potential, which are obtained by the Verlet algorithm and analytical methods, respectively.

❑ **Output data:** In the main work area of the JupyterLab interface, the readers can see the plot as shown in Fig. 4.16. We can see that the Hartree potential from the numerical method (the Verlet algorithm) reproduces the exact Hartree potential of the helium atom given by Eq. (4.39), too.

4.14.3 DFT-LDA for helium

❑ **Purpose:** Finally, we calculate the ground-state energy of the helium atom within LDA.

❑ **Background:** For the LDA, the exchange function $\epsilon_x(\boldsymbol{r})$ and the exchange potential $\mathcal{V}_x[n(\boldsymbol{r})]$ are defined by (see Sec. 4.11.1):

$$\epsilon_x(\boldsymbol{r}) = C_2 n^{1/3}(\boldsymbol{r}) \text{ and } \mathcal{V}_x[n(\boldsymbol{r})] = \frac{4}{3} C_2 n^{1/3}(\boldsymbol{r}), \tag{4.166}$$

respectively, where $C_2 = -0.738$.

We adopt the PZ correlation function $\epsilon_c(r_s)$, which is defined by (see Table 4.2):

$$\epsilon_c(r_s) = \begin{cases} A\ln(r_s) + B + Cr_s\ln(r_s) + Dr_s & \text{if } r_s < 1 \\ \gamma/(1+\beta_1\sqrt{r_s}+\beta_2 r_s) & \text{if } r_s \geq 1 \end{cases}, \tag{4.167}$$

where the parameters $A, B, C, D, \gamma, \beta_1, \beta_2$ are given in Table 4.2, too. Then, the correlation potential is given by

$$\mathcal{V}_c[n(\boldsymbol{r})] = \frac{\partial[\epsilon_c(\boldsymbol{r})n(\boldsymbol{r})]}{\partial n(\boldsymbol{r})}. \tag{4.168}$$

By substituting the electron density $n(\boldsymbol{r}) = 3/(4\pi r_s^3)$ from Eq. (4.112) into Eq. (4.168), we obtain:

$$\mathcal{V}_c(r_s) = \left(1 - \frac{r_s}{3}\frac{\mathrm{d}}{\mathrm{d}r_s}\right)\epsilon_c(r_s). \tag{4.169}$$

By substituting Eq. (4.167) into Eq. (4.169), we obtain [Thijssen (2007)]:

$$\mathcal{V}_c(r_s) = \begin{cases} A\ln(r_s) + B - \frac{A}{3} + \frac{2}{3}Cr_s\ln(r_s) + \frac{2D-C}{3}r_s, & \text{for } r_s < 1 \\ \gamma(1 + \frac{7}{6}\beta_1\sqrt{r_s} + \beta_2 r_s)/(1 + \beta_1\sqrt{r_s} + \beta_2 r_s)^2, & \text{for } r_s \geq 1 \end{cases}. \tag{4.170}$$

With the presence of the exchange-correlation potential, Eq. (4.156) becomes:

$$\left\{-\frac{1}{2}\frac{\mathrm{d}^2}{\mathrm{d}r^2} + \mathcal{V}_{en}(r) + \mathcal{V}_H(r) + \mathcal{V}_{xc}[n(\boldsymbol{r})]\right\}u(r) = \epsilon u(r), \tag{4.171}$$

where $\mathcal{V}_{xc}[n(\boldsymbol{r})] = \mathcal{V}_x[n(\boldsymbol{r})] + \mathcal{V}_c(r_s)$. Then, the total energy of the helium atom in Eq. (4.103) can be rewritten by

$$E = 2\epsilon - \int \mathcal{V}_H(r)u^2(r)\mathrm{d}r - \Delta\mathcal{E}_{\mathrm{xc}}[n(\boldsymbol{r})], \tag{4.172}$$

where $\Delta\mathcal{E}_{\mathrm{xc}}[n(\boldsymbol{r})]$ is expressed by

$$\begin{aligned} \Delta\mathcal{E}_{\mathrm{xc}}[n(\boldsymbol{r})] &= \int \mathcal{V}_{xc}[n(\boldsymbol{r})]n(\boldsymbol{r})\mathrm{d}\boldsymbol{r} - \mathcal{E}_{xc}[n(\boldsymbol{r})] \\ &= \int \{\mathcal{V}_{xc}[n(\boldsymbol{r})] - \epsilon_{xc}[n(\boldsymbol{r})]\}n(\boldsymbol{r})\mathrm{d}\boldsymbol{r} \\ &= 2\int \{\mathcal{V}_{xc}[n(\boldsymbol{r})] - \epsilon_{xc}[n(\boldsymbol{r})]\}u^2(\boldsymbol{r})\mathrm{d}r, \end{aligned} \tag{4.173}$$

where $\epsilon_{xc}[n(\boldsymbol{r})] = \epsilon_x[n(\boldsymbol{r})] + \epsilon_c(r_s)$. Note that we have to use the electron density $n(r)$ and $\mathcal{V}_H(r)$ for the helium atom.

❑ **How to run:** To run this tutorial, the readers will do the following command lines:

```
$ cd ~/SSP-QE/dft-he
$ jupyter-lab dft-lda-he.ipynb
```

- Line 1: Go to the `dft-he` directory that includes the input files.
- Line 2: Run `dft-lda-he.ipynb` with JupyterLab.

❑ **Input file:** The input files include two subroutine files (`schrodinger.py` and `poisson.py`) and a main file (`dft-lda-he.ipynb`). The main file calculates the total energy of the helium atom.

SSP-QE/dft-he/dft-lda-he.ipynb

```
# Import the necessary packages and modules
import numpy as np
from schrodinger import solve_schrodinger
from poisson import solve_poisson

# Set r from 0 to 15 bohr with 50000 steps
r, dr = np.linspace(0, 15, 50001, retstep=True)
r = r[1:]  # Skip r = 0
# Parameter for exchange potential Vx
C2 = -0.738;
# PZ parameters for correlation functional Vc
Aa = 0.0311; B= -0.048; C= 0.002; D= -0.0116; gamma=
       -0.1423; beta1= 1.0529; beta2= 0.3334;

# Define the exchange potential
def exc(n):
    V_x = (4/3)*C2*n**(1/3)
    e_x = C2*n**(1/3)
    return (V_x, e_x)

# Define the correlation potential
def cor(n):
    rs = (3/4/np.pi/n)**(1/3)
    V_c = np.piecewise(rs,[rs<1,rs>=1],[lambda rs:
        Aa*np.log(rs)+B-Aa/3+2/3*C*rs*np.log(rs)+(2*
        D-C)*rs/3, lambda rs: gamma/(1+beta1*rs
        **(1/2)+beta2*rs)*(1+7/6*beta1*rs**(1/2)+
        beta2*rs)/(1+beta1*rs**(1/2)+beta2*rs)])
```

```
    e_c = np.piecewise(rs,[rs<1,rs>=1],[lambda rs:
        Aa*np.log(rs)+B+C*rs*np.log(rs)+D*rs, lambda
         rs: gamma/(1+beta1*rs**(1/2)+beta2*rs)])
    return (V_c, e_c)

# The main DFT for helium with the LDA
def dft(V_en, maxiter=100, stop=0.001, correlation=
    False, verbose=False):
    V_H = np.zeros_like(V_en)
    V_x = np.zeros_like(V_en)
    V_c = np.zeros_like(V_en)
    e_x = np.zeros_like(V_en)
    e_c = np.zeros_like(V_en)
    eps = None
    for i in range(maxiter):
        eps_old = eps
        V = V_en + V_H + V_x + V_c
        V_xc = V_x + V_c
        e_xc = e_x + e_c
        # eigen energy and u(r)
        eps, u = solve_schrodinger(r, V)
        # normalize u(r)
        u /= np.linalg.norm(u)*np.sqrt(dr)
        # convergence of eigen energy
        if eps_old is not None:
            if verbose:
                print('Step %02d: eps = %f, |eps -
                    eps_old| = %f' %(i, eps, abs(eps
                     - eps_old)))
            if abs(eps - eps_old) < stop:
                return 2*eps - dr*np.dot(V_H, u**2)
                    - 2*dr*np.dot(V_xc - e_xc, u**2)
        elif verbose:
            print
        # update Hartree potential
        U_H = solve_poisson(r, u)
        V_H = 2*U_H/r
        # total electron density
        n = 2*(u**2/4/np.pi/r**2)
        # update exchange potential
        V_x, e_x = exc(n)
        # update correlation potential
        if correlation:
            V_c, e_c = cor(n)
        elif correlation:
            V_c = 0
            e_c = 0
```

```
    raise Exception('Not converged after %d
        iterations; |eps - eps_old| = %f' %(maxiter,
        abs(eps - eps_old)))

print('Total energy of He without correlation')
print('JOB DONE: total energy E = %f Ha' %(dft(-2/r,
    correlation=False,verbose=True)))
print('-----------------------------------------')
print('Total energy of He with correlation')
print('JOB DONE: total energy E = %f Ha' %(dft(-2/r,
    correlation=True, verbose=True)))
```

❑ **Output data:** In main work area of the JupyterLab interface, the readers can see the total energy of the helium atom as follows:

```
Total energy of He without correlation
Step 01: eps = -0.320312, |eps - eps_old| = 1.674937
Step 02: eps = -0.625862, |eps - eps_old| = 0.305550
Step 03: eps = -0.470276, |eps - eps_old| = 0.155586
Step 04: eps = -0.537201, |eps - eps_old| = 0.066925
Step 05: eps = -0.505615, |eps - eps_old| = 0.031586
Step 06: eps = -0.519958, |eps - eps_old| = 0.014343
Step 07: eps = -0.513306, |eps - eps_old| = 0.006653
Step 08: eps = -0.516357, |eps - eps_old| = 0.003052
Step 09: eps = -0.514954, |eps - eps_old| = 0.001404
Step 10: eps = -0.515625, |eps - eps_old| = 0.000671
JOB DONE: total energy E = -2.716883 Ha
----------------------------------------
Total energy of He with correlation
Step 01: eps = -0.376953, |eps - eps_old| = 1.618296
Step 02: eps = -0.661568, |eps - eps_old| = 0.284615
Step 03: eps = -0.528656, |eps - eps_old| = 0.132912
Step 04: eps = -0.582932, |eps - eps_old| = 0.054276
Step 05: eps = -0.559113, |eps - eps_old| = 0.023819
Step 06: eps = -0.569275, |eps - eps_old| = 0.010162
Step 07: eps = -0.564880, |eps - eps_old| = 0.004395
Step 08: eps = -0.566772, |eps - eps_old| = 0.001892
Step 09: eps = -0.565948, |eps - eps_old| = 0.000824
JOB DONE: total energy E = -2.826743 Ha
```

The total energy without the correlation energy is −2.716 Ha, which is close to the reported value (−2.72 Ha [Thijssen (2007)]). However, it is less accurate than the Hartree-Fock method (−2.85 Ha, see Sec. 4.6). This is because the exchange potential by the LDA is an approximation for the uniform-electron charge. Therefore, the total energy can be improved by considering the correlation potential. As a result, we obtain the total energy with the PZ correlation functional is −2.827 Ha, which is close to the reported value (−2.83 Ha [Thijssen (2007)]). Although −2.827 Ha is still worse than the

Hartree-Fock result, it is an important improvement with respect to −2.716 Ha. The experimental value of the total energy of the helium atom is −2.9 Ha [Sucher (1958)].

Chapter 5

Solid-State Physics for Quantum ESPRESSO

In order to understand the input and output files of Quantum ESPRESSO, the readers need to know the meaning of technical words in solid-state physics. Here we give minimum information of solid-state physics for Quantum ESPRESSO. The authors wish that the readers take interest in solid-state physics even though their major is not physics.

5.1 Unit cell and Brillouin zone

Unit cell: In Quantum ESPRESSO, we consider a crystal. A crystal is solid with a periodic structure on the position of atoms in the solid. Thus, in order to express the structure of a crystal, it is sufficient to show a minimum block of the periodicity, which is called a unit cell. The unit cell is a box or a parallelepiped,[1] which can be represented by three vectors, each of which is called a **unit vector,** $\boldsymbol{a}_i$ $(i = 1, 3)$. The angles between the unit vectors can be 90°, 120°, or any value

[1] A parallelepiped is a box in which consists of three pairs of parallel faces.

Quantum ESPRESSO Course for Solid-State Physics
Nguyen Tuan Hung, Ahmad R. T. Nugraha, and Riichiro Saito

ISBN 978-981-4968-37-9 (Hardcover), 978-981-4968-63-8 (Paperback), 978-1-003-29096-4 (eBook)
www.jennystanford.com

depending on the symmetry of the crystal. In three-dimensional materials, we have 230 distinct crystal shapes, which are defined in the space group. The 230 kinds of crystals are classified by (A) 5 crystal families, (B) 14 Bravais lattices, or (C) 32 point groups of the unit cell. Here the Bravais lattice is defined by the nonequivalent translational symmetry of the lattice. In the input file of Quantum ESPRESSO, we input the information of $\boldsymbol{a}_i$ $(i = 1, 3)$. The relative coordinate of the j-th atom in the unit cell, $\boldsymbol{r}_j$ is given by fractional numbers s_{ij} $(i = 1, 3, 0 \leq s_{ij} < 0)$ in the input file:

$$\boldsymbol{r}_j = \sum_i^3 s_{ij}\boldsymbol{a}_i. \tag{5.1}$$

It should be careful to set s_{ij} in the unit cell when the unit cell is not cubic or cuboid. A **lattice vector** which is defined by

$$\boldsymbol{R} = p\boldsymbol{a}_1 + q\boldsymbol{a}_2 + r\boldsymbol{a}_3, \qquad (p, q, r; \text{integers}), \tag{5.2}$$

where $\boldsymbol{R}$ represents a position of the unit cell in the crystal. Thus, the coordinate of any atom in a crystal is given by the sum of $\boldsymbol{R} + \boldsymbol{r}_j$.

Reciprocal lattice vector: Any wave in the crystal[2] that propagates in the direction of $\boldsymbol{k}$ is expressed by $f(\boldsymbol{k}, \boldsymbol{r}) \equiv \exp(i\boldsymbol{k} \cdot \boldsymbol{r})$ which we call a plane wave. The product of $\boldsymbol{k} \cdot \boldsymbol{r}$ in $\exp(i\boldsymbol{k} \cdot \boldsymbol{r})$ is called a phase of the wave that has a periodicity of 2π. If we take $\boldsymbol{r} = \boldsymbol{a}_i$ and $\boldsymbol{k} = \boldsymbol{b}_i$ so that $\boldsymbol{a}_i \cdot \boldsymbol{b}_i = 2\pi$, we get the same amplitude of f for $\boldsymbol{r}$ and $\boldsymbol{r} + \boldsymbol{a}_i$, that is $f(\boldsymbol{b}_i, \boldsymbol{r}) = f(\boldsymbol{b}_i, \boldsymbol{r} + \boldsymbol{a}_i)$. Further, if we take $\boldsymbol{r} = \boldsymbol{a}_i$, we get $f(\boldsymbol{k}, \boldsymbol{a}_i) = f(\boldsymbol{k} + \boldsymbol{b}_i, \boldsymbol{a}_i)$. It means that the wave has a periodicity not only in the real space but also in the $\boldsymbol{k}$ space, which we call **reciprocal lattice.**[3] The vector $\boldsymbol{b}_i$ is a reciprocal lattice vector. The unit box in the $\boldsymbol{k}$ space which consists of the three $\boldsymbol{b}_i$ is called the **Brillouin zone**. Similar to the lattice vector $\boldsymbol{R}$, we can define **reciprocal lattice vector, $\boldsymbol{G}$,**

$$\boldsymbol{G} = \ell\boldsymbol{b}_1 + m\boldsymbol{b}_2 + n\boldsymbol{b}_3 \equiv (\ell, n, m), (\ell, n, m; \text{integers}), \tag{5.3}$$

[2] In quantum mechanics, electron behaves as a wave. Lattice oscillation is a wave like earthquake in the earth. X-ray propagates in the lattice as electro-magnetic wave.

[3] If you know the Fourier transform, you can imagine that the $\boldsymbol{r}$ and $\boldsymbol{k}$ can be converted to each other. In physics, we call the extended 6 dimensional space of $\{\boldsymbol{r}, \boldsymbol{k}\}$ as a phase space. Any motion of a particle is expressed by a curve in the phase space.

where $\boldsymbol{G}$ represents the position of the Brillouin zone in the $\boldsymbol{k}$ space. The integers of (ℓ, n, m) are called **the Miller indices** which are used in X-ray analysis (Sec. 5.2).

The $\boldsymbol{b}_i$ is obtained in the output file of Quantum ESPRESSO since $\boldsymbol{b}_i$ can be calculated by the $\boldsymbol{a}_i$ in the input file as follows:

$$\boldsymbol{a}_i \cdot \boldsymbol{b}_j = 2\pi\delta_{ij}, \qquad \delta_{ij} \equiv \begin{cases} 1 & \text{if } i = j \\ 0 & \text{if } i \neq j \end{cases}. \tag{5.4}$$

In fact, we have the following formula for $\boldsymbol{b}_i$ which satisfies Eq. (5.4):

$$\boldsymbol{b}_1 = \frac{2\pi(\boldsymbol{a}_2 \times \boldsymbol{a}_3)}{\boldsymbol{a}_1 \cdot (\boldsymbol{a}_2 \times \boldsymbol{a}_3)}, \ \boldsymbol{b}_2 = \frac{2\pi(\boldsymbol{a}_3 \times \boldsymbol{a}_1)}{\boldsymbol{a}_2 \cdot (\boldsymbol{a}_3 \times \boldsymbol{a}_1)}, \ \boldsymbol{b}_3 = \frac{2\pi(\boldsymbol{a}_1 \times \boldsymbol{a}_2)}{\boldsymbol{a}_3 \cdot (\boldsymbol{a}_1 \times \boldsymbol{a}_2)}, \tag{5.5}$$

where $\times$ denotes a vector product or cross product. Because of the properties of the vector product that the vector $\boldsymbol{a}_2 \times \boldsymbol{a}_3$ is perpendicular to the both $\boldsymbol{a}_2$ and $\boldsymbol{a}_3$, we can check that Eq. (5.5) satisfies Eq. (5.4). The Brillouin zone is given by six vertical bisectors[4] of $\boldsymbol{b}_i$ and $-\boldsymbol{b}_i$, $(i = 1, 3)$.

5.2 X-ray analysis

Once we define the unit vectors $\boldsymbol{a}_i$ in the input file of Quantum ESPRESSO, we can plot the 3D picture of the unit cell and lattice, Brillouin zone (see Sec. 5.1) by XCrySDen (see Sec. 3.1.1). Experimentally the lattice structure of a crystal is observed by X-ray spectroscopy. The most popular X-ray analysis is the Debye-Scherrer method in which the incident X-ray is scattered in the direction of an angle so-called 2θ from the propagating direction of incident X-ray ($\boldsymbol{k}_i n$ in Fig. 5.1 (b)).[5] Thus, if we measure the intensity of the scattered X-ray as a function of 2θ for calculating the lattice, we can directly compare the calculated results with the experimental results, which can also be automatically calculated by XCrySDen. The X-ray spectra

[4] A vertical or perpendicular bisector for a given vector is a plane that is perpendicular to the vector and that intersects at the center of a vector.
[5] 2θ is called diffraction angle. The factor 2 of 2θ can be understood by Fig. 5.1.

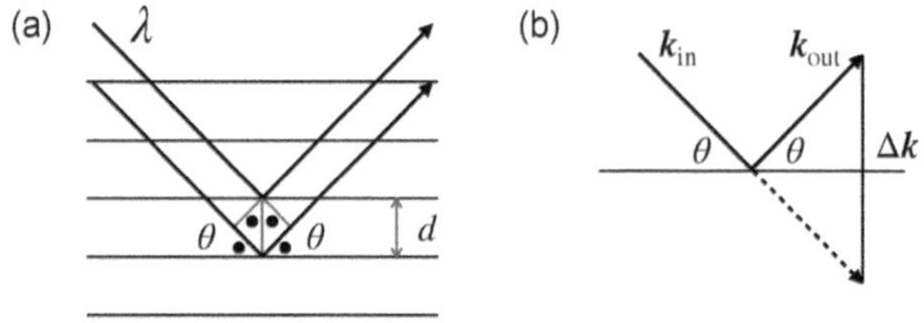

Figure 5.1 (a) Bragg's condition in the real space. (b) Bragg's condition is the $\boldsymbol{k}$ space.

have one peak at a 2θ, which we can label by a Miller index (ℓ, n, m) (see Sec. 5.1), which we explain below.

Bragg's condition: For a periodic lattice, we can define a **crystal plane** as is shown by thin lines in Fig. 5.1 (a) on which many atoms exist. Since the lattice is periodic, we have many equivalent crystal planes parallel to the crystal plane. In Fig. 5.1 (a), we show the side-view of the crystal planes as parallel lines. When the incident X-ray with the wavelength λ enters the crystal with an angle θ (solid dots in Fig. 5.1 (a)) measured from the plane, the diffracted X-ray that propagates in the direction of θ, too, interfere to each other from two adjacent planes as shown in Fig. 5.1 (a). The condition of the constructive interference of the diffracted light is given by

$$2d\sin\theta = n\lambda, \qquad \sin\theta = \frac{n\lambda}{2d}, \tag{5.6}$$

where d denotes the distance between two adjacent planes and $n = 1, 2, \ldots$ is an integer. Since d and λ are given, respectively, by the crystal and incident X-ray, only θ's for a given n are possible direction of diffracted light. It is noted that the constructive interference occurs not only for the two planes but also for all (say 10^8) parallel crystal planes in the crystal, which gives a strong X-ray intensity.[6] In Fig. 5.1 (b), we show that the diffraction angle of the scattered X-ray becomes 2θ which is the reason why we use 2θ in the horizontal axis of X-ray intensity plot. Since $|\boldsymbol{k}_{\rm in}| = |\boldsymbol{k}_{\rm out}| = k = 2\pi/\lambda$, the change of the

[6] If you know diffraction grating, the concept of the diffracting grating comes from the Bragg condition of one-dimensional periodicity.

wavevector Δk is given by

$$\Delta k = 2k \sin\theta = 2k \cdot \frac{n\lambda}{2d} = \frac{2\pi}{d} \cdot n, \tag{5.7}$$

where we use Eq. (5.6). The value of $2\pi/d$ corresponds to a reciprocal lattice vector for a given d, which should be one of $\boldsymbol{G}$ defined by Eq. (5.3) for being d as a distance of crystal plane. In fact, a crystal plane is defined by three points in the space, which is given by the end point of the three vectors of $\ell^{-1}\boldsymbol{a}_1$, $m^{-1}\boldsymbol{a}_2$, $n^{-1}\boldsymbol{a}_3$. The normal vector of the crystal plane is given by $\boldsymbol{G}$ with the Miller index (ℓ, m, n). Thus, d (2θ) for a given Miller index (ℓ, m, n) is expressed by

$$d = \frac{2\pi}{|\boldsymbol{G}|} = \frac{2\pi}{\sqrt{(\ell \boldsymbol{b}_1 + m\boldsymbol{b}_2 + n\boldsymbol{b}_3)^2}}. \tag{5.8}$$

When we plot X-ray intensity as a function of 2θ, we do the following procedure:

1. We assign the shape of the unit cell by $\boldsymbol{a}_i$, $(i = 1, 2, 3)$.
2. We get the reciprocal lattice vector $\boldsymbol{b}_i$, $(i = 1, 2, 3)$ by Eq. (5.5).
3. For a given Miller index, (ℓ, m, n), we calculate d by Eq. (5.8).
4. From the calculated d, we get θ by Eq. (5.7) by putting $n = 1$ and $\lambda = 1.54$ Å.[7]
5. Then we calculated X-ray intensity at 2θ.

For calculation of X-ray intensity, we need to know the concepts of "structure factor" and "atomic form factor". However, we will not explain in detail those words for simplicity. Since the software or database gives the X-ray intensity automatically, it is sufficient for the reader to know the meaning of 2θ. If you wish to know how to calculate the two factors, please take a look at the solid-state textbook.

Nevertheless, we should comment on an important concept of the "structure factor" for assigning the crystal symmetry. The scattering amplitude of solid is given by the sum of the scattering

[7] This wavelength of X-ray is called for Cu Kα line. If there is no specification of the wavelength in the experiment, you can use this value for comparison.

amplitude of atoms in the unit cell. The intensity of the X-ray is calculated by taking the square of the sum of scattering amplitudes of atoms in the unit cell. When the number of atoms in the unit cell is more than one, the destructive interface of the scattering amplitude occurs for two in-equivalent atoms in a unit cell. In this case, the X-ray intensity for some (ℓ, m, n) becomes zero, which is known as **X-ray extinction law**. The extinction law is calculated by calculating the structure factor. Since the extinct (ℓ, m, n) values depend on the lattice symmetry, the extinction law is important for identifying the symmetry of the lattice.

5.3 Plane wave expansion

The main results of Quantum ESPRESSO are energy dispersion or simply energy bands. The energy band is a function of electronic energy for an electron in the solid as a function of the wavevector (or crystal momentum) of the electron. The energy band is obtained by solving the Schrödinger equation in quantum mechanics (or more precisely the Kohn-Sham equation of the density-functional theory (Sec. 4.10.2)), $\mathcal{H}\Psi = E\Psi$, where $\mathcal{H}$ is the Hamiltonian operator, Ψ denote the wavefunction and E is energy. For a given $\mathcal{H}$, we solve $\mathcal{H}\Psi = E\Psi$ to obtain an eigenvalue E and an eigenfunction Ψ.[8]

When there is a periodicity of the unit cell in the real space, we have another periodicity of the reciprocal lattice vector $\boldsymbol{G}$ in the $\boldsymbol{k}$ space (Sec. 5.1). In this case, the wavevector $\boldsymbol{k}$ is a conserved variable and thus a good quantum number for expressing both $E(\boldsymbol{k})$ and $\Psi_{\boldsymbol{k}}(\boldsymbol{r})$. For the energy $E(\boldsymbol{k})$, we have the periodicity of $E(\boldsymbol{k}) = E(\boldsymbol{k} + \boldsymbol{G})$ for any $\boldsymbol{G}$. For the wavefunction Ψ, using the periodicity of $\boldsymbol{G}$, we can expand $\Psi_{\boldsymbol{k}}(\boldsymbol{r})$ as follows:

$$\Psi_{\boldsymbol{k}}(\boldsymbol{r}) = \sum_{\boldsymbol{G}} C_{\boldsymbol{G}}(\boldsymbol{k}) e^{i(\boldsymbol{k}+\boldsymbol{G})\boldsymbol{r}}, \tag{5.9}$$

where $C_{\boldsymbol{G}}(\boldsymbol{k})$ is the coefficient to be solved for each $\boldsymbol{k}$ and the summation on $\boldsymbol{G}$ is taken for many $\boldsymbol{G}$. This expansion is called **plane**

[8] The Schrödinger equation is given by the differential equation, and thus we solve the eigenvalue problem of the differential equation.

wave expansion. The plane expansion is one of the Fourier series expansion[9] for a periodic function and thus if we take more number of $\boldsymbol{G}$, we get more precise value of the function.

Let us explain how to solve the Schrödinger equation by using the plane wave expansion. The Hamiltonian $\mathcal{H}$ is given by

$$\mathcal{H}(\boldsymbol{r}) = -\frac{\hbar^2}{2m}\Delta + \mathcal{V}(\boldsymbol{r}), \tag{5.10}$$

where $-\frac{\hbar^2}{2m}\Delta$ and $\mathcal{V}(\boldsymbol{r})$ are kinetic and potential energy operators, respectively, that satisfy the periodicity of the unit cell.[10] Thus, we can expand $\mathcal{V}$ by the reciprocal lattice vector $\boldsymbol{G}$ as follows:

$$\mathcal{V}(\boldsymbol{r}) = \sum_{\boldsymbol{G}} \mathcal{V}_{\boldsymbol{G}} e^{i\boldsymbol{G}\boldsymbol{r}}, \text{ where } \mathcal{V}_{\boldsymbol{G}} = \frac{1}{N_{\mathrm{u}}} \int \mathrm{d}\boldsymbol{r}' \mathcal{V}(\boldsymbol{r}') e^{-i\boldsymbol{G}\boldsymbol{r}'}. \tag{5.11}$$

Here integration on $\boldsymbol{r}'$ is taken over the crystal and N_u denotes the number of the unit cells in the crystal. By substituting Eqs. (5.9), (5.10), and (5.11) into $\mathcal{H}\Psi = E\Psi$, we get

$$\begin{aligned} &\sum_{\boldsymbol{G}} \frac{\hbar^2}{2m}(\boldsymbol{k}+\boldsymbol{G})^2 C_{\boldsymbol{G}} e^{i(\boldsymbol{k}+\boldsymbol{G})\boldsymbol{r}} + \sum_{\boldsymbol{G}',\boldsymbol{G}''} \mathcal{V}_{\boldsymbol{G}'} C_{\boldsymbol{G}''} e^{i(\boldsymbol{k}+\boldsymbol{G}'+\boldsymbol{G}'')\boldsymbol{r}} \\ &\quad = E\sum_{\boldsymbol{G}} C_{\boldsymbol{G}} e^{i(\boldsymbol{k}+\boldsymbol{G})\boldsymbol{r}}. \end{aligned} \tag{5.12}$$

---

[9] For a periodic function $f(x)$ with a period L, we can express

$$f(x) = \sum_{p=0}^{\infty} f_p e^{i2\pi p x/L}$$

which is called the Fourier series expansion. Here $2\pi/L$ is the reciprocal lattice vector for L.

[10] The Laplacian Δ are invariant for transformation of $x' = x + a$ and $\mathcal{V}(\boldsymbol{r}+\boldsymbol{a}) = \mathcal{V}(\boldsymbol{r})$.

In the second term, we put $\boldsymbol{G}' + \boldsymbol{G}'' = \boldsymbol{G}$ and take the term of $e^{i(\boldsymbol{k}+\boldsymbol{G})\boldsymbol{r}}$, we get the following equation,[11]

$$\frac{\hbar^2}{2m}(\boldsymbol{k}+\boldsymbol{G})^2 C_{\boldsymbol{G}} + \sum_{\boldsymbol{G}''} \mathcal{V}_{\boldsymbol{G}-\boldsymbol{G}''} C_{\boldsymbol{G}''} = E C_{\boldsymbol{G}}. \tag{5.13}$$

Eq. (5.13) is simultaneous equation for obtaining $C_{\boldsymbol{G}}$. In the real calculation, when we define $N \times N$ Hamiltonian matrix

$$\mathcal{H}_{\boldsymbol{G}\boldsymbol{G}'} = \begin{cases} \frac{\hbar^2}{2m}(\boldsymbol{k}+\boldsymbol{G})^2, & \text{if } (\boldsymbol{G} = \boldsymbol{G}') \\ \mathcal{V}_{\boldsymbol{G}-\boldsymbol{G}'}, & \text{if } (\boldsymbol{G} \neq \boldsymbol{G}') \end{cases}. \tag{5.14}$$

Eq. (5.13) is expressed by a matrix equation $\mathcal{H}\boldsymbol{C} = E\boldsymbol{C}$ from which we can obtain eigenvalue E and eigenvector $\boldsymbol{C}$ by diagonalizing the matrix by numerical calculation.[12]

5.4 Cut-off energy and pseudopotential

The size of the Hamiltonian matrix, N, is given by the number of reciprocal lattice vectors $\boldsymbol{G}$. Although the accuracy of calculation monotonically increases with increasing N, the memory size and computational time of the computer quickly increases with increasing N. Thus, we must find a reasonable N by plotting the total energy and computational time as a function of N.

Cut-off energy: In the first-principles calculation, we do not usually compare N for different calculations but **cut-off energy** in units of atomic unit (see Table 4.1) which is defined by

$$E_{\text{cut-off}} = \frac{\hbar^2}{2m}\boldsymbol{G}^2_{\max}, \tag{5.15}$$

where $\boldsymbol{G}_{\max}$ is the largest reciprocal lattice vector. We use the cut-off energy because the values of N depend on the size of the unit

[11] Since $e^{i(\boldsymbol{k}+\boldsymbol{G})\boldsymbol{r}}$ are independent functions for different values of $\boldsymbol{G}$, the coefficients of the both-hand sides should be identical. Further, we change the summation on $\boldsymbol{G}'$ and $\boldsymbol{G}''$ to the sum on $\boldsymbol{G}$ and $\boldsymbol{G}''$. Note that $\boldsymbol{G}' = \boldsymbol{G} - \boldsymbol{G}''$.

[12] LAPACK is the name of a famous software package for linear algebra calculation including diagonalizing a matrix (zheev).

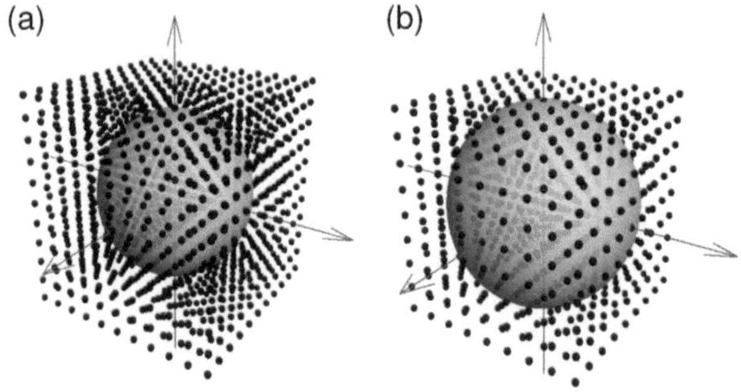

Figure 5.2 Cut-off energy. Dots in the k space represent the reciprocal lattice vectors $\boldsymbol{G}$. When we define a sphere whose diameter is $|\boldsymbol{G}_{\text{max}}|$, we will use $\boldsymbol{G}$'s in the sphere. Then cut-off energy is given by Eq. (5.15). (a) When the unit cell is large, $\boldsymbol{G}$ becomes small, and thus we have more $\boldsymbol{G}$'s in the sphere. (b) When the unit cell is relatively small, the number of $\boldsymbol{G}$'s becomes small even for the same cutting energy. Reprinted with permission from Asakura Publishing Co. Ltd.

cell. When the size of the unit cell is relatively large because of many atoms in the unit cell, the corresponding absolute value of $\boldsymbol{G}$ becomes small as shown in Fig. 5.2 (a). In order to obtain the same accuracy for calculating the atomic potentials in the smaller unit cell (Fig. 5.2 (b)), a larger $\boldsymbol{G}_{\text{max}}$ is needed. On the other hand, when we use the same cut-off energy for two different sizes of unit cells, we expect the accuracy of the calculation to be similar to each other.

The selection of cut-off energy also depends on the selection of **pseudopotential** for each atom, which we explain below. As shown in Eq. (5.12), the off-diagonal matrix element of the Hamiltonian matrix is the Fourier transform of the crystal potential V, which is given by the sum of atomic potentials. If we adopt the real atomic potential for an atom, we expect a large Coulomb potential, $-Ze^2/(4\pi\epsilon_0 r)$ (Z is the atomic number of the atom). Fourier transform of $1/r$ is $1/|\boldsymbol{G}|^2$ which is a slowly decreasing function of $|\boldsymbol{G}|$ compared with $\exp(-C|\boldsymbol{G}|)$ (C is a constant) in three dimensions. Thus, it is not a good idea that we do not adopt the real atomic potential in the plane wave expansion.

Pseudopotential: For discussing most of the solid-state properties, inner-core electrons in the $1s$, $2s$, $2p$ atomic orbitals for a heavy element are not so important. Only the valence electrons near the highest occupied orbitals contribute to the properties. The valence electrons feel screened Coulomb potentials by core electrons except for the core region of the atom. **Pseudopotential** is a potential in which the core region of an atom is artificially smoothed out near $\boldsymbol{r} = 0$ so that we do not need many $\boldsymbol{G}$ (or N) for obtaining $\mathcal{V}_{\boldsymbol{G}}$ with the same numerical accuracy.

In history, pseudopotentials have been proposed by many researchers, which we select from the list of pseudopotentials on the web pages as input files of Quantum ESPRESSO. The pseudopotential is given as a function of distance from the center of atom r and the density of electrons based on the density-functional theory (Sec. 4.12.3).

Norm-conserving pseudopotential: Although we will not go to in detail of the selection of pseudopotential, it is important to point out that the readers should select so-called **norm-conserving pseudopotential** if they want to use the wavefunction for calculating some matrix elements such as for calculating optical absorption spectra. For norm-conserving pestilential, (1) the pseudo wavefunction that is obtained by solving the Kohn-Sham (or the Schrödinger equation in the DFT) equation coincides with the real wavefunction for a larger r than a cut-off radius, $r_{\rm cut}$. (2) At $r_{\rm cut}$, the values and derivative of the pseudo wavefunction and real wavefunction should be the same. (3) For $r < r_{\rm cut}$, the norms of the pseudo wavefunction and real wavefunction are selected to be the same, though the pseudo wavefunction does not have a node as a function of r. (4) The eigenvalues or energies for pseudopotentials are the same as those by real potentials. The solid-state properties calculated by norm-conserving pseudopotentials can be compared with experimental results with reasonably high accuracy. On the other hand, **ultra-soft pseudopotential** is another direction of pseudopotentials in which the number of plane waves can be reduced significantly by neglecting the norm-conserving conditions. If one wants to obtain the energy band of many atoms in a large unit-cell, the selection of ultra-soft pseudopotential is important.

5.5 Energy bands and density of states

Energy bands: By selecting k points on high symmetry lines in the Brillouin zone in the input file of `bands.x` (see Sec. 3.2.2), we can obtain the energy as a function of k, $E(k)$, which we call **energy dispersion** or simply **energy bands**. The name of energy bands comes from the fact that the energy dispersion exists in a finite energy region. For a given energy dispersion, two electrons with the up- and down-spin of electron per unit cell can be occupied. As far as two energy dispersion does not cross each other, we can occupy two electrons per energy band from the lowest bands to the **valence band** which is the highest occupied energy band. Suppose two energy bands cross each other before filling the lower energy band fully. In that case, the electron starts to occupy the higher energy band with keeping the common highest occupied energies of the two energy bands, which is called the **Fermi energy**, E_F.

Density of states: For a given energy dispersion $E(k)$, when we denote $\mathrm{d}N$ for the number of states in the energy region from E to $E + \mathrm{d}E$, we define **density of states (DOS)**, $D(E)$, as $D(E) \equiv \mathrm{d}N/\mathrm{d}E$. The unit of DOS can be "states/eV/unit cell". DOS becomes large when $E(k)$ is flat in the energy dispersion.

Since the wavevector of a crystal in one direction is given by $k = 2\pi p/Na$, $(p = 0, 1, 2, ..., N - 1)$, where N is the number of the unit cell in the direction and a is the lattice constant. In a single crystal with a size of 1 cm, since N is large $N(\sim 10^8)$, the wavevector k exists almost continuously, which is why energy dispersion $E(k)$ is generally given a function of k. For each k state, we can put two electrons with up-spin and down-spin, and thus in total, we can put $2N$ electrons for N k states. Since there are N unit cells in the direction, we can say that we can occupy two electrons per unit cell, whose situation is the same as the case of three dimensions.

In Quantum ESPRESSO, DOS is calculated by making a mesh of k in the Brillouin zone. For a given E and for a given three-dimensional box made of mesh, if there is an equi-energy surface of $E(k) = E$, the area of the surface in the mesh is proportional to DOS. The number of mesh specified by the input file of DOS (see Sec. 3.2.3) is about $10 \times 10 \times 10$ since we can interpolate $E(k)$ in the box of the mesh. This method is called the "tetrahedron method", from which you can learn more about the tetrahedron method for obtaining DOS.

Metal, semimetal, semiconductor, insulator: Using the output values of the energy bands, the Fermi energy, and density of states, we classify the materials. When the Fermi energy is located in the middle energy of the energy band, electrons in the energy band can easily excite the unoccupied states within the same energy band. This situation is called "**metal**". We can say that metal is defined by a finite density of states at the Fermi energy, $D(E_F) \neq 0$. When the valence band is fully occupied, and the next higher energy band, that is **conduction band** is unoccupied, we usually have an **energy gap**. An energy gap, E_g, is defined by an energy difference between the highest energy of valence band and the lowest energy of conduction bands.[13] Since there are no states in the energy gap region, the occupied electron in the valence band needs at least E_g for exciting the electron. This situation corresponds to "semiconductor" or "insulator". The difference between "semiconductor" or "insulator" is the value of E_g. If $E_g < 3$eV, we can say that the material is semiconductor, while $E_g > 5$ eV, we call "insulator".[14]

When $E_g = 0$ or E_g is a small value compared with the thermal energy k_BT (k_B is the Boltzmann constant), part of electrons in the valence band can be excited at the finite temperature. In this situation, if the density of states at the Fermi energy is much smaller than $D(E_F)$ of three-dimensional metal, the number of carriers for flowing an electric current[15] becomes small, too. Such material is called **semimetal**. A typical example of semimetal is graphene in which $E_g = 0$ and $D(E_F) = 0$.

k sampling points: It is noted that Quantum ESPRESSO calculation of a metal takes more computational time than that of a semiconductor in the SCF calculation (Sec. 4.5). In the SCF calculation, we evaluate the charge density $\rho(r)$ by occupying electrons in the energy bands, which require more k points in the case of metal. When we evaluate $\rho(r)$ for a semiconductor, we can use a few numbers of sampling k points in the Brillouin zone. When all energy bands are fully occupied or unoccupied, we need only a few (even one) k points. However, in the case of metal, on the other hand, since we need to consider the occupation up to the Fermi energy, more k points are needed for obtaining the **Fermi surface** (a closed

---

[13] If you are a chemist, you can imagine that an energy gap is a kind of HOMO-LUMO gap of a molecule in solid.

[14] For $3 < E_g < 5$ eV, we can say "wide-gap semiconductor".

[15] Carrier is either an electron or a hole that conducts an electronic current in a solid. A hole is an unoccupied state in the almost occupied energy band. Like a going-up beer bubble, the unoccupied states carry a positive charge in the solid in the direction opposite to that of the electron.

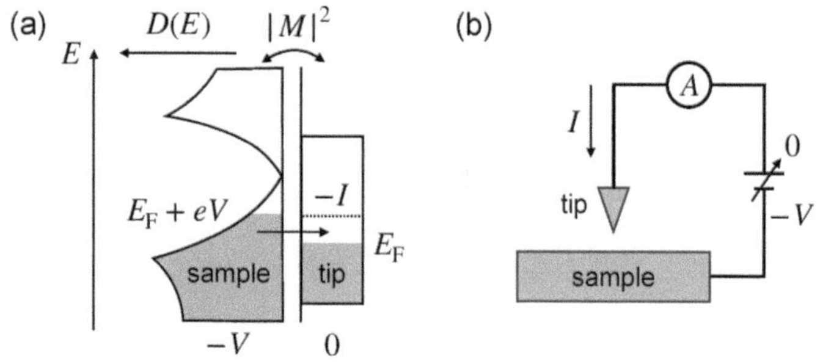

Figure 5.3 Scanning tunneling spectroscopy. (a) Tunneling current I (Eq. (5.16)) flows from the metallic materials (left) to a scanning metallic tip (right) in the energy region between E_F and $E_F + eV$, in which electrons occupied in the metallic side while not occupied in the tip side. (b) In the experiment, we fix the position of the tip, we change the voltage. Then the value dI/dV is proportional to the density of state $D(E_F + eV)$ (Eq. (5.17)).

surface in the k space on which the energy is the Fermi surface) which costs more CPU times. Total energy depends on the number of the sampling points, which should be checked by increasing the sampling points.

5.6 Experiments for $E(k)$ and DOS

The calculated $E(k)$ and DOS $D(E)$ by Quantum ESPRESSO can be directly compared with **ARPES** (angle-resolved photo-electron spectroscopy) and **STS** (scanning tunneling spectroscopy) measurements, respectively. In ARPES, we can measure a kinetic energy E of photo-excited electron which emitted from the materials as a function of the momentum direction of photo-electron $\boldsymbol{k}$ emitted into the vacuum by illuminated by light (photoelectric effect), which can be directly compared with the calculated electronic energy dispersion, $E(k)$. Since the electrons exist in the occupied energy bands, we can compare the $E(k)$ only for the occupied energy bands down to 10–20 eV below the Fermi energy. The range of energy that we can observe by ARPES depends on the light source.

In STS measurement (see Fig. 5.3 (b)), we measure the tunneling electric current I on a sharp metallic tip, which is electrically biased by an applied voltage. If we assumed that the tunneling probability

$|M|^2$ does not depend on the energy of electrons, the tunneling current I is given by

$$I = \int_{E_F}^{E_F+eV} |M|^2 D(E) \mathrm{d}E, \tag{5.16}$$

from which the differential conductance, $\mathrm{d}I/\mathrm{d}V$ is calculated as follows:

$$\frac{\mathrm{d}I}{\mathrm{d}V} = |M|^2 D(E_F + eV) \propto D(E_F + eV). \tag{5.17}$$

We get from Eq. (5.17) that the differential conductance defined by $\mathrm{d}I/\mathrm{d}V$ is proportional to the density of state at $E = E_F + eV$. Thus, we can measure $D(E)$ by changing V near the E_F. In the case of STS, the sample should be a conductor to have a finite I. Further, in order to avoid thermal broadening of the Fermi distribution function, the STS measurement is performed at a much lower temperature than the room temperature.

5.7 Phonon dispersion

Phonon dispersion relation is the phonon frequencies in units of cm^{-1} (1 eV = 8065 cm^{-1}) as a function of the wavevector of phonon, q. In Fig. 5.4, we show (a) phonon dispersion and phonon density of states of graphene. In Quantum ESPRESSO, we specify q by connecting symmetric points in the Brillouin zone, as is the case of the electronic energy band. When we have N atoms in a unit cell, we have $3N$ phonon dispersion in the Brillouin zone in which 3 of the $3N$ phonon dispersion are acoustic phonon modes whose frequency is zero at the zone-center (or the Γ point) of the Brillouin zone. The remaining $3N - 3$ phonon modes are optical phonon modes. For the three acoustic modes, one of the three acoustic modes is longitudinal acoustic (LA) phonon mode, in which the oscillational direction is parallel to the propagation direction of the wave of oscillation. The remaining two acoustic phonon modes are transverse acoustic (TA) phonon modes, in which the oscillational direction is perpendicular to the propagation direction. For optical phonon modes, $N - 1$ modes are longitudinal optic (LO) modes, and the remaining $2N - 2$ are

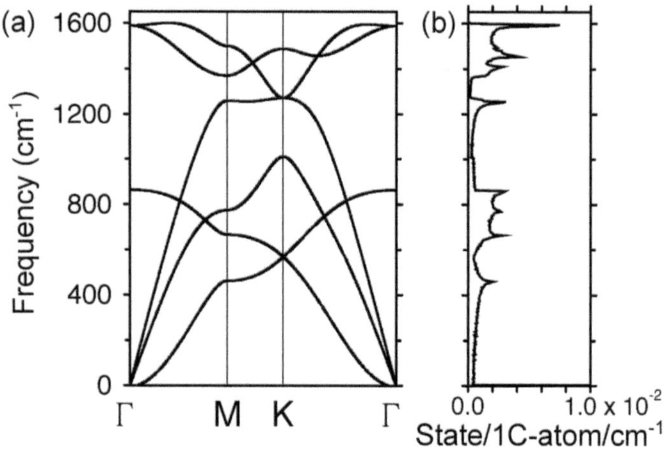

Figure 5.4 (a) Phonon dispersion of graphene. The unit of frequency is cm^{-1} (1 eV = 8065 cm^{-1}). (b) Phonon density of states.

transverse optic (TO) modes. The assignment of LA, TA, LO, and TO is possible by looking at the direction of oscillation of each atom at the Γ point (or near the Γ point for acoustic modes by the Material Cloud (see Sec. 3.3.1.)).

When the material is two-dimensional, such as graphene as shown in Fig. 5.4, TA modes are further classified by in-plane TA (iTA) and out-of-plane TA (oTA), and TO modes are classified by in-plane TO (iTO) and out-of-plane TO (oTO) phonon modes. In LA and LO modes, since the propagation direction is in-plane, we do not usually say iLA or iLO. On the other hand, we sometimes call oTA and oTO modes as ZA and ZO modes which shows that the oscillational direction is in the direction of z axis for xy plane of two-dimensional materials.

Phonon dispersion of a given material is calculated by solving the so-called dynamical matrix. In the dynamical matrix, we define force constants between two atoms in the crystal. In order to define the force constant, any atom should have a restoring force for all directions of the displacement of the atom. This situation corresponds to the ground state in which the total energy has a minimum at the optimized position as a function of displacement of each atom. The restoring force can be calculated by the gradient of the total energy by displacing an atom in the unit cell with keeping other

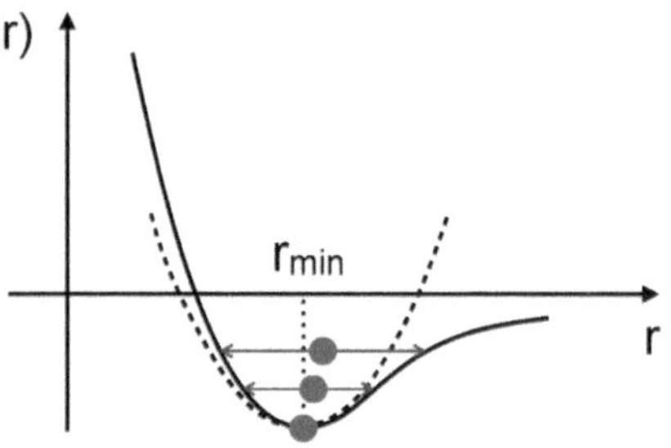

Figure 5.5 Anharmonicity of oscillation. When we expand the potential $V(r)$ as a displacement x measured from the minimum of $V(r)$, $r_{\min}$ the potential can be approximated by a harmonic potential ($\propto x^2$) as shown in the dashed line. When the amplitude of oscillation becomes relatively large, the centers of oscillation that are shown as dots shift to large values of r because of the deviation of $V(r)$ from the harmonic potential. The shift of dots corresponds to the thermal expansion of the materials. Thus, the anharmonicity of the potential is essential for thermal expansion.

atoms at the optimized position. If the optimization of the lattice structure is not sufficient, some restoring force might be negative, which gives an imaginary phonon frequency for the acoustic phonon modes at the Γ point.

In the phonon calculation by Quantum ESPRESSO, we usually adopt so-called "density-functional perturbation theory (DFPT)" [Baroni *et al.* (2001)] in which we need to optimize structure in relatively high precision by calculating the norm of the restoring forces of each atom, which takes a much more computational time than that of the electronic energy band.

When we get the minimum of the total energy by optimization, we can expand the total energy by a displacement x in which the quadratic function of the displacement, $kx^2/2$, corresponds to the harmonic oscillation with a spring constant k. However, when the displacement is relatively large (for example, 10% of the lattice constant, a), an anharmonic term which is proportional to x^3 can not be neglected anymore as shown in Fig. 5.5. When the amplitude of oscillation becomes relatively large, the center of oscillation that is shown a dot for each amplitude shift to a larger value of r than $r_{\min}$ because of the deviation of $V(r)$ from the harmonic potential (dashed line). The shift of dots corresponds to the thermal expansion of the materials. Thus, the anharmonicity of the potential is essential for thermal expansion.

5.8 Electron-phonon interaction

Since both electrons and phonons exist as a function of wavevector in solid, we expect the interaction between an electron and a phonon, that is, electron-phonon interaction, when the energy and momentum conservation satisfies in the interaction. The origin of electron-phonon interaction can be understood by the oscillation of atomic potential by lattice oscillation which modifies the electronic energy. In Quantum ESPRESSO, we obtain the electronic energy band by assuming that the atoms do not move. However, this assumption is an approximation in the case of finite temperature,[16] in which we have a finite amplitude of lattice oscillation, 0.01–0.1 Å, whose energy is the order of $k_{\mathrm{B}}T$ (k_{B} is the Boltzmann constant). Thus, the oscillation of the lattice accelerates (or decelerates) the velocity of an electron to have a thermal equilibrium state. Because of the law of action and reaction, the motion of an electron can modify the lattice oscillation, too. However, since the mass of an atom is $10^3 - 10^5$ times larger than that of an electron, like a small bird on the shoulder of a person, the energy of the electron depends only on the position but not on the velocity of the atom. This means that the electron can easily follow the motion of the atom without any delay of the motion of the electron. This approximation is called "**adiabatic approximation**". In this approximation, the motion of the lattice is not affected much by the motion of electrons as the motion of the human is not affected by the motion of the bird on the shoulder even though the person feels the bird.

When we denote distortion of an atom, $\delta\boldsymbol{R}$, from the original position of $\boldsymbol{R}$ as shown in Fig. 5.6, the distorted atomic potential $v(\boldsymbol{r} - \delta\boldsymbol{R} - \boldsymbol{R})$ can be expanded by

$$v(\boldsymbol{r} - \boldsymbol{R} - \delta\boldsymbol{R}) = v(\boldsymbol{r} - \boldsymbol{R}) + v'(\boldsymbol{r} - \boldsymbol{R})\delta\boldsymbol{R}, \tag{5.18}$$

in which the second term of the right-hand side of Eq. (5.18) is called deformation potential. When we adopt the adiabatic approximation

[16] Lattice does move even at $T = 0$ K, which is known as zero-point motion of the atom. An origin of the zero-point motion is the uncertainty principles of quantum mechanics. It is known that the zero-point motion gives a not-negligible contribution to thermal properties of materials.

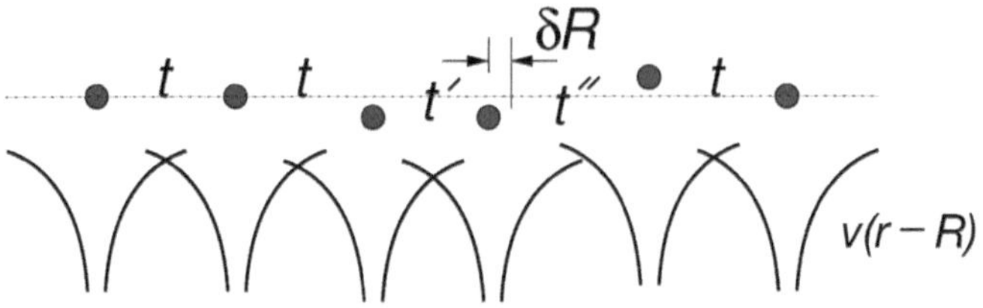

Figure 5.6 Deformation potential. Solid lines are atomic potentials. When an atom shifts position by $\delta\boldsymbol{R}$, the potential energy for an electron (showing solid circles) of the nearest neighbor atom increases (or decreases) by moving away (or approaching) the atom. Further the absolute value of the transfer integral $|t|$ decreases or increases by moving away (or approaching) the atom.

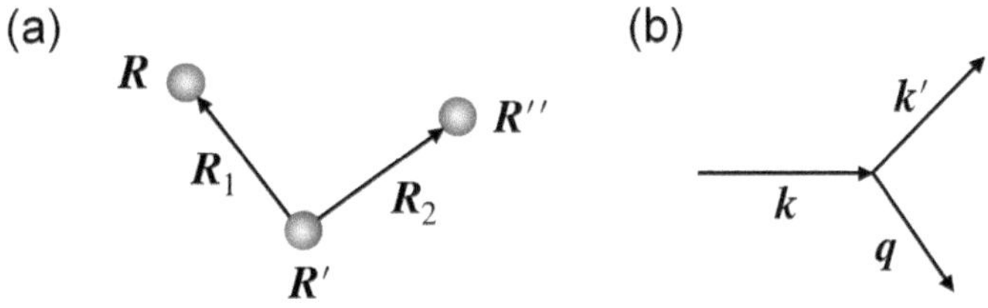

Figure 5.7 (a) Three-centers integral $M(\boldsymbol{R}', \boldsymbol{R}, \boldsymbol{R}'')$ in Eq. (5.20) is a function of the positions of three atoms, $\boldsymbol{R}', \boldsymbol{R}, \boldsymbol{R}''$. The integration is given as a function of the relative coordinate $\boldsymbol{R}_1 = \boldsymbol{R} - \boldsymbol{R}'$, $\boldsymbol{R}_2 = \boldsymbol{R}'' - \boldsymbol{R}'$. (b) The momentum of an electron $\hbar\boldsymbol{k}$ is scattered to $\hbar\boldsymbol{k}' = \hbar\boldsymbol{k}' - \hbar\boldsymbol{q}$ by emitting a phonon with the crystal momentum $\hbar\boldsymbol{q}$.

that the atomic wavefunction follows the position of the atom, the atomic wavefunction of the distorted atom is given by

$$\varphi(\boldsymbol{r} - \boldsymbol{R} - \delta\boldsymbol{R}) = \varphi(\boldsymbol{r} - \boldsymbol{R}) + \varphi'(\boldsymbol{r} - \boldsymbol{R})\delta\boldsymbol{R}. \tag{5.19}$$

When we consider the integration of the atomic potential by the atomic wavefunction, $\langle\varphi|v|\varphi\rangle$, the correction of $\langle\varphi|v|\varphi\rangle$ by putting Eqs. (5.18) and (5.19), up to the linear term of $\delta\boldsymbol{R}$, we get the expression of electron-phonon interaction. The correction of $\langle\varphi|v|\varphi\rangle$ appears in the diagonal and off-diagonal matrix element of the Hamiltonian matrix, which we call on-site and off-site electron-phonon interaction, respectively.

When we consider electron-phonon interaction in solid, an electron is scattered from $\boldsymbol{k}$ to $\boldsymbol{k}'$ states in the k space. When we consider the Bloch orbital $\Phi(\boldsymbol{k}, \boldsymbol{r})$ for electronic states and the lattice distortion of a phonon with the wavevector, $\boldsymbol{q}$, $\delta\boldsymbol{R} = A\exp(-i\boldsymbol{qR})$, the matrix element between the Bloch orbitals of $\boldsymbol{k}$ and $\boldsymbol{k}'$ is given by

$$V_{\boldsymbol{k}',\boldsymbol{k}} \equiv A\langle\Phi(\boldsymbol{k}',\boldsymbol{r})|v'(\boldsymbol{r})\exp(-i\boldsymbol{qR})|\Phi(\boldsymbol{k},\boldsymbol{r})\rangle = \frac{1}{N}\sum_{\boldsymbol{R},\boldsymbol{R}',\boldsymbol{R}''}\exp(-i\boldsymbol{k}'\boldsymbol{R} + i\boldsymbol{kR}'' - i\boldsymbol{qR}')M(\boldsymbol{R},\boldsymbol{R}',\boldsymbol{R}''), \quad (5.20)$$

where $M(\boldsymbol{R}', \boldsymbol{R}, \boldsymbol{R}'')$ is three-centers integral as a function of $\boldsymbol{R}', \boldsymbol{R}, \boldsymbol{R}''$, $M = A\langle\varphi(\boldsymbol{r}-\boldsymbol{R})|v'(\boldsymbol{r}-\boldsymbol{R}')|\varphi(\boldsymbol{r}-\boldsymbol{R}'')\rangle$. When we adopt the relative coordinates $\boldsymbol{R}_1 = \boldsymbol{R} - \boldsymbol{R}'$, $\boldsymbol{R}_2 = \boldsymbol{R}'' - \boldsymbol{R}'$ to the coordinate $\boldsymbol{R}'$, as shown in Fig. 5.7, M can be expressed as a function of $\boldsymbol{R}_1$ and $\boldsymbol{R}_2$, $M(\boldsymbol{R}_1, \boldsymbol{R}_2)$. Then we can take the summation on $\boldsymbol{R}'$ in Eq. (5.20), and we get

$$\begin{aligned} V_{\boldsymbol{k}',\boldsymbol{k}} &= \frac{1}{N}\sum_{\boldsymbol{R}_1,\boldsymbol{R}',\boldsymbol{R}_2}\exp(-i\boldsymbol{k}'\boldsymbol{R}_1 + i\boldsymbol{kR}_2 + i(-\boldsymbol{q} - \boldsymbol{k}' + \boldsymbol{k})\boldsymbol{R}')M(\boldsymbol{R}_1,\boldsymbol{R}_2) \\ &= \delta(-\boldsymbol{q} - \boldsymbol{k}' + \boldsymbol{k})\sum_{\boldsymbol{R}_1,\boldsymbol{R}_2}\exp(-i\boldsymbol{k}'\boldsymbol{R}_1 + i\boldsymbol{kR}_2)M(\boldsymbol{R}_1,\boldsymbol{R}_2) \end{aligned} \quad (5.21)$$

where $\delta(-\boldsymbol{q} - \boldsymbol{k}' + \boldsymbol{k})$ is the delta function that shows the conservation of momentum, $\boldsymbol{k} = \boldsymbol{k}' + \boldsymbol{q}$, before and after the scattering. In Fig. 5.7 (b), we show that the momentum of an electron $\hbar\boldsymbol{k}$ is scattered to $\hbar\boldsymbol{k}' = \hbar\boldsymbol{k} - \hbar\boldsymbol{q}$ by emitting a phonon with the crystal momentum $\hbar\boldsymbol{q}$. It is noted here that the delta function is satisfied even when $\boldsymbol{k} = \boldsymbol{k}' + \boldsymbol{q} + \boldsymbol{G}$ ($\boldsymbol{G}$ is the reciprocal lattice vector) since the exponential function $\exp(i(-\boldsymbol{q} - \boldsymbol{k}' + \boldsymbol{k})\boldsymbol{R}') = \exp(i\boldsymbol{GR}')$ becomes the unity. The scattering with $\boldsymbol{G} \neq 0$ in which additional momentum from the lattice is given (or released) to electron is called the Umklapp process while the scattering with $\boldsymbol{G} = 0$ is called normal process. In both cases the energy and momentum that include the crystal momentum conserved during the scattering processes. The electron-phonon interaction is essential for occurrence of electrical resistance, superconductivity, and Raman spectroscopy.

In Quantum ESPRESSO, the electron-phonon matrix element is calculated as a function of $\boldsymbol{k}$ and $\boldsymbol{q}$ (see Sec. 3.3.3). The calculated

Figure 5.8 When light comes to a material, the light is transparent (left), reflected (center), or scattered (right). Even for the case of transparent, the direction of light changes after the transmission. Reprinted with permission from Asakura Publishing Co. Ltd.

matrix elements are used for estimating the critical temperature of the superconductivity or resonant Raman intensity.

5.9 Optical properties of solid

When light hits a material (see Fig. 5.8), the light is transparent (left), reflected (center), or scattered (right). Even in the case of transparent properties, the direction of light changes after the transmission. It means that all phenomena are related to the interaction of a photon and materials.

Light or a photon interacts with an electron (or electrons) in the materials. The optical properties of the materials consist of (1) absorption, (2) emission, and (3) scattering of light. In the case of semiconductors, optical absorption (or emission) occurs for the photon whose energy is more than the energy gap of the semiconductor. Suppose the momentum and energy of a photon are matched to the difference of momentum and energy between the initial and final electronic states. In that case, the optical absorption occurs by annihilating the photon and by exciting the electron from the initial to the final states. This type of excitation for the electron is called single-particle excitation. On the other hand, for metal or heavily doped semiconductors, an electric field of a photon as an electro-magnetic wave can excite an electric current of metal. The optical absorption occurs by the Joule heat by the current. In this case, since the current is the collective motion of electrons, this type of excitation is called collective excitation. Usually, the single-particle excitation and collective excitation occur exclusively or simultaneously for a given energy of the photon.

Scattering of light in the semiconductor is defined as a sequential process of optical absorption and emission, in which the emitted light propagates in any direction of solid angle. The scattering amplitude depends on materials as a function of photon frequency, which determines the color of the materials. On the other hand, in metal, the reflection of light occurs by screening the electric field of the photon by the induced electric current, which is the origin of metallic luster.

For the region of the wavelength of the visible light, from 400 to 800 nm, there are two kinds of scattering of light: elastic and inelastic scattering of light, which we call the Rayleigh and Raman scattering, respectively. In the Rayleigh (Raman) scattering, the scattered light does not (does) change the energy from that of the incident light. The scattering amplitude becomes strong if the energy difference of the initial and the final electronic states is the same as the photon energy for either the incident or scattered photon. This effect is called resonant Rayleigh (Raman) scattering. The resonant Rayleigh scattering can be one of the reasons the materials have an intrinsic color.[17]

Optical absorption can be explained by the perturbation theory for electronic structure in which the perturbation Hamiltonian is given by the vector potential of the electromagnetic field as follows:

$$\begin{aligned}\mathcal{H} &= \frac{1}{2m}\{-i\hbar\nabla - e\boldsymbol{A}(t)\}^2 + V(\boldsymbol{r}) \\ &= \frac{1}{2m}\left\{-\hbar^2\Delta + ie\hbar\boldsymbol{A}(t)\cdot\nabla + ie\hbar\nabla\cdot\boldsymbol{A}(t) + e^2\boldsymbol{A}(t)^2\right\} + V(\boldsymbol{r}).\end{aligned} \tag{5.22}$$

Both electric field ($\boldsymbol{E} = i\omega\boldsymbol{A}$) and magnetic field ($\boldsymbol{B} = \nabla\times\boldsymbol{A} = i\boldsymbol{k}\times\boldsymbol{A}$) of the electro-magnetic wave can be expressed by the vector

[17] Other reasons are (1) absorption spectra by induced current, (2) structural color, and so on.

potential $\boldsymbol{A}(t)$. In Eq. (5.22), we adopt the Coulomb gauge (div$\boldsymbol{A} = 0$),[18] we get that $\nabla \cdot \boldsymbol{A} = \text{div}\boldsymbol{A} + \boldsymbol{A} \cdot \nabla = \boldsymbol{A} \cdot \nabla$. When we neglect the term of $e^2\boldsymbol{A}(t)^2/2m$ compared with the term $ie\hbar\boldsymbol{A} \cdot \nabla$, which is a good approximation for the conventional power of laser light, we get the perturbation Hamiltonian as follows:

$$\mathcal{H}' = \frac{ie\hbar}{m}\boldsymbol{A}(t) \cdot \nabla. \tag{5.23}$$

When we used the Fermi golden rule,[19] the transition probability per unit time, dP/dt from the initial state i to the final state f is given by

$$\frac{dP}{dt} = \frac{\pi}{\hbar^2} |\langle f|\mathcal{H}'|i\rangle|^2 \delta(E_f - E_i - \hbar\omega), \tag{5.24}$$

where δ is the Dirac delta function. The delta function tells us that the optical transition occurs when the energy difference between the initial state E_i and the final state E_f is matched to the energy of the phonon $\hbar\omega$, which we call resonant optical transition. In the mathematics, the delta function $\delta(x)$ describes a singular function at $x = 0$. However, in physics, because of the uncertainty principles for the energy, $\Delta E \Delta t \sim \hbar$, the strict condition of $x = 0$ can be relaxed for the fast optical transition with a small Δt. For example, if $\Delta t = 1$ ps (or 1 THz), ΔE becomes about 1 meV. $\Delta E = 1$ meV means that the optical absorption spectra have a spectral width of 1 meV. For the case of Raman scattering, a typical value of scattering is $\Delta t = 10$ fs, ΔE becomes 0.1 eV.[20]

$\langle f|\mathcal{H}'|i\rangle$ in Eq. (5.24), which is called transition dipole moment, is calculated by putting the wavefunction of valence and conduction

---

[18] The gauge is an additional condition for a vector potential combined with a scalar electro-static potential that does not change the value of electric and magnetic fields. If we adopt the Coulomb gauge, the scalar electro-static potential satisfies the Poisson equation, while if we adopt the Lorentz gauge, the scalar electro-static potential satisfies the wave equation.

[19] The Fermi golden rule is a general formula of the rate of transition from the initial to the final states for a given time-dependent perturbation that has a time dependence of exp $-i\omega t$. See detail in any textbook on quantum mechanics.

[20] $\Delta E = 0.1$ eV means that the resonance Raman scattering occurs within 0.1 eV energy width as a function of incident laser energy, which we call resonance Raman profile. It is noted that the spectral width of Raman spectra comes from the lifetime of a phonon.

bands for the initial and final states, respectively. Here we can express the vector potential of the light by polarization vector $\boldsymbol{P}$ that is the unit vector in the direction of $\boldsymbol{A}$ (or $\boldsymbol{E}$),

$$\boldsymbol{A} = \frac{i}{\omega}\sqrt{\frac{I_0}{c\varepsilon_0}} \exp\{i(\boldsymbol{k}_{\text{opt}} \cdot \boldsymbol{r} \pm \omega t)\}\boldsymbol{P}, \tag{5.25}$$

where I_0 is the intensity of incident laser light (W/m^2), ε_0, $\boldsymbol{k}_{\text{opt}}$ and ω denote, respectively, the permittivity of the vacuum, the wavevector and angular frequency of the light. When we define the dipole vector:

$$\boldsymbol{D}^{fi}(\boldsymbol{k}) = \langle f|\nabla|i\rangle, \tag{5.26}$$

we express the matrix element by an inner product of $\boldsymbol{D}^{fi}$ and $\boldsymbol{P}$, as follows:

$$\langle f|\mathcal{H}'|i\rangle = \frac{e\hbar}{m\omega}\sqrt{\frac{I}{c\varepsilon_0}} \exp\{i(\omega_f - \omega_i \pm \omega)t\}\boldsymbol{D}^{fi} \cdot \boldsymbol{P}. \tag{5.27}$$

In the Quantum ESPRESSO, since the wavefunction is expanded by the plane waves (Eq. (5.9)), by putting Eq. (5.9) to Eq. (5.26), $\boldsymbol{D}^{fi}(\boldsymbol{k})$ is expanded as follows:

$$\begin{aligned} \boldsymbol{D}^{fi}(\boldsymbol{k}) &= i\sum_{\boldsymbol{G}}\sum_{\boldsymbol{G}'} C_{\boldsymbol{G}'}^{f*}(\boldsymbol{k})C_{\boldsymbol{G}}^{i}(\boldsymbol{k})(\boldsymbol{k}+\boldsymbol{G}) \int \exp\{i(\boldsymbol{G}-\boldsymbol{G}')\boldsymbol{r}\}d\boldsymbol{r} \\ &= i\sum_{\boldsymbol{G}}\sum_{\boldsymbol{G}'} C_{\boldsymbol{G}'}^{f*}(\boldsymbol{k})C_{\boldsymbol{G}}^{i}(\boldsymbol{k})(\boldsymbol{k}+\boldsymbol{G})\delta_{\boldsymbol{G},\boldsymbol{G}'} \\ &= i\sum_{\boldsymbol{G}} C_{\boldsymbol{G}}^{f*}(\boldsymbol{k})C_{\boldsymbol{G}}^{i}(\boldsymbol{k})(\boldsymbol{k}+\boldsymbol{G}). \end{aligned} \tag{5.28}$$

Thus using the output of the coefficients of the wavefunction, we can calculate the optical absorption spectra.

The optical absorption spectra can be calculated in Quantum ESPRESSO once we calculate the dielectric functions as a function of ω, $\varepsilon(\omega)$ (Sec. 3.4.1). Using the calculated dielectric functions, the real and imaginary part of the refractive index, $n(\omega)$ and $\kappa(\omega)$ are, respectively, given by

$$n(\omega) = \sqrt{\frac{|\varepsilon(\omega)| + \text{Re}(\varepsilon(\omega))}{2}}, \tag{5.29}$$

and

$$\kappa(\omega) = \sqrt{\frac{|\varepsilon(\omega)| - \mathrm{Re}(\varepsilon(\omega))}{2}}. \tag{5.30}$$

The dimensionless reflectance, $R(\omega)$, and absorption coefficient, $\alpha(\omega)$ in units of 1/m are expressed by $n(\omega)$ and $\kappa(\omega)$ as follows:

$$R(\omega) = \frac{(n(\omega) - 1)^2 + \kappa(\omega)^2}{(n(\omega) + 1)^2 + \kappa(\omega)^2}, \tag{5.31}$$

and

$$\alpha(\omega) = \frac{2\omega\kappa(\omega)}{c}, \tag{5.32}$$

where c denotes the velocity of light. The absorption coefficient, $\alpha(\omega)$, is defined by **the Lambert-Beer law** in which the intensity of the transmitting light, $I(z)$, for a given incident intensity, I_0, is expressed by

$$I(z) = I_0 \exp(-i\alpha z). \tag{5.33}$$

The absorption coefficient becomes large when the number of states is large for a pair of the initial and final states for a given energy difference E is large. The number of states for the pair is calculated by Quantum ESPRESSO as **joint density of states** ($JDOS(E)$) which is defined by

$$JDOS(E) = \sum_{\sigma} \frac{V}{(2\pi)^3} \int d\boldsymbol{k} \delta \left\{ E_{c\sigma}(\boldsymbol{k}) - E_{v\sigma}(\boldsymbol{k}) - E \right\}, \tag{5.34}$$

where the summation is taken for the spin of electrons, V is the volume of the sample, $E_{c\sigma}(\boldsymbol{k})$ ($E_{v\sigma}(\boldsymbol{k})$) represents the energy dispersion of conduction (valence) band with the spin σ. Here, we assume that the photon wavevector is sufficiently small compared with the reciprocal lattice vector, and thus the photon wavevector is neglected for the momentum conservation. The assumption is valid for $\hbar\omega < 3$eV. In the case of X-ray, on the other hand, the wavevector of the photon can not be neglected in the calculation of X-ray absorption spectra [Chowdhury *et al.* (2012)]. In one-dimensional materials such as carbon nanotubes, it is known that the JDOS becomes singular $(1/\sqrt{E-E_0}$ at the energy bottom or top $(E=E_0)$, which is known as one-dimensional van Hove singularity [Saito *et al.* (1992)]. In the two-dimensional materials, van Hove singularity $(\log|E-E_0|)$ appears at the saddle point of energy dispersion. In the case of graphene, the saddle points exist at the M point $(E=E_0)$ in the two-dimensional Brillouin zone.

5.10 Transport properties of solid

When we apply the voltage to the electrodes at both ends of the materials, electric current flows. If you can control the current electronically, we can make a solid-state device such as a transistor. Further, applying a magnetic field to the devices can create an electric current with the aligned spin of an electron, which we call spin current. In this section, we discuss the electric current as a transport property of solid.

When we consider a cuboid with a width W and a length L, electrical resistance R in SI unit Ω (Ohm) is defined by

$$R=\rho\frac{L}{W^2}, \quad (\Omega) \tag{5.35}$$

where W^2 is a cross-section of the cuboid and ρ (Ωm) is called resistivity. The inverse of R is conductance G in SI unit S (Siemens):

$$G=\sigma\frac{W^2}{L}, \quad (S) \tag{5.36}$$

where $\sigma\equiv 1/\rho$ is conductivity (S/m). It is clear from Eqs. (5.35) and (5.36) that R and G depend on the shape of material such as W and L while ρ and σ do not depend on the shape but only on the materials.

When an electron (or a hole) feels a constant force of $-e\boldsymbol{E}$ ($e\boldsymbol{E}$) in the presence of the constant electric field $\boldsymbol{E}$, the velocity is proportional to the time t, as $e\boldsymbol{E}t/m$, where m is the effective mass of the electron. On the other hand, the electron loses velocity by scattering in the materials. If we assume that the electron loses the velocity after a time 2τ, where τ is called a relaxation time, the averaged velocity, $\bar{v}$ of the electron can be expressed by $e\boldsymbol{E}\tau/m$ since the velocity changes from 0 to $2e\boldsymbol{E}\tau/m$. The current density $\boldsymbol{J}$ in units of C/m^2s, is defined by the number of electrons (or holes) passing a unit area of 1 m^2 per unit time, which is given by

$$\boldsymbol{J} = n_c e\bar{v} = \frac{n_c e^2 \tau \boldsymbol{E}}{m}, \tag{5.37}$$

where n_c denotes the carrier[21] density in units of $1/m^3$. Since the conductivity is defined in electromagnetism by $\boldsymbol{J} = \sigma\boldsymbol{E}$, we get the formula for the conductivity by comparing with Eq. (5.37),

$$\sigma = \frac{n_c e^2 \tau}{m}. \tag{5.38}$$

Further, the conductivity is given by a product of (number of carriers) and mobility. Mobility is a physical quantity of how much velocity increases with respect to a given electric field. When we denote the number of electrons and hole per unit volume by n and p, respectively, the **mobility** of electron and hole, μ_e and μ_h are defined by

$$\sigma = -ne\mu_e + pe\mu_h. \tag{5.39}$$

The unit of mobility is $m^2/V{\cdot}s$. By comparing Eq. (5.38) and Eq. (5.39), the mobility can be expressed by

$$\mu_i = e\tau_i/m_i \qquad (i = \mathrm{e}, \mathrm{h}). \tag{5.40}$$

Eq. (5.40) means that the mobility is proportional to the recombination time, τ_i, and inversely proportional to the effective mass, m_i. In order to get a relatively large current density, $\boldsymbol{J}$, either a large carrier density or a large mobility are needed. In semiconductors, the number of carriers can be controlled by the number of impurities doped to the semiconductor. However, since the impurity can be a center of scattering for carriers, too much doping gives relatively low mobility.

In Quantum ESPRESSO, the relaxation time for the n-th band, $\tau_n(\boldsymbol{k})$, is given as a function of $\boldsymbol{k}$ by the inverse of the scattering rate,

[21] Carrier is either an electron or a hole that carries charges.

$\Gamma_{n\boldsymbol{k}}(\boldsymbol{k})$, as

$$\tau_n(\boldsymbol{k}) = \frac{1}{\Gamma_n(\boldsymbol{k})}, \tag{5.41}$$

in which $\Gamma_n(\boldsymbol{k})$ is calculated by EPW package [Poncé *et al.* (2016)] with using the electron-phonon interaction as

$$\begin{aligned} \Gamma_n(\boldsymbol{k}) = \frac{2\pi}{\hbar} \sum_{m\mu j} \int \frac{d\boldsymbol{q}}{\Omega_{\text{BZ}}} |M_{mn}(\boldsymbol{k}+\boldsymbol{q},\boldsymbol{k})|^2 \times \\ \left[\frac{1+j}{2} - jf^0_{m,\boldsymbol{k}+\boldsymbol{q}} + n_{\boldsymbol{q}.\mu}\right] \delta\left(\Delta E_{\boldsymbol{k}.\boldsymbol{q}} - j\hbar\omega_{\boldsymbol{q}\mu}\right), \end{aligned} \tag{5.42}$$

where $M_{mn}(\boldsymbol{k}+\boldsymbol{q},\boldsymbol{k})$ represents the electron-phonon matrix element for an electron to be scattered from the initial state $(n, \boldsymbol{k})$ to the final state $(m, \boldsymbol{k}+\boldsymbol{q})$ by the μ-th phonon at $\boldsymbol{q}$ with the phonon frequency $\omega_{\boldsymbol{q}\mu}$. The integer j represents either phonon emission $(j = +1)$ or phonon absorption $(j = -1)$. $f^0_{m,\boldsymbol{k}+\boldsymbol{q}}$ and $n_{\boldsymbol{q}.\mu}$ denote, respectively, the Fermi and Bose-Einstein distribution functions for the electron and the phonon. Ω_{BZ} denotes the volume of the Brillouin zone in the k space. The delta function shows the energy conservation in the Fermi golden rule for the time-dependent perturbation theory. The summation is taken over the possible intermediated states specified by m, μ, and j and the integration on $\boldsymbol{q}$ is taken over the Brillouin zone.

It is noted that the $M_{mn}(\boldsymbol{k}+\boldsymbol{q},\boldsymbol{k})$ for a given $\boldsymbol{q}$ is calculated by the so-called Wannier interpolation (see Sec. 5.14.3) in EPW (electron-phonon Wannier) package. The EPW package can usually install directly inside Quantum ESPRESSO by doing "`make epw`" after installing Wannier90 by "`make w90`". Even though the tutorials for the EPW package are beyond the scope of this book, several tutorials are available at `https://docs.epw-code.org/doc/Tutorial.html`. We also recommend that the readers practice with Quantum ESPRESSO and Wannier90 in Chapter 3 before running the tutorials for EPW.

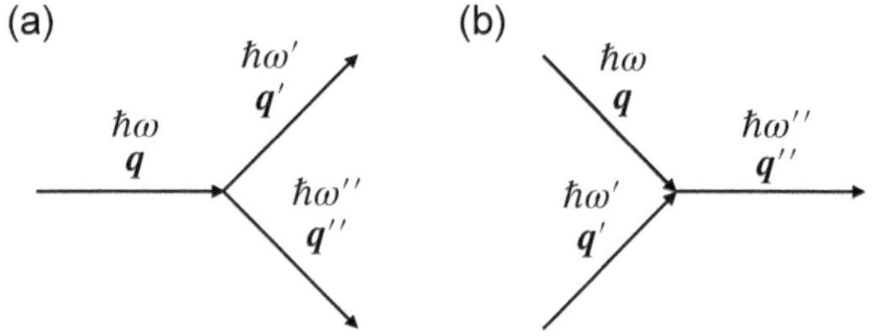

Figure 5.9 Phonon-phonon scattering by cubic term of the anharmonic potential, (a) phonon-emission and (b) phonon-absorption processes. The energy and momentum before and after the scattering conserve. It is noted that there is the Umklapp scattering in which the reciprocal lattice vector $\boldsymbol{G}$ can be added in the momentum after the scattering.

5.11 Phonon-phonon interaction

A phonon with the angular frequency ω and an energy $E_{\text{ph}} = \hbar\omega(n + 1/2)$ $(n = 0, 1, 2, ...)$ is an eigenstate of harmonic Hamiltonian in which the harmonic potential is given by $Kx^2/2$, where K is a spring constant and x is distortion of an atom from $x = 0$. Further, phonon in a crystal is given a propagating wave of oscillation as a function of the wavevector $\boldsymbol{q}$. For a given $\boldsymbol{q}$, we have $3N$ phonon modes in which N is the number of atoms in the unit cell. All the $3N$ modes are orthogonal to one another.

When we consider anharmonic terms of the potential, which is proportional to x^3 or x^4, as a perturbation Hamiltonian, a phonon state specified by an integer n in $E_{\text{ph}} = \hbar\omega(n + 1/2)$ is no more eigenstate but has a finite life-time. When we solve time-dependent perturbation theory for the cubic anharmonic potential that is proportional to x^3, a phonon is scattered into two phonons, or two phonons are scattered into a phonon by creating or annihilating a new phonon, respectively. In Fig. 5.9, we show the three phonon scattering processes for a phonon (a) emitting and (b) absorbing processes. This phonon scattering occurs by keeping the energy and momentum of the phonons before and after the scattering. In this scattering, not only the wavevector $\boldsymbol{q}$ but also the phonon modes are changed, which we call phonon-phonon interaction.

When the cubic anharmonic term, x^3 is considered, the potential is not symmetric around the minimum of $V(r)$, r_{min}, though the potential minimum still exists at $r_{\min}$ (see Fig. 5.5). As shown in Fig. 5.5, the averaged position of the oscillation is shifted from $x = 0$ whose shifted value increases with increasing the amplitude of the oscillation. This situation corresponds to the phenomena of thermal expansion (Sec. 5.7).

The x^4 term, on the other hand, does not contribute to the thermal expansion since the potential is symmetric around $x = 0$ but contributes to the energy transfer from one phonon mode to the other phonon modes. This situation corresponds to the diffusion of the phonon and thus is related to thermal conductivity. It is noted that the x^3 term also contributes to the thermal conductivity. Since the terms of x^3 and x^4 are given by Taylor expansion around $r_{\min}$, the x^4 term is generally smaller than the x^3 term.

Because of the phonon-phonon scattering, each phonon mode with the wavevector $\boldsymbol{q}$ at a temperature, T, has a finite relaxation time $\tau_j(\boldsymbol{q}, T)$ and a mean free path by multiply the $\tau_j(\boldsymbol{q}, T)$ and the group velocity $\partial\omega(\boldsymbol{q})/\partial\boldsymbol{q}$ of the phonon. It is important to note here that the mean free path we discuss here is not the same as the mean free path for calculating the thermal conductivity (see Sec. 5.12). In the thermal equilibrium at T, the probability to find the j-th phonon with the energy $\hbar\omega_j(\boldsymbol{q})$ is given by the Bose-Einstein distribution function, $n_{\boldsymbol{q}j}$, as a function of $\omega_j(\boldsymbol{q})$ and T as follows:

$$n_{\boldsymbol{q}j} \equiv \frac{1}{e^{\hbar\omega_j(\boldsymbol{q})/k_{\mathrm{B}}T} - 1}, \tag{5.43}$$

since a phonon is a boson. The values of $n_{\boldsymbol{q}j}$ should not be changed even if we consider the phonon-phonon scattering by anharmonicity of the potential because of the detailed valance of the scattering processes.

The scattering rate for the j-th phonon at the temperature T, $\gamma_j(\boldsymbol{q}, T)$, is inversely proportional of $\tau_j(\boldsymbol{q}, T)$ which is given by the Fermi golden rule as follows:

$$\begin{aligned}\gamma_j(\boldsymbol{q}, T) &\equiv \frac{1}{\tau_j(\boldsymbol{q}, T)} \\ &= \frac{\pi}{\hbar^2 N_q} \sum_{\boldsymbol{q}',\boldsymbol{q}'',j',j''} \left| V^{(3)}_{\boldsymbol{q}j,\boldsymbol{q}'j',\boldsymbol{q}''j''} \right|^2 \delta(-\boldsymbol{q} + \boldsymbol{q}' + \boldsymbol{q}'' + \boldsymbol{G}) \\ &\quad \times [(1 + n_{\boldsymbol{q}'j'} + n_{\boldsymbol{q}''j''})\delta(\omega_{\boldsymbol{q}j} - \omega_{\boldsymbol{q}'j'} - \omega_{\boldsymbol{q}''j''}) \\ &\quad + 2(n_{\boldsymbol{q}'j'} - n_{\boldsymbol{q}''j''})\delta(\omega_{\boldsymbol{q}j} + \omega_{\boldsymbol{q}'j'} - \omega_{\boldsymbol{q}''j''})],\end{aligned} \tag{5.44}$$

where $N_{\boldsymbol{q}}$ represents the number of $\boldsymbol{q}$ points in the Brillouin zone in the calculation. Three delta functions corresponds to the momentum and energy conservation in the phonon-phonon scattering processes as shown in Fig. 5.9. Here $\boldsymbol{G}$ denotes the reciprocal lattice vector. When $\boldsymbol{G} = 0$ ($\boldsymbol{G} \neq 0$), we call normal (the Umklapp) scattering process. The Umklapp process is important for thermal conductivity since the normal process does not give a net thermal flow but only contributes to $\tau_j(\boldsymbol{q}, T)$. The factors $(1 + n_{\boldsymbol{q}'j'} + n_{\boldsymbol{q}''j''})$ and $2(n_{\boldsymbol{q}'j'} - n_{\boldsymbol{q}''j''})$ in Eq. (5.44) are factors that the scattering rate for Fig. 5.9 (a) and (b), respectively, which is relevant to the $n_{\boldsymbol{q}j}$ before and after scattering. The derivation of these factors are long, and we do not discuss the origin more. See, for example, Eq. (6.68) in "The Physics of Phonons" [Srivastava (1990)]. It is noted that the two terms are equivalent to more meaningful terms by the identities

$$\begin{aligned} n_{\boldsymbol{q}'j'} - n_{\boldsymbol{q}''j''} &= \frac{(n_{\boldsymbol{q}'j'} + 1)n_{\boldsymbol{q}''j''}}{n_{\boldsymbol{q}j}}, \\ 1 + n_{\boldsymbol{q}'j'} + n_{\boldsymbol{q}''j''} &= \frac{n_{\boldsymbol{q}'j'}n_{\boldsymbol{q}''j''}}{n_{\boldsymbol{q}j}}, \end{aligned} \tag{5.45}$$

where we used Eq. (5.43) and the energy conservation. In the calculation of Eq. (5.44), we used so-called "single-mode relaxation-time approximation (SMTA)". In the SMTA, when we consider the deviation of $n_{\boldsymbol{q}j}$ for the j-th phonon mode at $\boldsymbol{q}$ from the thermal equilibrium, we assume that the other phonon-modes are in the thermal equilibrium.

Since the phonon has a finite lifetime, the uncertainty principle ($\Delta E \Delta t \geq \hbar$) tells us that the phonon energy can not be determined precisely. In fact, $\hbar\gamma_j(\boldsymbol{q}, T)$ corresponds an energy width of the energy dispersion of the j-th phonon modes. When the width is large, it means that the phonon has a short lifetime, and thus we can see $\hbar\gamma_j(\boldsymbol{q}, T)$ in the spectral width of the j-th phonon at $\boldsymbol{q}$ in the neutron scattering measurement of Raman spectra. This is important for us to discuss the resonant conditions of the scattering event. In Quantum ESPRESSO, the linewidth of the phonon energy dispersion is calculated by `d3.x`. However, since version 6.0 of Quantum ESPRESSO, the `d3.x` code has been superseded by the D3Q code (`https://anharmonic.github.io/d3q/`). Please be aware that each release of the D3Q code is developed and tested on top of a specific version of Quantum ESPRESSO, and it will probably not work on any other version. Therefore, the readers should choose the correct package.

It is noted that there are other origins of phonon scattering besides anharmonicity. For example, isotope or lattice defect breaks the periodicity of the lattice, which makes the phonon scattering. In the case of carbon, 1.1% of carbon is ^{13}C isotope. When a phonon propagates in two-dimension, a phonon meets a ^{13}C in 10×10 unit cells which can not be negligible. Electron-phonon interaction is another origin for phonon scattering indirectly in which a phonon is absorbed to or emitted from an electron, which gives a finite lifetime of the phonon.

5.12 Heat conduction in a solid

When there are hot and cold points in material, heat flows from the hot to the cold points. The heat flux $\boldsymbol{J}$ in units of W/m^2 at a point is proportional to the gradient of temperature ∇T, (K/m), that is,

$$\boldsymbol{J} = -\kappa \nabla T, \tag{5.46}$$

which is called **the Fourier law**, and the coefficient $\kappa > 0$ is called **thermal conductivity** (W/m/K). In Quantum ESPRESSO, we can calculate the thermal conductivity by Thermal2 code (`https://anharmonic.github.io/thermal2/`), which comes bundled with the D3Q code. Both Thermal2 and D3Q codes can usually install directly inside Quantum ESPRESSO by doing "`make d3q`". The thermal conductivity κ becomes large when the amplitude of the oscillation is large compared with a few % of the bond length. Diamond and carbon nanotubes are known to be materials with a large thermal conductivity since (1) the mass of carbon atoms is relatively small for a given spring constant and (2) the anharmonicity of the potential is large compared with other materials. It is noted here that the thermal conductivity of metal is generally large, too, because phonons are interacted by electron-phonons interactions. Thus, the phonon scattering process is relevant to the thermal conductivity.

When we use the equation of continuity, temporal change of a heat Q at $\boldsymbol{r}$ is given by

$$\frac{dQ}{dt} = -\text{div}\boldsymbol{J}, \tag{5.47}$$

where Q (J/m^3) per unit volume is expressed by the density ρ (kg/m^3) and the specific heat C (J/kg/K) as follows:

$$Q = \rho C T. \tag{5.48}$$

When we assume that ρ and C are constants for a given material, we put Eqs. (5.48) and (5.46) to Eqs. (5.47) and obtain the following equation:

$$\frac{dQ}{dt} = \rho C \frac{dT}{dt} = -\text{div}\boldsymbol{J} = \kappa \Delta T. \tag{5.49}$$

Eq. (5.49) is a diffusion equation for T, which we call **the heat equation**. The value of $\kappa/(\rho C)$ (m^2/s) is called the **thermal diffusivity**.

Temperature of the lattice in thermal equilibrium is defined by the Bose-Einstein distribution function, n_{qj} (Eq. (5.43)). Using the Boltzmann transport theory, the diffusive thermal conductivity κ is expressed by [Saito *et al.* (2018)]

$$\kappa = \frac{\hbar^2}{2Vk_BT^2} \sum_{qj} \omega_{qj}^2 \tau_j(\mathbf{q}, T) n_{qj}(n_{qj}+1)|v_{qj}|^2. \tag{5.50}$$

where V denotes the volume of material, $\tau_j(\mathbf{q}, T)$ is defined by Eq. (5.44), and v_{qj} represents the group velocity of the phonon ($\boldsymbol{v}_{qj} = \nabla_q \omega_{qj}$). In Eq. (5.50), we can define the specific heat per volume of the lattice for the j-th phonon at $\boldsymbol{q}$ and the specific heat C as follows:

$$C_{vqj} = \frac{\hbar^2}{Vk_BT^2}\omega_{qj}^2 n_{qj}(n_{qj}+1), \qquad C = \frac{V}{2N}\sum_{qj} C_{vqj}. \tag{5.51}$$

In Thermal2 code, we can select as an option either (1) single-mode approximation (SMA) or (2) full consideration by "conjugate gradient algorithm with precoding" (CGP) [Atkinson (1988)]. It is natural that it takes more computational time if we adopt (2). The product of the group velocity and the relaxation time corresponds to the phonon mean free path. When the sample size, L, is smaller than the phonon mean free path, thermal conductivity by the phonon becomes "ballistic" in which the phonon with a positive velocity directly contributes to the thermal conductivity as follows:

$$\kappa_{\text{ball}} = L \sum_{qj} C_{vqj} v_{qjx} \theta(v_{qjx}), \tag{5.52}$$

where θ is the Heaviside function. When part of phonon modes becomes ballistic, the diffusive thermal conductivity can not be used. Further, we must consider the thermal resistivity by isotope scattering [Saito *et al.* (2018)] and by electron-phonon interaction if we want to compare the calculated results with the experimental results.

When we decrease the temperature, the thermal conductivity increases monotonically up to the maximum and then becomes zero since the specific heat becomes zero at $T = 0$ K. It is not generally easy to calculate thermal conductivity by first-principles calculation at low temperatures since the mean free path becomes long compared with the size of the computation. Fitting the anharmonicity to the anharmonic results of Quantum ESPRESSO, the tight-binding method for thermal conductivity is a possible way to calculate thermal conductivity at low temperature [Saito *et al.* (2018)].

We note that the D3Q and Thermal2 codes are beyond the scope of this book. The readers can find the tutorials for graphene and silicon at `https://anharmonic.github.io/`. We recommended the readers to practice the phonon tutorials in Sec. 3.3 before running the tutorials of the D3Q and Thermal2 codes.

5.13 Non-resonant Raman scattering

Raman scattering is the inelastic scattering of light, in which energy of the incident light is partially consumed for exciting a phonon.[22]

[22] Any other elementary excitations such as a magnon can be an origin of Raman scattering.

The energy difference between the incident light and the scattered light is called the Raman shift, whose unit is cm^{-1} (1 eV = 8065 cm^{-1}). In the experiment, they observe the intensity of the scattered light as a function of the Raman shift, which is called Raman spectra. The peak positions of Raman spectra correspond to the phonon modes, which are Raman active modes. The Raman active modes are defined by the optical phonon modes in which the photo-excited electron emits a phonon by the electron-phonon interaction. The Raman active modes have the symmetry of quadratic form of x, y, z such as $x^2 + y^2$, $x^2 - y^2$, xz, etc. The quadratic function for each phonon mode is obtained as functions for the irreducible representation in the character table of point group for the unit cell if we know the information of the irreducible representation of the phonon mode in the point group.[23]

In Quantum ESPRESSO, we can calculate the so-called non-resonant Raman spectra to obtain the phonon mode's frequency and symmetry and the relative intensity in the Raman spectra, which we call density-functional perturbation theory (DFPT). The non-resonant Raman intensities are calculated within the DFPT [Lazzeri and Mauri (2003)] as

$$I(\nu) \propto |\vec{e}_s \cdot R(\nu) \cdot \vec{e}_i|^2 \, \frac{n_\nu + 1}{\omega_\nu}, \tag{5.53}$$

in which $\vec{e}_i(\vec{e}_s)$ is the polarization vector of incident (scattered) light. Here the polarization vector is defined by a unit vector of the direction of electric field of the incident (or scattered) light. n_ν $(= \frac{1}{e^{\hbar\omega_\nu/k_B T} - 1})$ is the occupation number of νth phonon at temperature T. $R(\nu)$ is so-called Raman tensor (a 3×3 matrix) for the ν-th phonon mode which connects the polarization vectors, $\vec{e}_i$ and $\vec{e}_s$. Depending on the symmetry of the phonon mode, Raman tensor does have non-zero matrix element. For example, the group theory tells us that the degenerate E_{2g} phonon modes with $x^2 - y^2$ and xy symmetry have

[23] Number of the Raman active modes and their irreducible representation are obtained by decomposing the reducible character of "atomic site character" $\times$ "irreducible representations of x, y, z" into irreducible representations [Dresselhaus *et al.* (2008)].

the following shapes

$$R(E_{2g}:x^2-y^2)=\begin{pmatrix} a & 0 & 0 \\ 0 & -a & 0 \\ 0 & 0 & 0 \end{pmatrix},\quad R(E_{2g}:xy)=\begin{pmatrix} 0 & b & 0 \\ b & 0 & 0 \\ 0 & 0 & 0 \end{pmatrix},\quad (5.54)$$

respectively, where a and b are non-zero constants at xx (yy) and xy (yx) components, which are calculated by Quantum ESPRESSO. The matrix elements of non-resonant Raman tensor are calculated by DFPT in Quantum ESPRESSO by differentiating the electric energy, E^{el}, with electric field components, $\vec{E}_\alpha$ or $\vec{E}_\beta$ ($\alpha, \beta = x, y, z$), and the lattice deformation of a phonon as follows:

$$R_{\alpha\beta}(\nu)=\sum_{\zeta\delta}\frac{\partial^3 E^{el}}{\partial E_\alpha \partial E_\beta \partial r_{\zeta\delta}}\frac{u^{\nu}_{\zeta\delta}}{\sqrt{M_\delta}}, \quad (5.55)$$

where M_δ the mass of the δth atom, $r_{\zeta\delta}$ the position of the δth atom along the direction ζ ($\zeta = x, y, z$), and $u_{\zeta\delta}$ the atomic displacement of the δth atom along the ζ direction for the eigenmode ν.

For the calculated Raman tensor and polarization vectors, we can calculate the relative Raman intensities by $|\vec{e}_s \cdot R(\nu) \cdot \vec{e}_i|^2$ in Eq. (5.53). For example, when the polarization of the incident light lies in the x direction, $\vec{e}_i =^t (1,0,0)$, we get a non-zero intensity for the x or y polarizations of the scattered light $\vec{e}_s = e_x \equiv^t (1,0,0)$, or $\vec{e}_s = e_y \equiv^t (0,1,0)$ for $R(E_{2g}:x^2-y^2)$ or $R(E_{2g}:xy)$, respectively, as follows:

$$^te_xR(E_{2g}:x^2-y^2)e_x=(1,0,0)\begin{pmatrix} a & 0 & 0 \\ 0 & -a & 0 \\ 0 & 0 & 0 \end{pmatrix}\begin{pmatrix} 1 \\ 0 \\ 0 \end{pmatrix}=a, \quad (5.56)$$

or

$$^te_yR(E_{2g}:xy)e_x=(0,1,0)\begin{pmatrix} 0 & b & 0 \\ b & 0 & 0 \\ 0 & 0 & 0 \end{pmatrix}\begin{pmatrix} 1 \\ 0 \\ 0 \end{pmatrix}=b. \quad (5.57)$$

From the calculated phonon dispersion relation as a function of q, we obtain the Raman shift frequency of each phonon at the optical phonon frequency at the Γ point in the Brillouin zone, which is called zone-center phonon mode. The reason why only the Γ point phonon is observed in the first-order Raman spectra is understood by the three Raman sub-processes: (1) optical absorption of an electron with the wavevector k in which the electron can excite almost "vertically" in the k space, (2) the photo-excited electron can emit a phonon with a phonon wavevector q, and (3) the scattered electron is recombined with a hole at the original k. In order for the scattered electron to recombine with a hole, the phonon wavevector should be $q = 0$.

If we can not find the corresponding phonon frequency in the observed Raman spectra, the Raman spectra might be two-phonon Raman spectra whose frequency is the sum of two-phonons. In the two-phonon Raman scattering, the restriction of $q = 0$ in the case of one-phonon scattering is relaxed, and a pair of $q \neq 0$ and $-q$ can be possible for the wavevectors of the emitted two phonons, which enables the photo-excited electron to recombine with the hole. Thus we expect that the two phonon spectra are generally broad and weak. When two of the three intermediate states are resonant to the electronic states, the Raman intensity of the two-phonon Raman scattering becomes comparable or even larger than the intensity of one-phonon Raman scattering. This situation is called "double-resonance Raman scattering" [Saito *et al.* (2002b), Saito *et al.* (2003), Saito *et al.* (2002a)]. The double resonance Raman occurs, too, when one of the two scattering processes with q and $-q$ is an elastic scattering of a photo-excited electron by impurity. One famous example of defect-oriented, double resonance Raman spectra is the D-band of graphite at 1350 cm^{-1} [Pimenta *et al.* (2007)] in which iTO phonon mode at the K point is relevant to the Raman spectra. The two-phonon Raman spectra of the iTO phonon mode at the K point appear at 2700 cm^{-1} which we call G' band (or 2D band). It is noted that the G' band is an intrinsic Raman spectra of graphite without any defects.

When the intermediate state of the photo-excited electrons is a real electronic state, the Raman intensity is enhanced significantly compared with non-resonant Raman spectra, which we call **resonant Raman spectra**. In order to calculate the resonant Raman intensity by first-principles calculation, we need to calculate electron-phonon and electron-photon matrix elements. See detail in the references [Tatsumi and Saito (2018), Tatsumi *et al.* (2018)].

5.14 Warnier functions

Wannier functions (WFs) [Wannier (1937)] are Fourier-transformed Bloch functions of DFT calculation into a smaller set that is localized in real space. While the Bloch function is oscillating and delocalized in the real spaces, as shown in Fig 5.10 (a), the WFs that are localized in the real space offer more microscopic insights for the chemical and physical properties. In particular, the WFs are useful for analyzing chemical bonding, building accurate tight-binding models, or calculating the material properties that require a dense integration in the Brillouin zone.

Let us start with the Bloch states of energy band n, $\psi_{n\boldsymbol{k}}(\boldsymbol{r})$, in terms of a Fourier series of $\boldsymbol{R}$ as

$$\psi_{n\boldsymbol{k}}(\boldsymbol{r}) = \sum_{\boldsymbol{R}} e^{i\boldsymbol{k}\cdot\boldsymbol{R}} w_{n\boldsymbol{R}}(\boldsymbol{r}), \tag{5.58}$$

where $\boldsymbol{R}$ is a lattice vector labeling a unit cell within a supercell that is conjugate to the $\boldsymbol{k}$-points grid. Here, $w_{n\boldsymbol{R}}(\boldsymbol{r})$ is the WF that is localized at each atomic position of $\boldsymbol{R}$, as shown in Fig. 5.10 (b). The Bloch functions of the 1D system can be plotted for $\boldsymbol{k} = 0, \boldsymbol{k}_1$ and $\boldsymbol{k}_2$ in Fig. 5.10 (a). In the case of $\boldsymbol{k} \neq 0$, we can see that the amplitude of $\psi_{n\boldsymbol{k}}(\boldsymbol{r})$ is oscillating with $\boldsymbol{k}_1$ (or $\boldsymbol{k}_2$). The inverse Fourier transformations of $\psi_{n\boldsymbol{k}}(\boldsymbol{r})$ on $\boldsymbol{k}$ for a given $\boldsymbol{R}$ gives a WF [Wannier (1937)]:

$$w_{n\boldsymbol{R}}(\boldsymbol{r}) = \frac{1}{N_{\boldsymbol{k}}} \sum_{\boldsymbol{k}} e^{-i\boldsymbol{k}\cdot\boldsymbol{R}} \psi_{n\boldsymbol{k}}(\boldsymbol{r}), \tag{5.59}$$

where $N_{\boldsymbol{k}}$ is the number of $\boldsymbol{k}$ points. The set of the WFs forms an orthogonal and complete basis set, respectively, as

$$\int w^*_{n\boldsymbol{R}}(\boldsymbol{r}) w_{n'\boldsymbol{R}'}(\boldsymbol{r}) \mathrm{d}\boldsymbol{r} = \delta(\boldsymbol{R} - \boldsymbol{R}')\delta_{nn'}, \tag{5.60}$$

and

$$\sum_{n\boldsymbol{R}} w^*_{n\boldsymbol{R}}(\boldsymbol{r}) w_{n\boldsymbol{R}}(\boldsymbol{r}') = \delta(\boldsymbol{r} - \boldsymbol{r}'). \tag{5.61}$$

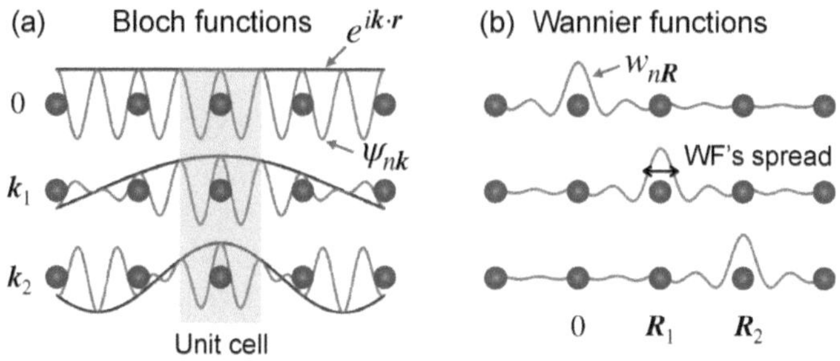

Figure 5.10 (a) The Bloch functions $\psi_{nk}(\boldsymbol{r})$ of a single band n in 1D system for three different values of the wave vector $\boldsymbol{k}$. (b) The Wannier functions $w_{n\boldsymbol{R}}(\boldsymbol{r})$ for the same band n for the three different values of the lattice vector $\boldsymbol{R}$.

According to the Bloch theorem [Kittel (1976)], $\psi_{n\boldsymbol{k}}(\boldsymbol{r})$ is expressed by

$$\psi_{n\boldsymbol{k}}(\boldsymbol{r}) = e^{i\boldsymbol{k}\cdot\boldsymbol{r}} u_{n\boldsymbol{k}}(\boldsymbol{r}), \tag{5.62}$$

where $u_{n\boldsymbol{k}}(\boldsymbol{r})$ is a periodic function ($u_{n\boldsymbol{k}}(\boldsymbol{r}) = u_{n\boldsymbol{k}}(\boldsymbol{r} - \boldsymbol{R})$). By substituting Eq. (5.62) into Eq. (5.59), we obtain:

$$\begin{aligned} w_{n\boldsymbol{R}}(\boldsymbol{r}) &= \frac{1}{N_{\boldsymbol{k}}} \sum_{\boldsymbol{k}} e^{i\boldsymbol{k}(\boldsymbol{r}-\boldsymbol{R})} u_{n\boldsymbol{k}}(\boldsymbol{r}) \\ &= \frac{1}{N_{\boldsymbol{k}}} \sum_{\boldsymbol{k}} \psi_{n\boldsymbol{k}}(\boldsymbol{r} - \boldsymbol{R}) \\ &= w_{n,0}(\boldsymbol{r} - \boldsymbol{R}). \end{aligned} \tag{5.63}$$

Therefore, $w_{n\boldsymbol{R}}(\boldsymbol{r})$ has the same periodicity as the atomic structure of the crystal, as shown in Fig. 5.10 (b).

It is important to note that the WFs are not unique because the Bloch functions are not unique. For each $\psi_{n\boldsymbol{k}}(\boldsymbol{r})$, we can obtain the same electron density by multiplying a different phase α in front of the Bloch function (i.e., $e^{i\alpha}\psi_{n\boldsymbol{k}}(\boldsymbol{r})$). However, by choosing a different phase of the Bloch functions, we get new Wannier functions that may exhibit very different shapes in real space. When we consider N_e energy bands, N_e Bloch functions at a given $\boldsymbol{k}$ can be rotated by the unitary operator as

$$\psi_{n\boldsymbol{k}}(\boldsymbol{r}) = \sum_{m=1}^{N_e} U_{mn}^{\boldsymbol{k}} \psi_{m\boldsymbol{k}}(\boldsymbol{r}), \tag{5.64}$$

where $U_{mn}^{\boldsymbol{k}}$ is an $N_e \times N_e$ unitary matrix at each $\boldsymbol{k}$. Then the WFs in Eq. (5.59) can be rewritten as

$$w_{n\boldsymbol{R}}(\boldsymbol{r}) = \frac{1}{N_{\boldsymbol{k}}} \sum_{\boldsymbol{k}} e^{-i\boldsymbol{k}\cdot\boldsymbol{R}} \left[\sum_{m=1}^{N_e} U_{mn}^{\boldsymbol{k}} \psi_{m\boldsymbol{k}}(\boldsymbol{r}) \right]. \tag{5.65}$$

For Eq. (5.65), the question is how to choose $U_{mn}^{\boldsymbol{k}}$ to obtain a maximally-localized WF in real space.

5.14.1 Maximally-localized Warnier functions

Among several methods, **maximally-localized Wannier functions** (MLWFs) [Marzari and Vanderbilt (1997)] are an effective method for minimizing the mean-square spread of the WFs, Ω, to choose $U_{mn}^{(\boldsymbol{k})}$. The MLWFs are employed in Quantum ESPRESSO by Wannier90 package. Ω is defined as [Marzari and Vanderbilt (1997)]

$$\Omega = \sum_{n=1}^{N_e} \left[\langle (\boldsymbol{r} - \bar{\boldsymbol{r}}_n)^2 \rangle_n \right] = \sum_{n=1}^{N_e} \left[\langle \boldsymbol{r}^2 \rangle_n - \bar{\boldsymbol{r}}_n^2 \right], \tag{5.66}$$

where $\bar{\boldsymbol{r}}_n$ and $\langle \boldsymbol{r}^2 \rangle_n$ are the center and the second moment of the average for WF at $\boldsymbol{R} = 0$ that are defined by

$$\bar{\boldsymbol{r}}_n = \langle w_{n,0} | \boldsymbol{r} | w_{n,0} \rangle \text{ and } \langle \boldsymbol{r}^2 \rangle_n = \langle w_{n,0} | \boldsymbol{r}^2 | w_{n,0} \rangle. \tag{5.67}$$

For the 1D system, the spread of the WFs is shown in Fig. 5.10 (b). Since $w_{n,0}$ is a function of $U_{mn}^{\boldsymbol{k}}$ (see Eq. (5.65)), Ω is a functional of $U_{mn}^{\boldsymbol{k}}$. Thus, it allows us to minimize the spread Ω by variation of these unitary matrices. To perform this minimization, steepest-descent or conjugate-gradient algorithms can be applied if $\mathrm{d}\Omega/\mathrm{d}U_{mn}^{\boldsymbol{k}}$ is known [Marzari *et al.* (2012b)].

5.14.2 Spread of the Wannier functions

As shown by Blount [Blount (1962)], the center of the WFs can be expressed in terms of the Bloch states as

$$\bar{\boldsymbol{r}}_n = \langle w_{n,0}|\boldsymbol{r}|w_{n,0}\rangle = \frac{1}{N_{\boldsymbol{k}}}\sum_{\boldsymbol{k}}\left\langle u_{n\boldsymbol{k}}\middle| i\frac{\partial}{\partial \boldsymbol{k}}\middle| u_{n\boldsymbol{k}}\right\rangle. \tag{5.68}$$

Thus, the second moment of the WFs can also be expressed as

$$\langle \boldsymbol{r}^2\rangle_n = \langle w_{n,0}|\boldsymbol{r}^2|w_{n,0}\rangle = -\frac{1}{N_{\boldsymbol{k}}}\sum_{\boldsymbol{k}}\left\langle u_{n\boldsymbol{k}}\middle|\left(\frac{\partial}{\partial \boldsymbol{k}}\right)^2\middle| u_{n\boldsymbol{k}}\right\rangle. \tag{5.69}$$

In a numerical approach, we can use the finite-difference expressions for $\partial/\partial\boldsymbol{k}$ and $(\partial/\partial\boldsymbol{k})^2$ [Marzari and Vanderbilt (1997)] in Eqs. (5.68) and (5.69), respectively, and we get:

$$\bar{\boldsymbol{r}}_n = \frac{i}{N_{\boldsymbol{k}}}\sum_{\boldsymbol{k},\boldsymbol{b}} w_{\boldsymbol{b}}\boldsymbol{b}[\langle u_{n\boldsymbol{k}}|u_{n,\boldsymbol{k}+\boldsymbol{b}}\rangle - 1], \tag{5.70}$$

and

$$\langle \boldsymbol{r}^2\rangle_n = \frac{1}{N_{\boldsymbol{k}}}\sum_{\boldsymbol{k},\boldsymbol{b}} w_{\boldsymbol{b}}[2 - 2\mathrm{Re}\langle u_{n\boldsymbol{k}}|u_{n,\boldsymbol{k}+\boldsymbol{b}}\rangle], \tag{5.71}$$

where $\boldsymbol{b}$ denotes the vectors connecting the points $\boldsymbol{k}$ to neighboring points $\boldsymbol{k}+\boldsymbol{b}$, and $w_{\boldsymbol{b}}$ denotes the associated weights that come from the finite difference representation.

Let us define an overlap matrix element $M_{mn}^{\boldsymbol{k},\boldsymbol{b}}$ as

$$M_{mn}^{\boldsymbol{k},\boldsymbol{b}} = \langle u_{m\boldsymbol{k}}|u_{n,\boldsymbol{k}+\boldsymbol{b}}\rangle. \tag{5.72}$$

It is proved that $M_{mn}^{\boldsymbol{k},\boldsymbol{b}}$ satisfies the following conditions [Marzari and Vanderbilt (1997)]:

$$M_{nn}^{\boldsymbol{k},\boldsymbol{b}} - 1 = i\mathrm{Im}\left(\ln M_{nn}^{\boldsymbol{k},\boldsymbol{b}}\right), \tag{5.73}$$

and

$$2 - 2\mathrm{Re}M_{nn}^{\boldsymbol{k},\boldsymbol{b}} = 1 - \left|M_{nn}^{\boldsymbol{k},\boldsymbol{b}}\right|^2 + \left[\mathrm{Im}\left(\ln M_{nn}^{\boldsymbol{k},\boldsymbol{b}}\right)\right]^2. \tag{5.74}$$

By substituting Eqs. (5.73) and (5.74) into Eqs. (5.70) and (5.71), we obtain:

$$\bar{\boldsymbol{r}}_n = -\frac{1}{N_{\boldsymbol{k}}} \sum_{\boldsymbol{k},\boldsymbol{b}} w_{\boldsymbol{b}} \boldsymbol{b} \operatorname{Im} \left(\ln M_{nn}^{\boldsymbol{k},\boldsymbol{b}} \right), \tag{5.75}$$

and

$$\langle \boldsymbol{r}^2 \rangle_n = \frac{1}{N_{\boldsymbol{k}}} \sum_{\boldsymbol{k},\boldsymbol{b}} w_{\boldsymbol{b}} \left(1 - \left| M_{nn}^{\boldsymbol{k},\boldsymbol{b}} \right|^2 + \left[\operatorname{Im} \left(\ln M_{nn}^{\boldsymbol{k},\boldsymbol{b}} \right) \right]^2 \right). \tag{5.76}$$

From Eqs. (5.66), (5.75), and (5.76), we can see that the spread Ω with respect to the unitary matrix $U_{mn}^{\boldsymbol{k}}$ can be expressed as a function of the overlap matrix element $M_{mn}^{\boldsymbol{k},\boldsymbol{b}}$. Therefore, $M_{mn}^{\boldsymbol{k},\boldsymbol{b}}$ is needed to minimize the spread Ω to obtain the optimized choice of $U_{mn}^{\boldsymbol{k}}$. On the other hand, the iterative minimization of Ω will starts with an initial guess orbitals as

$$|\tilde{w}_{n,0}\rangle = \frac{1}{N_{\boldsymbol{k}}} \sum_{m\boldsymbol{k}} A_{mn}^{\boldsymbol{k}} |\psi_{m\boldsymbol{k}}\rangle, \tag{5.77}$$

where $A_{mn}^{\boldsymbol{k}}$ is projected matrix element, which is defined by

$$A_{mn}^{\boldsymbol{k}} = \langle \psi_{m\boldsymbol{k}} | g_n \rangle, \tag{5.78}$$

where g_n are projections such as s, p, or hybrids. In Quantum ESPRESSO, $M_{mn}^{\boldsymbol{k},\boldsymbol{b}}$ and $A_{mn}^{\boldsymbol{k}}$ are calculated from `pw2wannier90.x` after the Bloch states calculated by `pw.x`.

5.14.3 Tight-binding model and Wannier interpolation

The unique set of the MLWFs constitutes an orthogonal and complete basis, which enables us to use these orbitals for setting up the **effective model Hamiltonian** in real space for the tight-binding model as

$$\mathcal{H} = \sum_{\boldsymbol{R}\boldsymbol{R}'} \sum_{nn'} t_{nn'}(\boldsymbol{R}, \boldsymbol{R}') |w_{n\boldsymbol{R}}\rangle \langle w_{n'\boldsymbol{R}'}|, \tag{5.79}$$

where $t_{nn'}(\boldsymbol{R}, \boldsymbol{R}')$ are the hopping parameters that depend on only the distance vector connecting the MLWFs as

$$\begin{aligned} t_{nn'}(\boldsymbol{R}, \boldsymbol{R}') &= \langle w_{n\boldsymbol{R}} | \mathcal{H} | w_{n'\boldsymbol{R}'} \rangle \\ &= \langle w_{n,0} | \mathcal{H} | w_{n',\boldsymbol{R}'-\boldsymbol{R}} \rangle \\ &= t_{nn'}(\boldsymbol{R}' - \boldsymbol{R}). \end{aligned} \tag{5.80}$$

Using a Fourier transformation of $\mathcal{H}$ in the basis of the MLWFs, we can obtain $\mathcal{H}$ in momentum space as

$$\begin{aligned} \mathcal{H}_{nn'}(\boldsymbol{q}) &= \sum_{\boldsymbol{R}} e^{i\boldsymbol{q}\cdot\boldsymbol{R}} t_{nn'}(\boldsymbol{R}) \\ &= \sum_{\boldsymbol{R}} e^{i\boldsymbol{q}\cdot\boldsymbol{R}} \left(\frac{1}{N_k} \sum_{\boldsymbol{k}} e^{-i\boldsymbol{k}\cdot\boldsymbol{R}} \left[(U^{\boldsymbol{k}})^{\dagger} \epsilon(\boldsymbol{k}) U^{\boldsymbol{k}} \right]_{nn'} \right), \end{aligned} \tag{5.81}$$

where $\epsilon(\boldsymbol{k})$ are the Kohn-Sham eigenvalues by given a $\boldsymbol{k}$-grid in the DFT calculation. By diagonalizing the tight-binding Hamiltonian $\mathcal{H}_{nn'}(\boldsymbol{q})$ in Eq. (5.81), we can obtain the energy dispersion, $\epsilon(\boldsymbol{q})$, in which $\boldsymbol{q}$-grid can be much denser than $\boldsymbol{k}$-grid. This calculation is known as the **Wannier interpolation**, which is useful to calculate transport properties (e.g., electrical conductivity) that require a dense $\boldsymbol{k}$-grid in the Brillouin zone.

Chapter 6

Productivity Tools

In practical use of first-principles calculation software, we will need to manage and visualize our data in an attractive manner. Furthermore, when we research some unknown materials, we have to know the starting point for generating the Quantum ESPRESSO input files. In most cases, we should repeat similar procedures to check the consistency of the first-principles calculations. Such situations require us to supplement our skills of using Quantum ESPRESSO with other utilities. In particular, understanding the way of plotting figures and using a few Linux commands along with basic shell-scripting will be useful for us to be more productive when performing research. In this chapter, we will learn these extra tools, from the preparation of our own Quantum ESPRESSO files, some useful Linux commands/scripts, plotting with Matplotlib in the Python environment, to additional packages online.

6.1 Quantum ESPRESSO input generators

Recent advances in artificial intelligence and machine learning have allowed us to obtain pre-defined input files for Quantum ESPRESSO calculations from input generator tools available online. Two out-

Quantum ESPRESSO Course for Solid-State Physics
Nguyen Tuan Hung, Ahmad R. T. Nugraha, and Riichiro Saito

ISBN 978-981-4968-37-9 (Hardcover), 978-981-4968-63-8 (Paperback), 978-1-003-29096-4 (eBook)
www.jennystanford.com

standing examples are AFLOW (`aflowlib.org`) and MaterialsCloud (`materialscloud.org`). We can generate Quantum ESPRESSO input files based on the information provided in those websites for a particular material. The two websites also provide us with the structure visualization tools that can be used to confirm the structure of the material under study and to select the possible *k*-path for the band structure plot, which is helpful to accelerate our work.

As an example of this section, we will learn how to make the Quantum ESPRESSO input files for bulk germanium telluride (GeTe) so that we can obtain its density of states (DOS) and energy dispersion. The input files will be named as follows:

1. `gete.scf.in` for the SCF calculation.
2. `gete.nscfdos.in` for the NSCF DOS calculation.
3. `gete.dos.in` for post-processing the DOS data.
4. `gete.nscfbands.in` for the energy dispersion calculation.
5. `gete.bands.in` for post-processing the dispersion data.

Let us see how we can make all the input files.

6.1.1 Obtaining a structural CIF file from AFLOW

The first step to creating the Quantum ESPRESSO input files is obtaining a structure definition for the material that we are studying. We will use AFLOW to get the structure information that will define the material in all Quantum ESPRESSO input files. We can click the address of the AFLOW website (`aflowlib.org`) in the web browser, and we will see its front page as shown in Fig. 6.1. There are several menus available in the top panel of the website. What we need is the `Search` menu, highlighted by the box on the top right of Fig. 6.1.

After clicking the `Search` menu, we can see the search page as shown in Fig. 6.2. We highlight four steps in the case of specifying GeTe for obtaining its structural information:

1. Click atom Ge.
2. Click atom Te.
3. Make sure "Ge, Te" appear in the search box.

Figure 6.1 The front page of the AFLOW website.

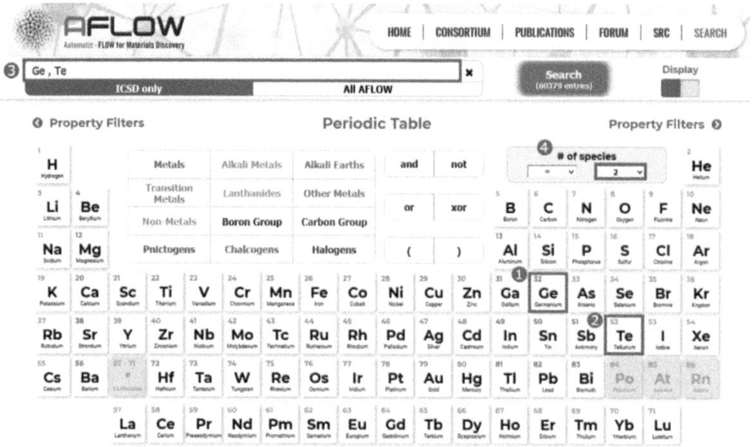

Figure 6.2 AFLOW search page with some highlighted steps for obtaining the GeTe information. (1) Push Ge, (2) push Te and then we get Ge, Te in (3) and we select "2" for the number of species.

4. Set the number of species to "2" because we want to focus on finding compounds that strictly consist of Ge and Te atoms only, *without* other atoms.

It is also possible to skip Steps 1–2 and jump directly to Step 3 by manually writing "Ge, Te" in the search box.

After specifying the Ge and Te atoms on the AFLOW search page, as shown in Fig. 6.2, we can click the `Search` button beside the search box to let AFLOW find all relevant information about GeTe compounds. A screenshot of the search results is shown in Fig. 6.3. At the time of writing this book, there are 29 entries found related to our

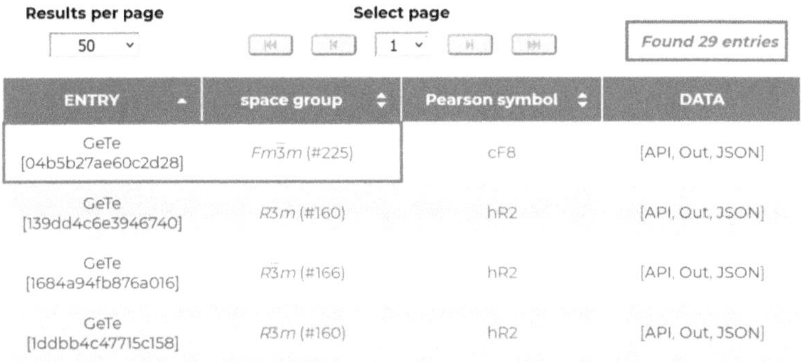

Figure 6.3 Sample search results for GeTe. We click the first entry to get the information of bulk GeTe in its cubic phase with the $Fm\bar{3}m$ (#225) space group.

specified "Ge, Te" keyword. The different entries mean that various structures and parameters for the GeTe calculations are reported to be explored.

For example, in the "space group" column, we can find $Fm\bar{3}m$ (#225) and $R3m$ (#160) groups that correspond to the cubic and rhombohedral lattice systems, respectively, as shown in the first and the second rows of Fig. 6.3. The same space groups may appear several times, indicating the other details of the parameters used in the structures. Here we take the first row of the search results as our example and click the corresponding GeTe entry as highlighted in Fig. 6.3. This GeTe entry gives information about the bulk GeTe in its cubic phase.

By opening the page after clicking the first entry of GeTe, we will see several structural details and physical properties of the bulk cubic GeTe. The first information available on that page is the structure visualization. The screenshot is given in Fig. 6.4, from which we can obtain the crystallographic information file (CIF) of the material. It is recommended to set the relaxed structure as calculated in AFLOW by clicking the corresponding button indicated by number 1 in Fig. 6.4. Then, we can save it in the CIF format by clicking the button number 2 in Fig. 6.4. We may save it as any filename we like. In this example, we use the default one given by AFLOW:

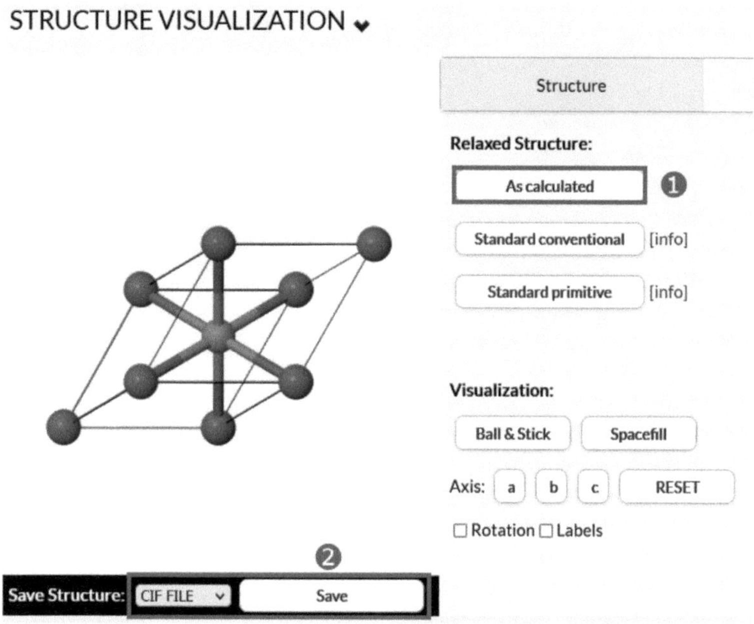

Figure 6.4 Structure visualization of bulk cubic GeTe in AFLOW. Button 1 should be clicked to set the relaxed structure as calculated originally in AFLOW. Button 2 is to save as the CIF format.

```
Ge1Te1_ICSD_638014.cif
```

Note that AFLOW, in this case, already gives us the relaxed geometry of GeTe. If we are not satisfied with the relaxed structure, we may repeat the structural optimization process in Quantum ESPRESSO by following the related hands-on in Sec. 3.1.5. However, for simplicity, in this chapter, we will not do such a process again. Instead, we rely on the relaxed structure given by AFLOW.

6.1.2 Generating SCF input file from MaterialsCloud

On the same AFLOW page of Fig. 6.4 below the structure visualization, we can explore physical properties such as thermodynamic, magnetic, and electronic properties calculated by AFLOW. When we further scroll down the page, there are also AFLOW calculation

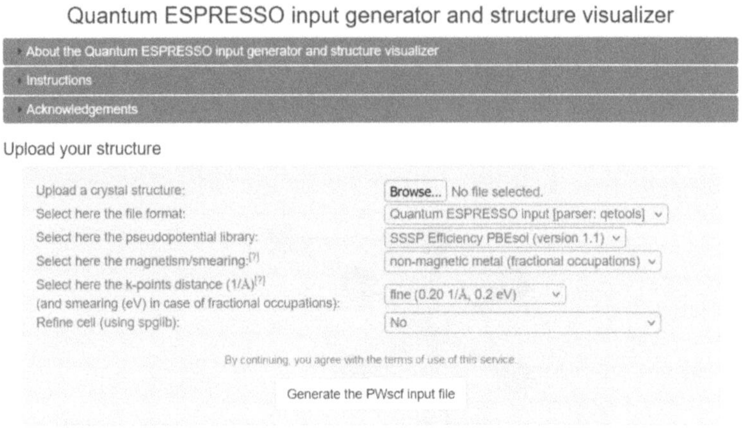

Figure 6.5 The default view of Quantum ESPRESSO input generator and structure visualizer in MaterialsCloud (`materialscloud.org`).

details such as cutoff energy, *k*-point grid, and pseudopotential version used for the material under investigation, but all the results are given there are not calculated by using Quantum ESPRESSO. Therefore, we turn to the other website, i.e., MaterialsCloud, for obtaining the Quantum ESPRESSO inputs. The relevant tool for generating the SCF input file in MaterialsCloud can be accessed at the following address: `https://materialscloud.org/work/tools/qeinputgenerator`. The web address written above may change in the future depending on the MaterialsCloud developers. However, we should be able to access it from the main page of MaterialsCloud through the following order of menus: `Work` → `Tools` → `Quantum ESPRESSO input generator and structure visualizer`. The default view of this tool is shown in Fig. 6.5

What we need now is only a structure file, which has been obtained for the GeTe example as explained in Sec. 6.1.1 in the CIF format. Provided the CIF file, we can do the following steps:

1. Click the `Browse` button in the Quantum ESPRESSO input generator on the MaterialsCloud website as shown in Fig. 6.5 and find the CIF file that we want to calculate.
2. Select the CIF file format in MaterialsCloud (either by `ASE` or `pymatgen` parser is fine; *do not use* the `qetools` parser).

3. Choose the PBE or PBEsol SSSP pseudopotential libraries. For each library, there are two versions: (i) a more accurate one, called "Precision" and (ii) a general-purpose one, called "Efficiency". Later, we may change the pseudopotential files manually if we do not want to use those recommended by MaterialsCloud.
4. Decide if we want to treat our system as (i) a non-magnetic metal (i.e., using smearing/fractional occupations), (ii) as a non-magnetic insulator (i.e., using fixed occupations equal to 2), or (iii) as a magnetic system (in this case, it is safer to use smearing/fractional occupations). When we are not sure about the metallicity and magnetism, just leave them and choose the default.
5. Decide the density of the k-points sampling. The default value is 0.2 $Å^{-1}$, which is a conservative, all-purpose choice. We call this "fine" sampling, and the appropriate smearing ("degauss") value is 0.2 eV. The two other options are "very fine" sampling (0.15 $Å^{-1}$, 0.1 eV), and "normal" (0.3 $Å^{-1}$, 0.3 eV). Note that we change the "degauss" parameter since a coarser sampling benefits from more smearing. See Sec. 3.1.1 for the other details about the smearing parameter.
6. Finally, we can choose whether or not we want to refine the cell. Normally we do not need to do it, so we can directly click the "`Generate`" button.

In the above steps, beware that there is no point in sampling the Brillouin Zone in a direction where there should not be any dispersion, i.e., in the case of a 0D, 1D, or 2D system in which 3, 2, or 1 direction(s) should not be sampled, we should update the input file manually. We may also want to use a shifted Monkhorst-Pack mesh instead of the unshifted one by changing `0 0 0` to `1 1 1` in the last three numbers of the k-point entry from the input file generated by Materials Cloud (c.f. Sec. 3.1.3). All these manual modifications can be done after we obtain the PWscf input file generated by the MaterialsCloud tool.

For the GeTe example, we show our selected options as highlighted in Fig. 6.6. We mostly select the default options, except the file format, which is CIF, and the magnetism, which we select as

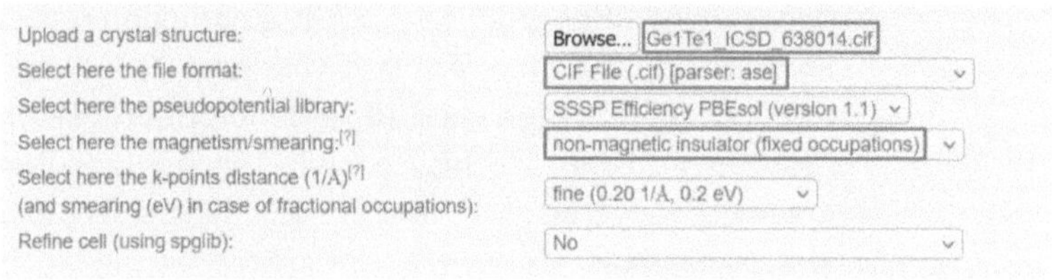

Figure 6.6 Options selected to generate the SCF input file for the bulk cubic GeTe in MaterialsCloud.

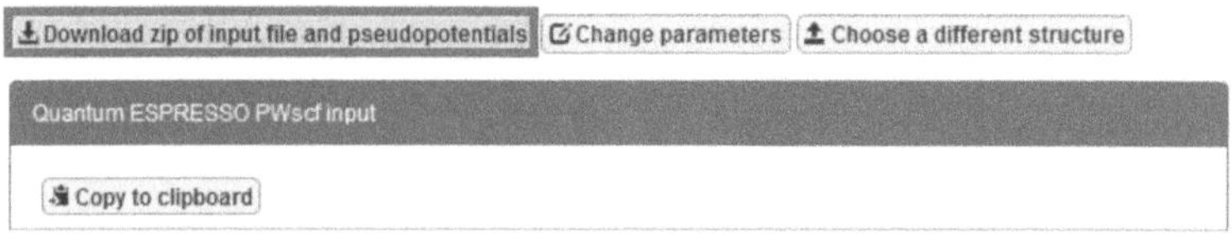

Figure 6.7 Downloading the SCF input file generated by MaterialsCloud.

"non-magnetic insulator" because we already expect GeTe to have a band gap. After we click the "`Generate`" button, we will see the resulting SCF input file given by MaterialsCloud. Note that it may take a few seconds until we can see the input file fully. It is recommended to download the SCF input file along with the pseudopotentials that are zipped in a file called `PWscf.zip`. We can click the download button as shown in Fig. 6.7. It is recommended to save the zip file under the `QE-SSP` folder following all the examples in this book.

The zipped file can be unzipped in the Linux terminal by the following command:

```
$ unzip PWscf.zip
```

After unzipping the file, we will see that there is a folder named `PWscf` containing a `PWscf.in` file and a `pseudo` directory with the suggested pseudopotential files from MaterialsCloud. To match with the material's name (GeTe), it is better if we rename the folder and the SCF input file by the following set of command lines:

```
$ mv PWscf gete
$ cd gete
$ mv PWscf.in gete.scf.in
```

The first line means that we change the name of the PWscf folder to gete. The second line is to enter the gete folder (which was previously named PWscf). In this folder, by using the third command line, we rename one more file, i.e., PWscf.in to gete.scf.in. Now we can see the contents of gete.scf.in by either opening it with a text editor or by typing the cat command in the terminal:

```
$ cat gete.scf.in
```

QE-SSP/gete/gete.scf.in

```
1  &CONTROL
2    calculation = 'scf'
3    outdir      = './out/'
4    prefix      = 'gete' # originally 'aiida'
5    pseudo_dir  = './pseudo/'
6    verbosity   = 'high'
7  /
8  &SYSTEM
9    ecutrho     = 3.2000000000d+02
10   ecutwfc     = 4.0000000000d+01
11   ibrav       = 0
12   nat         = 2
13   ntyp        = 2
14 /
15 &ELECTRONS
16   conv_thr    = 4.0000000000d-10
17   mixing_beta = 4.0000000000d-01
18 /
19 ATOMIC_SPECIES
20 Ge 72.64 ge_pbesol_v1.4.uspp.F.UPF
21 Te 127.6 te_pbesol_v1.uspp.F.UPF
22 ATOMIC_POSITIONS crystal
23 Ge  0.0000000000   0.0000000000   0.0000000000
24 Te  0.5000000000   0.5000000000   0.5000000000
25 CELL_PARAMETERS angstrom
26     4.2603675143   0.0000000000   0.0000000000
27     2.1301837572   3.6895864968   0.0000000000
28     2.1301837572   1.2298621656   3.4785755089
29 K_POINTS automatic
30 13 13 13 1 1 1 # originally 10 10 10 0 0 0
```

Note in gete.scf.in above, we need to change the prefix variable, which originally read 'aiida' in the MaterialsCloud input generator, to gete in our computer (at line 4). We also use slightly denser *k*-points and shifted Monkhorst-Pack mesh (`13 13 13 1 1 1`) and remove some variables, such as `etot_conv_thr`, `force_conv_thr`, `tprnfor`, `tstress`, `nosym`, `electron_maxstep`, and `occupations`, which are too detailed to be specified in the SCF file. In the next section, for the DOS calculation, we will need to set the `occupations` variable back in the NSCF (non-SCF) input file.

6.1.3 Preparing DOS and band structure inputs

Next, let us prepare the input files for DOS and band structure calculations from the information already available in the previous subsection combined with the new information we will gather here. In the following example, we use `gete.scf.in` as the starting point for the DOS and band structure calculations, which are both categorized as the NSCF calculations. We can copy the SCF input file to make new input files for the NSCF calculations of DOS and band structure:

```
$ cp gete.scf.in gete.nscfdos.in
$ cp gete.scf.in gete.nscfbands.in
```

Then, we should edit `gete.nscfdos.in` and `gete.nscfbands.in` according to our needs.

DOS input files

As explained in Sec. 3.2.3, the main differences between the SCF and NSCF inputs for the DOS calculation are the appearance of `occupations` parameter (= `'tetrahedra_opt'`) and setting much denser *k*-points mesh in the NSCF input file. In this GeTe example, we set the number of *k*-points twice larger than that in the SCF input file (line 28). It is also recommended to make the `nbnd` parameter same as in the NSCF input for the band structure calculation. The input file for the DOS calculation is shown as below:

QE-SSP/gete/gete.nscfdos.in

```
1 &CONTROL
2   calculation = 'nscf'
```

```
  outdir       = './out/'
  prefix       = 'gete'
  pseudo_dir   = './pseudo/'
  verbosity    = 'high'
/
&SYSTEM
  ecutrho      = 3.2000000000d+02
  ecutwfc      = 4.0000000000d+01
  ibrav        = 0
  nat          = 2
  ntyp         = 2
  nbnd         = 30 # same as in *.nscfbands.in
  occupations = 'tetrahedra_opt' # only for DOS
/
&ELECTRONS
  conv_thr     = 4.0000000000d-10
  mixing_beta = 4.0000000000d-01
/
ATOMIC_SPECIES
Ge 72.64 ge_pbesol_v1.4.uspp.F.UPF
Te 127.6 te_pbesol_v1.uspp.F.UPF
ATOMIC_POSITIONS crystal
Ge   0.0000000000   0.0000000000   0.0000000000
Te   0.5000000000   0.5000000000   0.5000000000
K_POINTS automatic
26 26 26 1 1 1       # twice the SCF's case
CELL_PARAMETERS angstrom
    4.2603675143   0.0000000000   0.0000000000
    2.1301837572   3.6895864968   0.0000000000
    2.1301837572   1.2298621656   3.4785755089
```

Besides gete.nscfdos.in, we should create a post-processing input file to gather the DOS data. In this case, we do not have to copy the SCF input file, just make a new file with a few lines of information for the DOS post-process using the &DOS namelist. The specifications of outdir (line 2) and prefix (line 3) should be consistent with the *.nscfdos.in file. We also add parameters emin and emax in eV (lines 5 and 6) in the following DOS post-processing file:

QE-SSP/gete/gete.dos.in

```
1 &DOS
2  outdir = './out/'
3  prefix = 'gete'
4  fildos = 'gete.dos'
5  emin   = -5
```

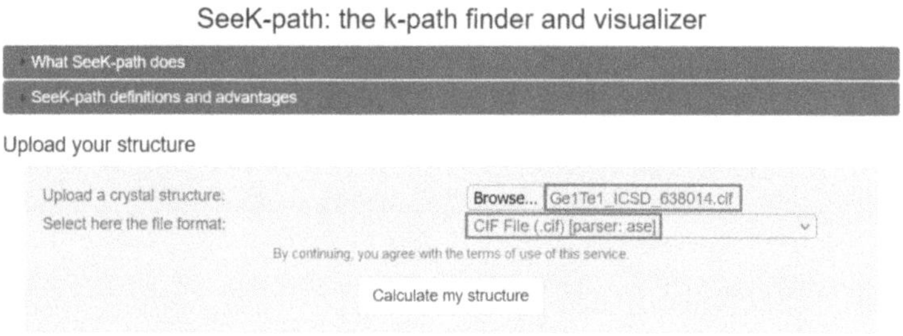

Figure 6.8 Obtaining the k-path and high-symmetry points of bulk cubic GeTe in MaterialsCloud.

```
emax    = 25
/
```

Band structure input file

The most important point when creating the band input file is the k-path for plotting the energy dispersion. MaterialsCloud has an additional tool called "SeeK-path" that can accept a CIF format, which is used for finding and visualizing a possible k-path, as shown in Fig. 6.8. The SeeK-path tool will show the coordinates of the high-symmetry points that we can copy to the band input file. The SeeK-path tool can be accessed at the following address: `https://materialscloud.org/work/tools/seekpath` or through the following order of MaterialsCloud menus: `Work` → `Tools` → `SeeK-path: the k-path finder and visualizer`.

In Fig. 6.8, we show how to obtain the k-path and high-symmetry points of bulk cubic GeTe in MaterialsCloud. We only need to browse/upload the CIF file of GeTe obtained earlier (from Sec. 6.1.1) and specify, for example, the `ase` parser for the CIF format. After pressing the button to "Calculate my structure" in the SeeK-path tool, we will be able to see the high-symmetry points of bulk cubic GeTe, as shown in Fig. 6.9.

Now suppose that we want to calculate and plot the energy dispersion along the path of W–L–U–X–Γ–K high-symmetry points. We just need to change the automatic `K_points` setup into `crystal_b` in the `gete.nscfbands.in` file and list all the corresponding high-

High-symmetry points (scaled units)

Label	**k_1**	**k_2**	**k_3**
Γ	0.0000000000	0.0000000000	0.0000000000
K	0.3750000000	0.3750000000	0.7500000000
L	0.5000000000	0.5000000000	0.5000000000
U	0.6250000000	0.2500000000	0.6250000000
W	0.5000000000	0.2500000000	0.7500000000
W_2	0.7500000000	0.2500000000	0.5000000000
X	0.5000000000	0.0000000000	0.5000000000

Figure 6.9 The high-symmetry points of bulk cubic GeTe in the scaled units or `crystal_b` coordinates given by the SeeK-path tool in MaterialsCloud.

symmetry points along which we want to obtain the dispersion data (lines 31–37). In this GeTe example, we list the 6 high-symmetry points for the specified path using the coordinate information from the SeeK-path tool. Do not forget that in `gete.nscfbands.in` we should set the `nbnd` parameter (line 14), which was not written in the `gete.scf.in` file. It is recommended to use the same value of `nbnd` in both `gete.nscfbands.in` and `gete.nscfdos.in` files.

QE-SSP/gete/gete.nscfbands.in

```
1  &CONTROL
2    calculation = 'bands'
3    outdir      = './out/'
4    prefix      = 'gete'
5    pseudo_dir  = './pseudo/'
6    verbosity   = 'high'
7  /
8  &SYSTEM
9    ecutrho     = 3.2000000000d+02
10   ecutwfc     = 4.0000000000d+01
11   ibrav       = 0
12   nat         = 2
13   ntyp        = 2
14   nbnd        = 30 # same as in *.nscfdos.in
15 /
16 &ELECTRONS
17   conv_thr    = 4.0000000000d-10
18   mixing_beta = 4.0000000000d-01
19 /
```

```
ATOMIC_SPECIES
Ge 72.64 ge_pbesol_v1.4.uspp.F.UPF
Te 127.6 te_pbesol_v1.uspp.F.UPF
ATOMIC_POSITIONS crystal
Ge  0.0000000000   0.0000000000   0.0000000000
Te  0.5000000000   0.5000000000   0.5000000000
CELL_PARAMETERS angstrom
    4.2603675143   0.0000000000   0.0000000000
    2.1301837572   3.6895864968   0.0000000000
    2.1301837572   1.2298621656   3.4785755089
K_POINTS crystal_b # was automatic in SCF file
6
  0.5000000000  0.2500000000  0.7500000000  50 !W
  0.5000000000  0.5000000000  0.5000000000  30 !L
  0.6250000000  0.2500000000  0.6250000000  20 !U
  0.5000000000  0.0000000000  0.5000000000  50 !X
  0.0000000000  0.0000000000  0.0000000000  50 !G
  0.3750000000  0.3750000000  0.7500000000  1  !K
```

Similar to the DOS part, we create a post-processing input file to gather the energy dispersion data. We can make a new file with a few lines of information for post-processing the band structure using the &BANDS namelist. The specifications of `outdir` and `prefix` should be consistent with the `*.nscfbands.in` file.

QE-SSP/gete/gete.bands.in

```
&BANDS
 outdir  = './out/'
 prefix  = 'gete'
 filband = 'gete.bands'
/
```

Having created the above file, we have completed all the steps in the preparation of Quantum ESPRESSO input files for the DOS and band structure calculations.

6.1.4 Wannier90 input generator from CIF file

In addition to the SCF, NSCF, and DOS or band processing input files, we may want to generate the Wannier90 input file directly from the provided CIF file. For this purpose, there is a useful tool in the forms of Python script named as cif2qewan, which can be downloaded by the following command:

```
$  git clone https://github.com/wannier-utils-dev/
     cif2qewan.git
```

The above command will create a cif2qewan directory in the current working directory. We should also execute the following command if we do not have some packages needed by cif2qewan tool:

```
$ pip install docopt toml cif2cell
```

Now, move to the cif2qewan directory and copy the Ge1Te1_ICSD_638014.cif file to this directory. Suppose that the CIF file was located in ~/QE-SSP/gete directory.

```
$ cd cif2qewan
$ cp ~/QE-SSP/Ge1Te1_ICSD_638014.cif .
$ python cif2qewan.py Ge1Te1_ICSD_638014.cif
     cif2qewan.toml
```

With the above set of command lines, we can obtain the Quantum ESPRESSO and Wannier90 input files automatically (scf.in, nscf.in, pw2wan.in, and pwscf.win). In order to understand and run these files, please check Sec. 3.6.1. Make sure that the pseudo_dir in \cif2qewan.toml file has been edited properly to be the location of "PS Library" pseudopotentials as explained in Sec. 3.1.6, and the name of the pseudopotential of each atom in pp_psl_rrkj.csv file is correct.

6.2 Linux commands

The most basic way that we have explained in Chapters 2 and 3 for running the Quantum ESPRESSO calculations is using the following Linux commands typed in the terminal:

```
$ QEcommand < input.in > output.out
```

where `QEcommand` is one of any Quantum ESPRESSO binary (application) file such as `pw.x`, `bands.x`, `dos.x`, etc. For obtaining the DOS and band structure of bulk cubic GeTe, we might want to execute the following commands one by one in the terminal:

```
$ pw.x < gete.scf.in > gete.scf.out
$ pw.x < gete.nscfdos.in > gete.nscfdos.out
$ dos.x < gete.dos.in > gete.dos.out
$ pw.x < gete.nscfbands.in > gete.nscfbands.out
$ bands.x < gete.bands.in > gete.bands.out
```

However, as we become more experienced with Quantum ESPRESSO and when we want to perform many different calculations (usually at the same time), we will realize that typing commands one by one is not efficient. Furthermore, if we have parallel computing resources, we will need to type an additional command, such as `mpirun -np 4`, before the `QEcommand` for specifying the parallel calculations. This situation will force us to type longer command lines repeatedly.

Imagine also when the calculation takes a lot of hours (or even days and weeks), we surely do not want to wait in front of our computer without doing other productive works just to see that the calculation is still running. Therefore, it is better for us to learn some Linux commands and so-called **scripts** that will run even after we log out. We also need to manage Quantum ESPRESSO calculations as well as for the administration of files and folders in our working directory.

We expect that the workflow or calculation routine developed by using sufficient ("not so much") knowledge of Linux commands and scripts should be more efficient than repeating long commands. We may notice in Chapter 3 there is a script file like `run.sh`, which consists of several Linux commands in a single file. This file is also known as a "batch" file because it is useful for running many jobs one-by-one by a single execution of the file. In this section, we review some commonly-used Linux commands. If the readers are familiar with Linux commands, the readers can go to in Sec. 6.3, in which we explain how to make shell scripts.

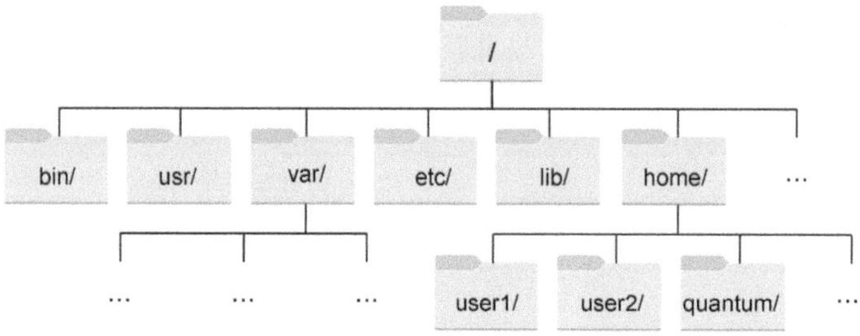

Figure 6.10 Linux directory structure.

6.2.1 File- and directory-related commands

In Linux (also in macOS or other Unix-like systems), every file and directory is put under the root directory, which is the first or topmost directory in the directory tree. The root directory is referred to by a single leading slash (/). Then, as shown in Fig. 6.10, under the root directory, there can be several default directories, such as `bin/`, `usr/`, `var/`, `etc/`, `lib/`, and `home/`, depending on the Linux distribution we install in our computer.

When navigating the file and directory structure, we can use either the absolute path or the relative path to the resource. For example, a user in Linux with a specific `username` usually has their main working directory as `/home/username/` in its absolute path that starts by "/". If our `username` is `quantum`, the home directory will be `/home/quantum/` (unless we specified another directory during the Linux installation). Suppose we are already inside `/home/quantum/` and we have a folder named `QE-SSP`, we can refer to that folder as `./QE-SSP` or simply `QE-SSP` in its relative path (while the absolute path is `/home/quantum/QE-SSP`).

Now let us learn some frequently-used commands that will be useful for the file and directory management as shown in Table 6.1.

`pwd`: **Show present working directory**

The present working directory is the directory in which the user is currently working. Each time we interact with the command prompt, we are working within a directory. We can use the `pwd` command to find out what directory we are currently working in:

Table 6.1 File- and directory-related commands in Linux, where file1 or dir1 are the name of file or directory, respectively.

Command	Purpose
`pwd`	show present directory
`cd dirname`	change directory to `dirname`
`cd`	change to home directory
`ls`	list the current directory
`ls -al`	list the current directory with hidden files
`touch filename`	create or update `filename`
`ln -s filename linkname`	create a symbolic link called `linkname` to `filename`; it can also works for directory
`mkdir dirname`	create a directory called `dirname`
`rm filename`	delete a file called `filename`
`rm -r dirname`	delete `dirname`
`rm -f filename`	force remove `filename`
`rm -rf dirname`	force remove `dirname`
`cp filename1 filename2`	copy `filename1` to `filename2`
`cp -r dirname1 dirname2`	copy `dirname1` to `dirname2`; create `dirname2` if it does not exist
`mv filename1 filename2`	rename or move `filename1` to `filename2`; however, if `filename2` is an existing directory, moves file1 into directory `filename2`
`cat > filename`	place standard input into `filename`
`more filename`	output the contents of `filename`
`head filename`	output the first 10 lines of `filename`
`tail filename`	output the last 10 lines of `filename`
`tail -f filename`	output the contents of `filename` as it grows, starting with the last 10 lines
`chmod octal filename`	change the permissions of `filename` (or `dirname`) to `octal` (see the main text for the detailed explanation)

```
$ pwd
```

The pwd command displays the absolute (full) path of our present working directory. Below is an example of pwd execution along with its output:

```
$ pwd
/home/quantum/QE-SSP
```

cd: Change directory

The cd ("change directory") command is used to move from the present working directory to another directory. Without any argument, cd will take us to our home folder:

```
$ cd
```

We can check its output by the pwd command

```
$ pwd
/home/quantum
```

Now to change to a particular directory, we can use its absolute path or relative path. Assuming that the directory QE-SSP exists in the home folder, we can navigate to QE-SSP by using the relative path as follows

```
$ cd QE-SSP
```

or

```
$ cd QE-SSP/
```

For better clarity of the relative path, we may also add a period character (.) before the directory name we want to enter since the period character emphasizes the path of the current working directory. The following command will give the same output as before:

```
$ cd ./QE-SSP/
```

If we prefer to navigate to the directory using its absolute path, we should specify the full path of the directory starting from the root "/" until the folder name we want to enter. Using the same directory example as before, we should type the `cd` command as follows in its absolute path:

```
$ cd /home/quantum/QE-SSP
```

Note that the home directory in Linux (in this example: `/home/quantum`) also has an alias denoted by a tilde symbol (~). Therefore, the following command:

```
$ cd ~/QE-SSP
```

is equivalent to the previous one.

What if we want to move back from QE-SSP to `/home/quantum` and to `/home`? Two dots (`..`) represent the parent directory, which is the directory immediately above the current one. It means the parent directory of QE-SSP in this case is `/home/quantum`, while the parent directory of `/home/quantum` is `/home` (see Fig. 6.10 for the illustration). Therefore, to switch to `/home/quantum` from the current working directory QE-SSP, we would type:

```
$ cd ../
```

If we want to directly move two levels up from QE-SSP to `/home`, it is possible to type the following command

```
$ cd ../../
```

Note that if the directory that we want to change has spaces in its name, we should either surround the path with quotes (single or double quotes):

```
$ cd "dirname with space"
```

or use the backslash (\) character to escape the space:

```
$ cd dirname\ with\ space
```

`ls`: List directory contents

The `ls` command will list information about files and directories within a directory. If we use it without any options and arguments, `ls` displays a list in alphabetical order of the names of all files in the current working directory:

```
$ ls
```

If we are already in `/home/quantum/QE-SSP/gete`, the output of the `ls` command execution would be similar to the following:

```
gete.bands.in      gete.nscfdos.in
gete.dos.in        gete.scf.in
gete.nscfbands.in  pseudo
```

To list files in a specific directory, we can pass the path to the directory as an argument. For example, if we are in `/home/quantum/QE-SSP/`, the following command will give exactly the same output as before:

```
$ ls gete
gete.bands.in      gete.nscfdos.in
gete.dos.in        gete.scf.in
gete.nscfbands.in  pseudo
```

We see that the default output of the `ls` command shows only the names of the files and directories. We can use the `-l` option to print files in a "long" listing format:

```
$ ls -l
```

The command output includes file/directory size, file/directory type, permissions, number of files/directories therein, owner, and timestamps (of latest modifications). An example of the output is given below:

```
$ ls -l
total 24K
-rw-r--r-- 1 user group   74 Jul 10 06:39 file1
-rw-r--r-- 1 user group  100 Jul 10 06:39 file2
-rw------- 1 user group 1.2K Jul 10 06:39 file3
-rw------- 1 user group  892 Jul 10 06:39 file4
-rw------- 1 user group  829 Jul 10 06:39 file5
drwxr-xr-x 2 user group 4.0K Jul 10 06:39 dir1
drwxr-xr-x 1 user group 4.0K Jul 10 06:39 dir2
```

The first line of the output is the total size of files (and directory) in the directory. Then, the next lines list the full information of the files and directories. The leftmost character in each line will be either a dash (-), which represents a file or a letter (`d`), which represents a directory. The set of following letters (`rwxrwxrwx`) indicate the permissions given to the particular file or directory.

The permissions are broken into three roles, i.e. `rwx`, `rwx`, and `rwx` for a user, a group, and others, respectively. The name of the user and the group are listed in `user` and `group` columns in the output of `ls -al`. Then, each role has specified permission for the file or directory in the following order: read (r), write (w), and execute (x). A dash character (-) is shown if there is no permission to read, write, or execute. Therefore, the following line,

```
rw-r--r-- ... file1
```

means that `file1` is

- `rw-`: readable, writable, and unexecutable by the user;

- r--: readable, unwritable, and unexecutable by the group;
- r--: readable, unwritable, and unexecutable by the others (beyond the user and the group).

Similarly,

```
rwxr-xr-x ... dir2
```

means that dir2 is

- rw-: readable, writable, and executable by the user;
- r--: readable, unwritable, and executable by the group;
- r--: readable, unwritable, and executable by the others.

Note that the ls command does not list the hidden files by default. A hidden file is any file that begins with a period character. To display all files, including the hidden files, use the -a option:

```
ls -al
```

We can then combine different options in one line, whether with an argument or not.

```
$ ls -al
$ ls -al /home/quantum/QE-SSP/gete
```

The first command above will list all files and folders in the current working directory, including the hidden files therein in the long listing format. Meanwhile, the second command will output the list for the specified directory. An example of the output is given below:

```
$ ls -al
total 36K
drwxr-xr-x 3 user group 4.0K Jul 10 06:39 .
drwxr-xr-x 5 user group 4.0K Jul 10 06:39 ..
-rw-r--r-- 1 user group   74 Jul 10 06:39 file1
-rw-r--r-- 1 user group  100 Jul 10 06:39 file2
-rw------- 1 user group 1.2K Jul 10 06:39 file3
-rw------- 1 user group  892 Jul 10 06:39 file4
```

```
-rw------- 1 user group  829 Jul 10 06:39 file5
-rw------- 1 user group  829 Jul 10 06:39 .file6
drwxr-xr-x 2 user group 4.0K Jul 10 06:39 dir1
drwxr-xr-x 1 user group 4.0K Jul 10 06:39 dir2
drwxr-xr-x 2 user group 4.0K Jul 10 06:39 .dir3
```

In this example, `.file6` and `.dir3`, which were previously not shown, are now listed in the output. You can practice the `ls -al` command to the other directories that contain hidden files.

`touch`: Create files

The `touch` command is used to update the timestamps on existing files and directories as well as to create new and empty files, which previously do not exist.

To create a file, specify the file name as an argument:

```
$ touch new-file.txt
```

If the file already exists, the `touch` command will change the file's last access and modification times to the current time. If the file does not exist yet, the command creates an empty `new-file.txt` with a zero-byte size.

`ln`: Create symbolic links

A symbolic link (or symlink) is a special type of file that points to another file or directory. The concept is similar to making a shortcut in Windows operating system (OS). In Linux-based OS, to create a symbolic link to a given file, we use the `ln` command with the `-s` option, the name of the file as the first argument, and the name of the symbolic link as the second argument:

```
$ ln -s sourcefile symboliclink
```

Suppose our current working directory is the home folder. We want to create a symbolic link to `~/QE-SSP/gete` and put it under the home directory as `GeTe`. We should type

```
$ cd ~
$ ln -s ~/QE-SSP/gete gete
```

The above commands will give a symlink named as `gete` in the home directory so we can access `gete folder` directly from the home directory.

```
$ cd ~/gete
```

If only one file/directory is given as an argument, `ln` creates a link to that file/directory in the current working directory with the same name as the file/directory it points to. The following command executed in the home directory will give the same result as the previous symlink:

```
$ ln -s ~/QE-SSP/gete
```

`mkdir`: **Create a new directory**

We can create a new directory using the `mkdir` command. To create a directory, pass the name of the directory as the argument to the `mkdir` command. For example,

```
$ mkdir /tmp/newdirectory
```

`mkdir` can take one or more directory names as its arguments. If the argument is a directory name, without the full path, the new directory is created in the present working directory.

To create directories along with the parent directory, we should use the `-p` option. For example,

```
$ mkdir -p QE-SSP/gete/out
```

creates the whole directory structure in the present working directory, from `QE-SSP` going to `gete`, and finally `out`. Note that when `mkdir` is invoked with the `-p` option, the command creates the directory only if it does not exist.

`rm`: Remove files or directories

To remove files or directories in Linux, use the `rm` command. We should be very careful when using this command because we cannot (in most cases) recover the files after removing them. By default (and also for safety), the `rm` command will not work when it is executed without any option.

To delete a file or a symbolic link, use the `rm` command followed by the file name as an argument:

```
$ rm file.txt
```

The `rm` command accepts one or more files or directories as its arguments. Note that when we delete a symlink, the original `sourcefile` will not be removed by the `rm` command.

The `-i` option tells `rm` to prompt the user for each given file before removing it:

```
$ rm -i file.txt
...
rm: remove regular empty file 'file.txt'?
```

Only when we answer the above question with "y" (without quotes), the `rm` command will perform.
Use the -d option to remove one or more empty directories:

```
$ rm -d dirname
```

To remove non-empty directories and all the files within them recursively, use the `-r` (recursive) option:

```
$ rm -rf dirname
```

The `-f` option does not prompt the user, and `rm -rf` command deletes all files and arguments under the directory without asking remove or not. It should be carefully noted that the `rm -rf ./*` command is considered as the **most dangerous command** in Linux since all files and folders under the current working directory will completely disappear. Imagine if we execute this command in our home directory, all our works will be wiped out, and if we execute the command in the root (/), our OS can be broken. Therefore, the readers should not use this command or use it after to back up the contents of the directory before performing the command.

`cp`: **Copy files and directories**

The `cp` command allows us to copy files and directories. To copy a file in the present working directory, use the source file as a first argument and the new file as the second:

```
$ cp file.txt filebackup.txt
```

To copy a file to another directory, we should specify the absolute or the relative path to the destination directory. When only the directory name is specified as a destination, the copied file under the destination will have the same name as the original file. For example, the following command

```
$ cp file.txt /home/quantum/QE-SSP
```

will copy `file.txt` in the present working directory to the `/home/quantum/QE-SSP` (or `~/QE-SSP`) folder with the same filename (`file.txt`). Note that by default, if the destination file exists, it will be overwritten without asking you unless you do not use `cp -i`.

To copy a directory, including all its files and subdirectories, use the `-R` or `-r` option, as in the following example:

```
$ cp -r QE-SSP QE-SSP-backup
```

`mv`: **Move and rename files/directories**

The `mv` command (short from move) is used to rename and move files/directories from one location to another. For example, to move a file named `file.txt` to the `/tmp` folder, we execute:

```
$ mv file.txt /tmp
```

To rename the file, we need to specify the destination file name:

```
$ mv file.txt filerenamed.txt
```

The command order for moving directories is the same as when moving files. To move multiple files and directories at once, specify the destination directory as the last argument:

```
$ mv file1.txt file2.txt /tmp
```

`cat`: Show file contents

The `cat` command shows the contents of one or more files and merge (concatenate) files by appending one file's contents to the end of another file. To display the contents of a file on the screen, pass the file name to cat as an argument. For example, executing

```
$ cat gete.scf.in
```

will output the contents of `gete.scf.in` (Sec. 6.1.1) to the terminal's screen. Adding an operator of standard output (>) after the `cat` command can make the command act as the cp. The following command line:

```
$ cat gete.scf.in > gete.scf.in.backup
```

will give an exact copy of the `gete.scf.in` file in the `gete.scf.in.backup` file.

`more` and `less`: Show file contents interactively

The `more` or `less` command is used to view the text files in the command prompt, displaying one screen at a time in case the file has many lines. The two commands are very similar, so we will consider that they are just the same command for our purpose of navigating large output files of Quantum ESPRESSO calculations. The merit of the commands is that they allow us to scroll up and down the file page by page. For example, we can check the usage of the `more` command on the `gete.scf.out` file:

```
$ more gete.scf.out
```

After executing the above command, we will be prompted with an output page (starting from the first line down to a certain line number) waiting for instruction. Press the space bar (or letter `f`) key to go to the next page. If we want to go back to the previous page, we can press letter `b` or `p` key. Use letter `q` key to quit the prompt.

`head` (or `tail`): Display the beginning (or end) of the file

The `head` and `tail` commands can be used to display the beginning and end of the file, respectively. By default, the number of lines shown by the two commands is limited to 10. If we want to change the number of lines printed on the screen, we can use the `-n` option followed by the desired number. We can try performing the `head` and `tail` commands on the `gete.scf.out` file. For the `head` command, the example execution along with the resulting output is as follows:

```
$ head -n 13 gete.scf.out

Program PWSCF 6.7 starts on 13Aug2021 at 6:53:15

This program is part Quantum ESPRESSO suite
for quantum simulation of materials; please cite
    "P. Giannozzi et al., JPCM 21 395502 (2009);
    "P. Giannozzi et al., JPCM 29 465901 (2017);
    URL http://www.quantum-espresso.org",
in publications or presentations.
More details at http://www.quantum-espresso.org/

Parallel version (MPI), running on 8 processors
```

For the `tail` command (this is useful to see `JOB DONE` message):

```
$ tail -n 8 gete.scf.out
  PWSCF         :  10m33.89s CPU  16m 8.73s WALL


  This run was terminated on: 7: 9:24  13Aug2021

=------------------------------------------------------=
  JOB DONE.
```

```
=--------------------------------------------------=
```

Similar to the cat command, we can combine either head or tail command with the standard output operation.

```
$ tail -n 8 gete.scf.out > log.txt
```

The above command will print the last 8 lines of gete.scf.out into the log.txt file, which can be opened later by a text editor.

The tail command also has a special option: -f, which is useful for displaying the contents of the file as it grows, starting with the last 10 lines. We can benefit from such an option, in particular, when monitoring a job running on the background. For example, when we execute

```
$ pw.x < gete.scf.in > gete.scf.out
```

we may want to observe the evolution of gete.scf.out during the running process of pw.x, which can take several minutes or even hours depending on the size of the calculated system. After executing the above command, we can follow it by the tail -f command to see the output (gete.scf.out) in real time when the tail -f command is entered:

```
$ tail -f gete.scf.out

  highest occupied level (ev):      7.1721

! total energy              =    -239.63636259 Ry
  estimated scf accuracy    <          3.6E-11 Ry

  The total energy is the sum of following terms:
  one-electron contribution =    -147.05152059 Ry
  hartree contribution      =      88.78504266 Ry
  xc contribution           =     -71.86764050 Ry
```

Note that, depending on the time of execution, it is normal if the output obtained by the readers is different from the above example. If pw.x calculation is successfully completed, the same "JOB DONE"

Table 6.2 Process management and system information commands.

Command	Purpose
`whoami`	display who we are logged in as
`w`	display who is online on the computer
`df`	show disk usage
`du`	show directory space usage
`whereis app`	show possible locations of an application named `app`
`which app`	show which `app` will be run by default
`bg`	lists stopped or background jobs; resume a stopped job in the background
`fg`	bring the most recent job to foreground
`top` `kill pid`	display all running processes kill a process whose "id" number is `pid`
`killall proc`	kill all processes named `proc`
`lscpu`	show cpu information, a shortened version of `cat /proc/cpuinfo` command
`free`	show memory and swap usage, a shortened version of `cat /proc/meminfo` command
`mpirun -np N app`	run `app` in parallel with N processors
`nohup command`	keep `command` (or `app`) process running even if we exit the shell (but not when the computer is shutdown)
`exit`	logout from the current session

message for `gete.scf.out` (and other kinds of Quantum ESPRESSO calculation outputs) should appear.

6.2.2 System information and process management

Linux and other Unix-based systems allow multiple processes to operate simultaneously without interfering with each other. Due to this nature, there are many commands dedicated to managing the processes and for displaying system information in Linux. We will not discuss all the commands, but we will list some of them that can be used along with the Quantum ESPRESSO calculations. The summary of Linux commands for process management and for displaying system information is given in Table. 6.2. We briefly describe each command one by one below.

`whoami`: Ensuring we know ourselves

The `whoami` command is used to check our own username. Suppose our username is `quantum`, the output of this command will also be `quantum`.

```
$ whoami
quantum
```

At a glance, we may consider this command not useful if we only have a simple username that is easy to memorize. However, in multi-user environments such as high-performance computing (HPC) clusters and supercomputers, we usually cannot set the username by ourselves because the system administrator will create the account for us. It is also possible that we obtain several usernames for one HPC system. In that case, the usernames can be a combination of letters and numbers that could be difficult to memorize. When we log into one username and switch to another username, we can forget who we are. In such a situation, the `whoami` command is a handy command to rescue us.

`w`: Display online users

The `w` command is used to know who is online on the computer we are logged in to. This command also shows other details such as the load average of the computer, the terminal number ("tty") we are connected to, how long the computer is already on, and how long the users have been logged in (or oppositely, have been idle). We simply type the letter `w` to run this command.

```
$ w
```

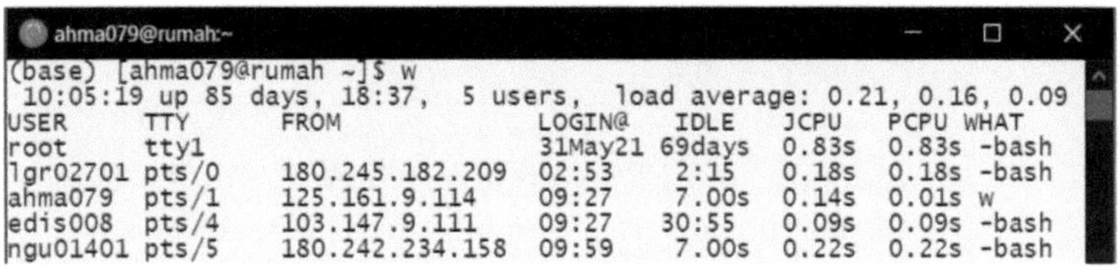

Figure 6.11 Example output of the `w` command in a HPC.

An example output of this command performed on a HPC is shown in Fig. 6.11. The username of one of the book authors is `ahma079`, running the `w` command, shown with other users who are online at that time. When more than two users use a computer, it would be nice to check it other users run Quantum ESPRESSO by `w` command.

`df` and `du`: Disk and directory space usage

Sometimes we want to know the disk (or file system in Linux) and directory space usage. For that purpose, we can use the `df` and `du` commands. In particular, if we are a user of a computer with limited space, we might be concerned with the directory space usage (the du command). Suppose we are in the directory where the input and output of the GeTe calculation files are located. We can use the du command to check the space used by all these files. An example execution and output of the du command is given below.

```
$ du -h
392K    ./out
1.0M    ./pseudo
6.7M    .
```

From the output above, we can see the usage of that directory (denoted by the period symbol) is around 6.7 MB. The output obtained by the readers may be different with this example depending on the readers' activities in the directory.

`whereis` and `which`: Understanding `app` locations

To know possible locations of an application named app, we can use the `whereis` command:

```
$ whereis app
```

For example, in a HPC system at which one of the book authors is registered, the output of `whereis pw.x` is as follows.

```
$ whereis pw.x
pw:
/apps/tools/qe-6.4.1/bin/pw.x
/home/ahma079/qe-6.7/bin/pw.x
```

The output indicates that there are two locations of `pw.x`. The first one is `/apps/tools/qe-6.4.1/bin/pw.x` and the second one is `/home/ahma079/qe-6.7/bin/pw.x`. This output can vary depending on the system used by the readers.

The question now is, "Which app is used by default when we call `pw.x`?" To get the answer, we can execute the `which` command with the following format:

```
$ which app
```

For example, in the HPC system same as before:

```
$ which pw.x
/apps/tools/qe-6.4.1/bin/pw.x
```

The output above informs us that the `pw.x` command from `/apps/tools/qe-6.4.1/bin/` folder will be used when we execute the `pw.x` command without its absolute path. If we want to use `pw.x` from `/home/ahma079/qe-6.7/bin/` instead, we must supply its full absolute path to `pw.x`, i.e.,

```
$ /home/username/qe-6.7/bin/pw.x < input > output
```

bg and fg: Background and foreground jobs

The bg command is used to resume a stopped job in the background. On the other hand, the `fg` command is used to bring the most recent job to the foreground. To understand the concepts of background and foreground in the Linux terminal, we can try running the SCF calculation of GeTe on our computer.

```
$ pw.x < gete.scf.in > gete.scf.out
```

By entering the above command only, we should notice that the cursor in the terminal does not return to the original command prompt. The cursor waits until the system finishes running `pw.x` as a foreground job, which actually takes a long time.

If we want to stop the calculation temporarily, we can press `CTRL+z` keystroke combination and continue writing other commands in the terminal as we wish. An example output message that appears after we press `CTRL+z` is given below.

```
^Z
[1]+  Stopped  pw.x < gete.scf.in > gete.scf.out
```

Now the command prompt returns to the original state with the cursor waiting for our action. What happens to the `pw.x` calculation above is that it stops until we enter either the `fg` command or `bg` command. First, let us try the `fg` command. Its execution with the example output is as follows.

```
$ fg
pw.x < gete.scf.in > gete.scf.out
```

In the above case, after executing the `fg` command, we can realize that `pw.x` continues running as a foreground job, indicated by the appearance of the second line (`pw.x < gete.scf.in > gete.scf.out`) on the screen, but then the cursor gets stuck until the calculation finishes. If the calculation is fast enough, we can do this kind of `fg` command and just wait for a moment with the inability of typing further commands in the terminal. However, we might not want to wait for the calculation, especially when the calculation takes a very long time. Instead, we want to continue other works in the terminal. In this case, it is a good idea to send the job to the background. We can press `CTRL+z` again, followed with the `bg` command.

```
$ bg
[1]+ pw.x < gete.scf.in > gete.scf.out &
```

The output message appearing on the second line indicates that `pw.x` is running as a background job, and the calculation will continue running while at the same time we can type other commands on the terminal. The ampersand character (&) in the example above tells the job to be run in the background. Therefore, instead of pressing `CTRL+z` and the `bg` command, we can also accompany the `pw.x` command line with the ampersand character if we decide to run the calculation on the background since the beginning of the calculation.

```
$ pw.x < gete.scf.in > gete.scf.out &
```

This command will take several minutes or hours depending on the computing resources we have in our PC. It is then natural to ask how we can monitor the process and how we can stop it when we wish to do so. The `top` and `kill` command will be helpful now.

`top`: Displaying all running processes

The `top` command is used to show the Linux processes. The command provides a dynamic real-time view of the running system, including the summary information of the system and the list of processes. As soon as we run this command, it will open an interactive command mode where the top half portion will contain the statistics of processes and resource usage. The lower half contains a list of the currently running processes. Pressing the letter q will exit the `top` command mode.

After we run the SCF calculation in our PC, we can check its status by the `top` command. Below is an example of the `top` command execution with some parts of its output.

```
$ top
Tasks: 16 total, ...
...
 PID USER    ... COMMAND
 850 quantum ... pw.x
...
```

There are actually further details (...) of the output but we just show the number of tasks, the process number (PID), the user who has the process, and the command related to the process. Note that the output of PID and USER may vary. However, if we run the same pw.x command for the SCF calculation of GeTe, we can see it somewhere below the COMMAND information of the top output. If we think that the calculation was run wrongly or if we just want to stop it, we should write down its PID, press q to exit the top command mode, and run the kill command as will be explained next.

kill: Termination of jobs in the middle

Suppose we want to stop the SCF calculation with PID already known from the previous example of top command, we can execute

```
$ kill pid
```

In our example of GeTe above, we should perform

```
$ kill 850
```

Is the process killed completely? Sometimes, no! We should confirm it by executing the top command again. If we still see the pw.x command with the same PID, we have two ways to completely kill the process. The first way is by using -9 option:

```
$ kill -9 850
```

The second way is by using the killall with the process (or command) name as its argument.

```
$ killall pw.x
```

lscpu and free: CPU and memory information

We can use the lscpu and free commands to check the processor (CPU) and memory information of the computer we are logged in. The lscpu command can be considered as a shortened version of cat /proc/cpuinfo command, while the free command is a shortened version of `cat /proc/meminfo`. The example execution of `lscpu` and `free` are shown below with some parts of their outputs.

```
$ lscpu
Architecture:          x86_64
CPU op-mode(s):        32-bit, 64-bit
Byte Order:            Little Endian
CPU(s):                8
On-line CPU(s) list:   0-7
Thread(s) per core:    2
Core(s) per socket:    4
Socket(s):             1
...

$ free
              total        used        free     ...
Mem:       11337068      143932    10964292     ...
Swap:       3145728           0     3145728     ...
```

`nohup`: No hang up

When exiting the terminal or shell of a Linux System, all running processes are usually terminated or hang up. If we want to keep the processes running when exiting the terminal, we can use the `nohup` command. It should be noted, however, that if the computer (not only a physical computer but also a virtual machine or WSL) is shut down, the `nohup` command and the processes to which it is assigned will also stop. Therefore, if we want to run very long Quantum ESPRESSO jobs, in addition to using the `nohup` command, we should make sure that the computer can survive running within the calculation time frame.

The `nohup` command can be combined with the background operator & as follows

```
$ nohup command &
```

where `command` can also be an app or other processes. For example, if we want to keep running the GeTe calculation in the background and safely exit our terminal, we can execute

```
$ nohup pw.x < gete.scf.in > gete.scf.out &
```

When we re-login to the terminal (as long as the computer is not shutdown), we can find by using the `top` command that the calculation is not interrupted. The `nohup` command is often used with a batch script as we will explain in Sec. 6.3.

`exit`: Quitting the session

As the name clearly tells us, the `exit` command is used to quit or logout from the current session of terminal/shell. We can simply type this command without any argument

```
$ exit
```

We can practice the `exit` command after the `nohup` command executed previously, and then login again to check (by using the `top` command) that the Quantum ESPRESSO calculations we submitted to the computer are still running.

6.2.3 Running parallel calculations

Recently, with the availability of "multicore" and "multithread" processors, most personal computers already allow us to perform parallel calculations with the `mpirun` command. However, to use this command, we should make sure that `mpirun` is installed in our system. If it is not installed in a Debian/Ubuntu Linux-based distribution, we can select one of few messages passing interfaces (MPI) implementations such as Open MPI, MPICH, and Intel MPI. The following is an example installation for the Open MPI library:

```
$ sudo apt install libopenmpi-dev
```

Note that we should have already done this step in Chapter 2 if we compile Quantum ESPRESSO with the parallel processing option.

A generic way to use the `mpirun` command is by using the `-np` option followed by the number of cores (more precisely threads) in a CPU (or CPUs) (`N`) for parallel processing:

```
mpirun -np N app
```

In simple HPC systems, the number of cores (threads) given by `lscpu` command is usually the maximum limit that we can assign to `mpirun` when running Quantum ESPRESSO in parallel. For example, if the number of CPUs from the `lscpu` output is 8, we can put 8 as the value of `N` for the `-np` option in the `mpirun` command, combined with `pw.x` or other Quantum ESPRESSO apps. However, we recommend not to use the maximum number for the -np option since OS is also running at least one core.

```
$ mpirun -np 8 pw.x < gete.scf.in > gete.scf.out
```

The majority of the codes that come with the Quantum ESPRESSO package can run in parallel similarly to the above way.

In a more sophisticated HPC system (let us say a supercomputer), the result of `lscpu` should not be taken into consideration for deciding the value of the `-np` option because when we login to that supercomputer, we are only brought to the "control node" that will distribute our job to the other nodes dedicated for parallel processing. In this case, we should consult the system administrator or the manual of the supercomputer for the best setting of the CPU number we can use for our calculations. On the other hand, if we are just running Quantum ESPRESSO calculations on our own laptop, it is safe to use a smaller CPU number than what is indicated by `lscpu`.

We should also be aware that plane-wave DFT calculations (like that in Quantum ESPRESSO) do not scale linearly. Our calculation will get faster to a certain value of `N`, after which the calculation will become slow if we add more parallel processes. The situation can vary depending on the PC hardware (chipset, memory size, etc.), software, and type of calculation we are doing, but usually we will see a reduction in the speedup after around 8–32 processes.

If we do the generic way of `mpirun` above (such as `mpirun -np 8 pw.x < gete.scf.in > gete.scf.out`), we accept the default "strategy" for parallelizing the calculation. Using this default strategy for the `pw.x` command, in most cases, is already a good choice. However, if we are not satisfied with the speedup given by the default strategy, we can use other parallelization schemes as discussed in detail in the Quantum ESPRESSO documentation, especially for `pw.x` ("PWscf") and `ph.x` ("PHonon") applications, in the sections of performances and parallelism, respectively.

6.2.4 Parallelization in Quantum ESPRESSO

In the following, we just briefly discuss some types of parallelization that can be implemented in Quantum ESPRESSO because advanced users may want to play by themselves with the possible parallelization strategies beyond the default one.

Parallelization for several independent calculations. If the calculation we have asked for involves running several similar calculations, the computers can break up these calculations between parallel processes. Quantum ESPRESSO offers this functionality for some applications, particularly `ph.x`), which refers to these sets of calculations as "images". We can set how many of these images are used for a parallel calculation with `-nimage` or `-ni`. If we run with 20 processors and specify `-ni 2`, each image will use 10 processors. The default for `-ni` is just 1.

Parallelization of k-points. This parallelization strategy needs very little communication between processes and thus can offer good scaling, as each k-point can be treated as effectively an independent calculation where results are added together at the end. This strategy will be effective when the number of k-points in our calculation is quite large (typically larger than $10 \times 10 \times 10$ k-point grid in the Monkhorst-Pack scheme). We can set how many parallel groups of k-points for our calculation with the `-npool` or `-nk` flag in Quantum ESPRESSO (e.g., add `-nb 2` option next to the `pw.x` command). The default `-nk` is 1.

Parallelization over bands. The calculation for each energy band can also be parallelized. This strategy may cut down the amount of memory used by each process but requires a bit more communication between processes. We can set the number of bands to be grouped in the parallel calculation by using the `-nband` or `-nb` flag (e.g., add `-nb 2` option next to the `pw.x` command). The default `-nb` is 1.

Parallelization for plane waves. For the sake of completeness of the information, we should note that Quantum ESPRESSO by

Table 6.3 Search in Linux using the grep commands.

Command	Purpose
`grep pattern files`	search for pattern in files
`grep - r pattern dirname`	search recursively for pattern in dirname
`command | grep pattern`	search for pattern in the output of command

default already distributes the plane wave basis sets efficiently to the parallel processors. This strategy is always turned on if Quantum ESPRESSO is compiled and run in parallel.

6.2.5 Searching

For searching files and patterns, Linux has some powerful commands such as `grep`, `find`, and `locate`. For our purpose of analyzing and understanding Quantum ESPRESSO output files, it is sufficient for us to focus on the `grep` command. The usage summary of the `grep` command is given in Table. 6.3. Let us practice this command for obtaining some important information about our calculation.

The most basic syntax for `grep` is

```
$ grep pattern files
```

where the `pattern` can be a character, a word, a sentence, or any string we want to find in files. Suppose we already have a `gete.scf.out` file in the `~/QE-SSP/gete/` directory, we can search for all occurrences of "total energy" in `gete.scf.out` as follows:

```
$ grep "total energy" gete.scf.out
```

The output will be similar to as follows:

```
    total energy = -239.63505730 Ry
    total energy = -239.63624073 Ry
    total energy = -239.63633447 Ry
    total energy = -239.63635918 Ry
    total energy = -239.63636081 Ry
    total energy = -239.63636155 Ry
    total energy = -239.63636227 Ry
    total energy = -239.63636260 Ry
    total energy = -239.63636259 Ry
    total energy = -239.63636258 Ry
!   total energy = -239.63636259 Ry
    The total energy is the sum of the following terms
          :
```

In this case, since the SCF calculation performed iterations for finding the minimum total energy, we see the words "total energy" several times according to the number of iterations. The one with the exclamation mark (!) is the final result of total energy. Therefore, if we just want to know this final result, we can change the search pattern to the exclamation mark. The `grep` execution with its output in this case is

```
$ grep ! gete.scf.out
!   total energy = -239.63636259 Ry
```

The `grep` command can be used recursively to find the pattern in a specified directory. The syntax is

```
$ grep - r pattern dirname
```

We can practice this syntax for finding the Fermi energy in the gete working directory (`~/QE-SSP/gete/`).

```
$ grep -r Fermi .
```

Remember that the period (`.`) sign denotes the current directory, so that the above syntax will look for all occurrences of "Fermi" in all files of the current directory. The example of the output is given below:

```
./gete.nscfdos.out: the Fermi energy is 7.2526 ev
./gete.dos        : ... EFermi =        7.253  eV
```

The output tells us that the grep command could find the word "Fermi" in the gete.nscfdos.out and gete.dos files in the current directory.

The grep command can also be combined with the other Linux commands through the "piping" method with the pipe (|) character. The syntax is

```
$ command | grep pattern
```

where command is any command or app. For example, we can list the current directory and find any file or directory which contains "out" pattern. The example execution and some parts of its output are given below.

```
$ ls -al | grep out
-rw-r--r-- 1 ... gete.bands.out
-rw-r--r-- 1 ... gete.dos.out
-rw-r--r-- 1 ... gete.nscfbands.out
-rw-r--r-- 1 ... gete.nscfdos.out
-rw-r--r-- 1 ... gete.scf.out
drwxr-xr-x 2 ... out
```

6.2.6 Keyboard shortcuts

There are some useful shortcuts available in the Linux terminal. The summary of these keyboard shortcuts is given in Table. 6.4. You can try them one by one following the table.

Self-contained "help" of the command

Usually, a Linux command can have many options that are specified by a single hyphen (-) or double hyphen (--). If we do not use the command frequently, we might forget its options. However, memorizing command options is not necessary because almost all Linux commands have a --help option, which prints a short message

Table 6.4 Keyboard shortcuts in the terminal.

Shortcut	Purpose
`CTRL+c`	halt the current command
`CTRL+z`	stop the current command, resume with `fg` in the foreground or `bg` in the background
`CTRL+d`	logout from the current session, similar to `exit`
`CTRL+w`	erase one word before the cursor in the current line
`CTRL+k`	erase any characters *after* the cursor
`CTRL+u`	erase the whole line
`CTRL+p`	move to a command backward; browse/recover previous command(s) according to the terminal history
`CTRL+n`	move to a command forward; in combination with `CTRL+p`, browse/recover previous command(s) according to the terminal history
`ALT+f`	move the cursor forward to the end of the next word in the current line
`ALT+b`	move the cursor backward to the beginning of the previous word in the current line
`CTRL+e`	move the cursor to the end of the current line
`CTRL+a`	move the cursor to the beginning of the current line line

about how to use the command. The way we write it in the terminal is as follows.

```
$ command --help
```

For example,

```
$ cp --help
```

will give us the details of the cp command as well as its possible options.

The Linux commands are also often distributed together with manual (man) pages. A man or manual page is a form of documentation that explains what the command does, examples of how you run the command, and what arguments it accepts. The man command is used to display the manual page of a given command.

```
$ man command_name
```

For example, to open the man page of the command to change the directory (cd), you would type

```
$ man cd
```

To navigate the man pages, use the Arrow, Page Up, and Page Down keys. We can also press the Enter key to move one line at a time, the Space bar to move to the next screen, and the letter b key to go one screen back. To exit the man page, press the letter q key.

6.3 Shell scripts and batch jobs

Shell scripts are useful for automating processes and running many jobs simultaneously because shell scripts allow us to string together sets of commands. With the shell scripts, we could write a script that runs calculations, parses the important results to another file, and even generates a plot. In this section, we will learn basic shell scripting to be able to automate Quantum ESPRESSO calculations and to submit batch jobs as a background job with nohup command.

There are often several different shells installed on a Linux system, and bash is typically the default one in most Linux distributions. The bash shell is running in the terminal, and it interprets and executes the commands we type in.

6.3.1 Environment

On startup, such as when a new terminal is opened, or when we connect to a remote system over the network, bash will read several configuration files.

- For login shells (the shells where we are prompted to login when we start it), bash will read configuration options in ~/.bash_profile or ~/.profile if this file exist. Note ~ is interpreted by the shell as the users home directory.
- For non-login shells (such as when we open the terminal emulator), bash will read configuration options in ~/.bashrc.
- For simplicity, we can put all our configuration options in ~/.bashrc and leave the ~/.profile file empty except for a command to read the ~/.bashrc file.

These files can be used to configure many aspects of the shell, such as how the prompt looks in the terminal and also how specific apps such as Quantum ESPRESSO can be found by defining the so-called environment variables.

Many important aspects of how bash behaves are governed by environment variables. These variables can be modified to our preference in the configuration files. To see the current value of one of these variables, e.g., the PATH variable, we can do the following

```
$ echo $PATH
```

The $ symbol in front of the variable (e.g., PATH variable) is used when referring to the data stored in a variable, and echo is a built-in bash command that outputs something to the terminal.

The PATH variable tells bash which directories to look in for executables so they can be run without typing in their full path. For example, we can just type in nano in the terminal to launch the program, as the PATH variable contains the directory holding an executable of this name. Different directories are separated by ":".

To add the directory ~/.bin to our path, we can add the line export PATH="~/bin:$PATH" to the .bashrc file. Here the export statement is used to allow the variable to be usable by any child processes rather than local to the script. Note when we are naming the variable we want to assign a value to we do not use a $. Also we just want to add a directory to the existing PATH so we add :$PATH to keep all the existing directories.

We can practice updating the PATH variable to include not only the ~/bin directory but also the Quantum ESPRESSO binary directory. Suppose we install Quantum ESPRESSO latest version (q-e) under ~/opt directory, we can add the export statements for the PATH variable as follows:

~/.bashrc

```
export PATH="~/bin:$PATH"
export PATH="~/opt/q-e/bin:$PATH"
```

We need to execute the following command in the terminal to ensure that the updated variables will be read by the shell without logout.

```
$ source ~/.bashrc
```

When you login the next time, you do not need this.

6.3.2 Scripting

This part covers basic scripting. If readers are interested in developing more advanced scripts, we suggest referring to the http://tldp.org/LDP/abs/html/index.html.

Let us make a simple example script to output "Hello World!". We can use the nano text editor to create it.

```
$ nano helloworld.sh
```

The contents are as follows:

QE-SSP/scripts/helloworld.sh

```
#!/bin/bash
# Simple script that outputs "Hello World!"
```

```
echo "Hello World!"
```

The first line above tells the system what command is used to interpret the contents of the script. This line should always begin with `#!` and then the path to the executable. For `bash` the executable will almost always be `/bin/bash`. In the script, comments are represented by #. Linux commands can be typed the same as we would enter them to the terminal. After writing the script, we will need to make the script executable with `chmod u+x hello.sh`.

```
$ chmod u+x helloworld.sh
```

Then, we can run it by typing `./helloworld.sh`. The execution and example output are given below.

```
$ ./helloworld.sh
Hello World!
```

Note that if we move the `helloworld.sh` file to `~/bin/` that has been added to the `PATH` variable, we can execute the script from any directory without specifying the current working directory (`./`).

```
$ mv helloworld.sh ~/bin/
$ helloworld.sh
Hello World!
```

Variables in shell scripts

In the bash shell scripts, variables need not be declared in any way, we can assign a value and start using them. For example:

QE-SSP/scripts/helloworld-rev.sh

```
#!/bin/bash
# Example using variables.

var1="Hello"
var2="World!"
```

```
echo $var1 $var2
# Note the use of the $ symbol when
# we want to use the value stored in the variable.

# Any command can use these variables
dirname="tmpdir"
mkdir $dirname
```

We can also read a value from standard input using the `read` command as follows:

QE-SSP/scripts/enterinput.sh

```
#!/bin/bash
# Example showing how the read command is used.

echo "Please enter some text:"
read user_text

echo "The text you entered was:" $user_text
```

Variable names are case sensitive in bash. The usual convention used is that environment variables (e.g., `PATH`), and internal shell variables (e.g., `BASH_VERSION`) are capitalized, while other variables are in lowercase. Adopting this convention will ensure that we do not accidentally overwrite any important variables in our scripts.

Command substitution

Command substitution allows the result of a command to replace the command itself, acting much like a variable in practice. For example:

QE-SSP/scripts/substitute.sh

```
#!/bin/bash
# Example of command substitution.

# This can be done by
# enclosing the command in $( )
echo "It is now $(date)."
# Note: the "date" command outputs
# the current date and time.

# This can also be done by
# enclosing the command in backtics ` `
```

```
echo "The files in this directory are:" `ls`
```

Conditional statements

The conditional `if` statements can be used in bash, and many types of conditional tests are possible. We can test if the value stored in a variable equals something as follows:

QE-SSP/scripts/if1.sh

```
#!/bin/bash
# Example using if statements

echo "Please enter a yes or no: "
read user_response

# Note the spaces
# following `[` and before `]` are important.
if [ $user\_response = yes ]
then
    echo "You entered yes."
elif [ $user\_response = no ]
# "elif" is the same as "else if"
then
    echo "You entered no."
else
    echo "You didn't enter yes or no."
fi
# "if" statements are always ended with "fi".
```

We can also check, e.g., if a file or directory exists:

QE-SSP/scripts/if2.sh

```
#!/bin/bash
# Check if the directory "tmpdir" exists, and if not
    , create it, then check
# if the file "tmpdir/testfile" exists, and if not,
    create it.

dirname="tmpdir"
filename="testfile"

if [ ! -d "$dirname" ]
# The "!" is logical negation
# -d tests that a file exists and is a directory.
```

```
then
    mkdir $dirname
fi

if [ ! -f "$dirname/$filename" ]
# -d tests that a file exists and is a regular file
    (i.e., not a directory).
then
    touch "$dirname/$filename"
fi
```

Arithmetic and testing

To test numeric values, the `(( ... ))` construction can be used. For example:

QE-SSP/scripts/arithmetictest.sh

```
#!/bin/bash
# Example showing the use of the (( )) construction

var1=4
var2=5
var3=8

if (( var1 + var2 > var3 ))
# Note within the (( ))
# don't use $ symbols before variables.
then
    echo "$var1 + $var2 > $var3"
else
    echo "$var1 + $var2 <= $var3"
fi

# We can also use the construction
# to perform basic arithmetic
var4=$(( var1  var3 ))echo "\$var1  $var3 = $var4"
```

Looping

In bash shell, we can use the `for` loops to iterate a variable over a range of values. For example:

QE-SSP/scripts/loop1.sh

```
#!/bin/bash

# Simple for loop example.
# This construction loops
# over a space separated list of values.

# A variable whose contents contains spaces
# would work in the same way.

for i in 1 2 3
do
    echo "Iteration number $i"
done
```

The `for` loop can be used to apply a command repeatedly to a set of files. For example:

QE-SSP/scripts/loop2.sh

```
#!/bin/bash

# Example showing a simple for loop
# over a list of arguments used to convert
# a set of data files with comma separated columns
# to tab separated columns.

for csvfile in "$(ls .csv)"dodatfile="\$(basename
    \$csvfile .csv).dat"sed 's/,/\t/g' \$csvfile >
    \$datfiledone
```

In the above example, basename is a tool to strip the suffix from a file. Here we use it to construct a new filename with the `.dat` extension.

Bash also supports numeric loop ranges using the syntax `{start..finish..increment)`. For example:

```
#!/bin/bash
# Example of for loop with numeric increments

for i in {0..6..2}
do
    echo "The value of i is $i"
done
```

In addition to the `for` loops, the bash shell also supports `while` loops, where a set of commands are repeated continuously when a given condition is true. For example:

QE-SSP/scripts/while1.sh

```
#!/bin/bash
# Example of while loop in bash

counter=0

# Here -lt means less than
while [ $counter -lt 10 ]
do
  echo $counter
  counter=$((counter + 1))
done
```

It is often convenient to use `while` loops to operate on every line of a file. For example:

QE-SSP/scripts/while2.sh

```
#!/bin/bash
# Example of while loop reading lines from a file

linenumber=0

while read line
do
  linenumber=$((linenumber + 1))
  echo "$linenumber: $line"
  # This will prepend each line of a file with its
      line number.
done < test.txt
```

Arguments

We can pass arguments to bash scripts from the command line. For example:

QE-SSP/scripts/argument1.sh

```
#!/bin/bash
# Example using command line arguments.
```

```
# Call this script with several arguments.
echo "The first argument is $1"
echo "The second argument is $2"
echo "The number of arguments is $#"
echo "The full list of arguments is $@"
```

We can loop over arguments using a `for` loop as follows:

QE-SSP/scripts/argument2.sh

```
#!/bin/bash
# Example using a for loop to iterate over command
    line arguments.

for arg in $@
do
  echo $arg
done
```

6.3.3 Quantum ESPRESSO job script

Now let us make a simple script for submitting batch jobs of Quantum ESPRESSO calculations. The script `runEnDOS.sh` includes Quantum ESPRESSO calculations of SCF, NSCF for obtaining DOS, and NSCF for obtaining the energy dispersion.

QE-SSP/gete/runEnDOS.sh

```
#!/bin/bash
# Simple QE job script
# This script can be re-used for other materials
# Later we just need to change
# the following 3 variables:
# first is the Quantum ESPRESSO binary location
export QEBIN=/home/quantum/opt/q-e/bin/
# second is the number of CPUs
export NCPU=4
# last is the prefix of the calculations
export PREFIX=gete
# QE binaries that we want to use
export PW=$QEBIN/pw.x
export DOS=$QEBIN/dos.x
```

```
export BND=$QEBIN/bands.x
# define variables for calculations of
# SCF, DOS, and energy dispersion
export SCF=$PREFIX.scf
export NSCFD=$PREFIX.nscfdos
export POSTD=$PREFIX.dos
export NSCFB=$PREFIX.nscfbands
export POSTB=$PREFIX.bands
echo "start SCF"
mpirun -np $NCPU $PW < $SCF.in > $SCF.out
echo "SCF done"
echo "start NSCF DOS"
mpirun -np $NCPU $PW < $NSCFD.in > $NSCFD.out
echo "NSCF DOS done"
echo "start Postprocessing DOS"
mpirun -np $NCPU $DOS < $POSTD.in > $SCF.out
echo "Postprocessing DOS done"
echo "start NSCF BANDS"
mpirun -np $NCPU $PW < $NSCFB.in > $NSCFB.out
echo "NSCF BANDS done"
echo "start Postprocessing BANDS"
mpirun -np $NCPU $BND < $POSTB.in > $POSB.out
echo "Postprocessing BANDS done"
echo "All jobs are finished"
```

The script can be run in the background using the nohup command. This method will ensure that we can leave our terminal safely and continue with other works.

```
$ nohup runEnDOS.sh &
```

6.4 Plotting and visualization tools

Plotting a graph is essential for the readers of the scientific paper to understand the numerical output values in a consistent and scientific way. There are several libraries in the Python programming language that can be used for plotting. Matplotlib is one of the most well-known and free Python libraries that can be used to produce a variety of plots and graphics in many styles and can be interacted with in flexible ways. Therefore, in this book, we choose Matplotlib in the Python environment as the platform to produce graphical outputs from Quantum ESPRESSO data. The Matplotlib official site at `https://matplotlib.org/users/tutorials.html` gives us a full list of tutorials and various options we can adopt for any visualization project. We also use the JupyterLab (https://jupyter.org/), which is the next-generation web-based user interface for Python and Matplotlib. The installation of Python, Matplotlib, JupyterLab, along with other important Python libraries, is already explained in Sec. 2.2. This section will show you the most direct and relevant practice of using Matplotlib that will give us a reasonable publication quality, which is good enough to appear in a scientific paper.

Hereafter, we will explain how to write a Python plotting script file with the *.ipynb extension, which can be opened and run directly by JupyterLab as

```
$ jupyter-lab plot.ipynb
```

In the Python script, we have to put `import` statements to load some libraries related to the plotting and numerical computations. We will load Matplotlib along with NumPy libraries for our purpose. It is also recommended to create a "short form" of the library (e.g., `mpl` for `matplotlib`) so that we can refer to its functions using the this short form instead of typing out `matplotlib` every single time. The # sign is used for comments.

QE-SSP/scripts/plot.ipynb

```
# Import required packages
import matplotlib as mpl
import matplotlib.pyplot as plt
import numpy as np
```

Now suppose we have a data file called "gr.dos", which can be found in `QE-SSP/scripts/gr.dos`. The readers can also obtain the gr.dos file by running the tutorial in Sec. 3.3.2. The gr.dos file is the output file of the phonon density of states (phonon DOS) of graphene, which contains four columns separated by space. The 1st column corresponds to the frequency ω in cm^{-1}, the 2nd column is the total phonon DOS in state/cm^{-1}/unit-cell, and 3rd and 4th columns are the partial phonon DOS of the C atom number 1 and number 2, respectively. We can load this data using the `loadtxt` function from the NumPy package that we imported as `np`.

QE-SSP/scripts/plot.ipynb

```
1 # Use numpy.loadtxt to import our data
2 omega, ph_tot, ph_c1, ph_c2 = np.loadtxt('gr.dos',
      unpack=True)
```

In the above blocks, we read the variable set `omega`, `ph_tot`, `ph_c1`, and `ph_c2` by the `np.loadtxt` function. By setting `unpack=True` (line 2), we transpose each column into an array.

6.4.1 Plain plotting of the data

Once we have loaded the data, we can make a default plot by `plt.figure()` (the short form of `matplotlib.pyplot.figure()`) and inspect the dataset with the following code:

QE-SSP/scripts/plot.ipynb

```
# Create figure and add axes object
fig = plt.figure()
ax = fig.add_axes([0, 0, 1, 1])

# Plot our data
ax.plot(omega, ph_tot)
ax.plot(omega, ph_c1)

# Save and show our data
plt.savefig('plot.pdf', bbox_inches='tight')
plt.show()
```

Some brief explanations for this block of code are as follows:

- `fig = plt.figure()`: assigns a figure object with the default setting of Matplotlib.
- `ax = fig.add_axes([0, 0, 1, 1])`: defines default axes of the `fig` object with the axis origin (the corner of the axis at the bottom left) located at (0, 0), width = 1, and height = 1 in the scaled units.

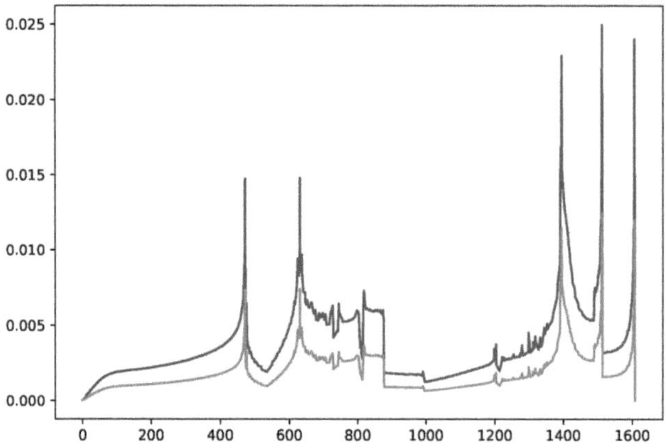

Figure 6.12 A default plot of gr.dos, which is not suitable for a scientific paper.

- `ax.plot(omega, ph_tot)`: calls the `plot` function for the `ax` axes defined earlier, for which we plot `ph_tot` as a function of `omega`.
- `ax.plot(omega, ph_c1)`: on the same `ax` axes, calls the `plot` function again to plot `ph_c1` as a function of `omega`.
- `plt.savefig('plot.pdf', bbox_inches='tight')`: saves the figure to "plot.pdf", in which the setting `bbox_inches='tight'` will remove all of the extra white space around the figure.
- `plt.show()`: displays the figure on the web-based user interface of JupyterLab.

We can see that the plot shown in Fig. 6.12 has a few meanings already, but we notice that the default Matplotlib settings do not give the figure with sufficient publication quality. For example, there is no axis label, the font size is too small, and we may also not like the tick marks outside the main frame. As we change some of the parameters described in the next subsections, we can get a better looking plot that is good enough for publication as shown in Fig. 6.13.

It is safe to say that the appearance of Fig. 6.13 is better than Fig. 6.12 in many ways. In particular, the plot now has axis labels for variables along with their units, readable fonts with a reasonable size, standard colors, clear tick marks for both the major and minor

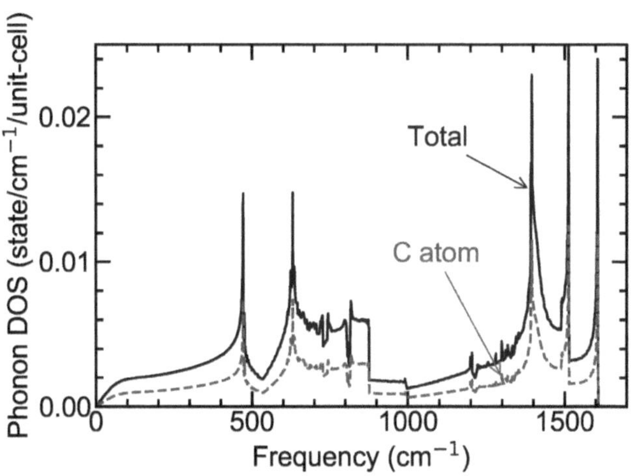

Figure 6.13 An improved, polished version of Fig. 6.12.

ticks, and also well-placed annotations or legends to distinguish two plots of different samples. Note that to clearly distinguish the two plots, if we use the only color, it would be troublesome for black-white printing or for people who have difficulties in recognizing colors. We should use both different colors and line styles at the same time, such as "blue-solid line" and "red-dashed line". All such details can be precisely managed through some Matplotlib settings that will be discussed below.

6.4.2 Changing general plot parameters

There are at least three general parameters that we need to set at the beginning of the Python plotting script: (1) font, (2) font size, and (3) axis line width. These three are essentially global parameters defined at the beginning of the Python script that we do not change later, i.e., we do not have to explicitly set font/sizes for each label down the line). We can add the following code before we generate any figures:

QE-SSP/scripts/plot.ipynb

```
# Edit the font, font size, and axes width
mpl.rcParams['font.family'] = 'Arial'
```

```
plt.rcParams['font.size'] = 22
plt.rcParams['axes.linewidth'] = 2
```

This block of code is typically put after importing packages.

You may change the font with other fonts you like, but it is recommended to use the “Sans” font family, such as “Helvetica”, “Arial”, and “Verdana”. The “Sans” font family is suitable for reading the font clearly even when we reduce the size of the graph. If you are not sure whether or not the font is available, you can type the following command in Python and see the output that lists all the font names available to Matplotlib in your computer.

QE-SSP/scripts/plot.ipynb

```
import matplotlib.font_manager as fm

# Collect all the available fonts
font_names = [f.name for f in fm.fontManager.ttflist]
print(font_names)
```

6.4.3 Setting axes and ticks

We should not forget to add a label to each axis. The tick widths and lengths should be edited to match our axis parameters. If we have minor ticks, we can also specify by a parameter `which='mirror'`.

QE-SSP/scripts/plot.ipynb

```
# Add the x and y-axis labels
ax.set_xlabel(r'Frequency (cm$^{-1}$)')
ax.set_ylabel(r'Phonon DOS (state/cm$^{-1}$/unit-cell)')
# Set the axis limits
ax.set_xlim(0, 1700)
ax.set_ylim(0, 0.025)
# Edit the major and minor ticks
ax.xaxis.set_tick_params(which='major', size=10,
    width=2, direction='in', top='on')
ax.xaxis.set_tick_params(which='minor', size=6,
    width=2, direction='in', top='on')
ax.yaxis.set_tick_params(which='major', size=10,
```

```
    width=2, direction='in', right='on')
ax.yaxis.set_tick_params(which='minor', size=6,
    width=2, direction='in', right='on')
# Edit the major and minor ticks of x and y axes
ax.xaxis.set_major_locator(mpl.ticker.MultipleLocator(500))
ax.xaxis.set_minor_locator(mpl.ticker.MultipleLocator(100))
ax.yaxis.set_major_locator(mpl.ticker.MultipleLocator(0.01))
ax.yaxis.set_minor_locator(mpl.ticker.MultipleLocator(0.002))
```

Note that Matplotlib can accept special mathematical formatting from the LaTeX such as `cm$^{-1}$`.

6.4.4 Annotations and saving the plots

After all basic settings are ready, we can plot the dataset and give annotations or legends to the plots.

QE-SSP/scripts/plot.ipynb

```
# Plot our data
ax.plot(omega, ph_tot, linewidth=2, color='blue', linestyle='solid')
ax.plot(omega, ph_c1, linewidth=2, color='red', linestyle='dashed')

# Add annotate
ax.annotate('Total', color='blue',
            xy=(1390, 0.015), xycoords='data',
            xytext=(1000, 0.018), textcoords='data',
            arrowprops=dict(arrowstyle="->", color='blue')
            )
ax.annotate('C atom', color='red',
            xy=(1300, 0.002), xycoords='data',
            xytext=(950, 0.01), textcoords='data',
            arrowprops=dict(arrowstyle="->", color='red')
            )

# Save and show figure
plt.savefig('plot-rev.pdf', bbox_inches='tight')
```

```
plt.show()
```

At the very last, do not forget to save the plot. The PDF format is recommended because it is a vector image that will not lose the resolution.

6.4.5 Creating and using your Matplotlib style

When the readers want to plot many figures with the same style, making your Matplotlib style is helpful to save time. You do not need to copy/paste settings in Matplotlib whenever you create a new figure. By importing a style file, you can ensure consistency while still maintaining the ability to override settings as you wish within the individual scripts.

The readers can find a style file for this book at `QE-SSP/matplotlib/sci.mplstyle`. The `sci.mplstyle` file can open and edit by any text editor for example, nano:

```
$ nano sci.mplstyle
```

QE-SSP/matplotlib/sci.mplstyle

```
# Figure properties
figure.figsize      : 6.5, 5

# Font properties
font.family         : Arial
font.size           : 22

#### LaTeX
mathtext.default    : regular

# Axes properties
axes.titlesize      : medium   # fontsize of the axes title
axes.titlepad       : 10       # pad between axes and title in points
axes.titleweight    : normal   # font weight for axes title
axes.linewidth      : 2        # edge linewidth
```

```
axes.labelpad       : 10      # space between label and axis
axes.prop_cycle     : cycler(color=['1f77b4', 'd62728', '2ca02c', 'ff7f0e', '9467bd', '8c564b', 'e377c2', '7f7f7f', 'bcbd22', '17becf'])

# Tick properties
# x-axis
xtick.top           : True
xtick.direction     : in
xtick.major.size    : 9
xtick.major.width   : 2
xtick.major.pad     : 4
xtick.minor.visible: True
xtick.minor.size    : 6
xtick.minor.width   : 2
xtick.minor.pad     : 4

# y-axis
ytick.right         : True
ytick.direction     : in
ytick.major.size    : 9
ytick.major.width   : 2
ytick.major.pad     : 4
ytick.minor.visible: True
ytick.minor.size    : 6
ytick.minor.width   : 2
ytick.minor.pad     : 4

# Line properties
lines.linewidth     : 2
lines.markersize    : 10

# Legend properties
legend.framealpha   : 1
legend.frameon      : False
legend.fontsize     : 19

# Increase the default DPI, and change the file type from png to pdf
savefig.dpi         : 600     # figure dots per inch
savefig.format      : pdf     # png, ps, pdf, svg
savefig.bbox        : tight   # {tight, standard}
savefig.transparent: True     # transparent background
```

The readers can modify the file freely. The details for each setting can be found at `https://matplotlib.org/stable/tutorials/`

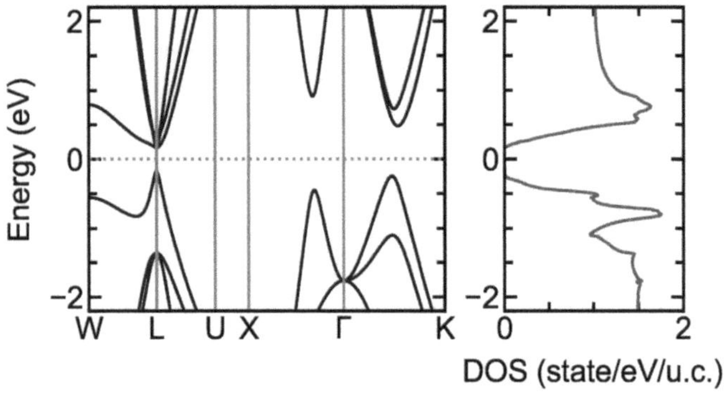

Figure 6.14 Energy dispersion (left) and density of state (DOS) (right) of GeTe.

`introductory/customizing.html`. Now, when we want to use this style, we need to add the following line in our Python script:

```
plt.style.use('~/QE-SSP/matplotlib/sci.mplstyle')
```

It is important to note that the directory path of `sci.mplstyle` must be correct.

6.4.6 Plotting DOS and energy dispersion

With the knowledge of Python and Matplotlib we have obtained before, we can create our own script for plotting DOS and band structure of Quantum ESPRESSO calculation outputs. We show an example for GeTe in Fig. 6.14 with the following `plot-gete.ipynb` script.

QE-SSP/gete/plot-gete.ipynb

```
# Import the necessary packages and modules
import matplotlib.pyplot as plt
plt.style.use('../matplotlib/sci.mplstyle')
import numpy as np

# Load data from gete.bands.gnu
data = np.loadtxt('gete.bands.gnu')
```

```
k = np.unique(data[:, 0])
bands = np.reshape(data[:, 1], (-1, len(k)))
# Load data from gete.dos
ener, dos, idos = np.loadtxt('gete.dos', unpack=True)

# Fermi energy from ./out/*.xml :
HOMO  = 2.635710373417968e-1
LUMO  = 2.758546355831248e-1
unitE = 27.2114
E_F   = (HOMO+LUMO)*unitE/2

# Set high-symmetry points from *.nscf.in file
rW = k[0]; rL = k[50]; rU = k[80]; rX = k[100]; rG = k[150]; rK = k[200]

# Create figure object
fig = plt.figure(figsize=(6, 3))
# Add x and y-axes
axE = fig.add_axes([0.00, 0.0, 0.60, 1])
axD = fig.add_axes([0.70, 0.0, 0.30, 1])

# Plot band structure
for band in range(len(bands)):
    axE.plot(k, bands[band, :]-E_F, c='b')
# Plot dotted line at Fermi energy
axE.axhline(0, c='gray', ls=':')
# Plot dotted lines at high-symmetry points
axE.axvline(rL, c='gray')
axE.axvline(rU, c='gray')
axE.axvline(rX, c='gray')
axE.axvline(rG, c='gray')
# Add labels for high-symmetry points
axE.set_xticks([rW, rL, rU, rX, rG, rK])
axE.set_xticklabels(['W', 'L', 'U', 'X', '$\Gamma$', 'K'])
# Hide x-axis minor ticks
axE.tick_params(axis='x', which='both', length=0)
# Set the axis limits
axE.set_xlabel('')
axE.set_ylabel('Energy (eV)')
# Set the axis limits
axE.set_xlim(rW, rK)
axE.set_ylim (-2.2, 2.2)

# Plot the DOS
axD.plot(dos, ener-E_F, c='r')
# Set the axis limits
axD.set_xlim(0, 2)
```

```
axD.set_ylim(-2.2, 2.2)
# Add the x label
axD.set_xlabel('DOS (state/eV/u.c.)')

# Save figure to the pdf file
plt.savefig('GeTeEnDOS.pdf')
# Show figure
plt.show()
```

Bibliography

Adamo, C. and Barone, V. (1999). Toward reliable density functional methods without adjustable parameters: The pbe0 model, *J. Chem. Phys.* **110**, pp. 6158–6170.

Ashcroft, N. W. (1966). Electron-ion pseudopotentials in metals, *Phys. Lett.* **23**, pp. 48–50.

Atkinson, K. E. (1988). *An Introduction to Numerical Aanalysis* (John Wiley & Sons).

Bardeen, J. (1936). Theory of the work function. II. The surface double layer, *Phys. Rev.* **49**, p. 653.

Baroni, S., De Gironcoli, S., Dal Corso, A. and Giannozzi, P. (2001). Phonons and related crystal properties from density-functional perturbation theory, *Rev. Mod. Phys.* **73**, p. 515.

Becke, A. D. (1988). Density-functional exchange-energy approximation with correct asymptotic behavior, *Phys. Rev. A* **38**, p. 3098.

Becke, A. D. (1993). Density-functional thermochemistry. III. The role of exact exchange, *J. Phys. Chem.* **98**, p. 5648–5652.

Becke, A. D. and Johnson, E. R. (2007). Exchange-hole dipole moment and the dispersion interaction revisited, *J. Chem. Phys.* **127**, p. 154108.

Bengtsson, L. (1999). Dipole correction for surface supercell calculations, *Phys. R. B* **59**, p. 12301.

Berland, K., Chakraborty, D. and Thonhauser, T. (2019). van der Waals density functional with corrected c6 coefficients, *Physical Review B* **99**, 19, p. 195418.

Blount, E. I. (1962). *Formalisms of Band Theory* (Academic Press).

Born, M. and Oppenheimer, R. (1927). Zur quantentheorie der molekeln, *Annalen der Physik* **389**, 20, pp. 457–484.

Brack, M. (1993). The physics of simple metal clusters: self-consistent jellium model and semiclassical approaches, *Rev. Mod. Phys.* **65**, p. 677.

Broyden, C. G. (1970). The convergence of a class of double-rank minimization algorithms 1. General considerations, *J. Inst. Math. Appl.* **6**, pp. 76–90.

Cao, Y., Fatemi, V., Fang, S., Watanabe, K., Taniguchi, T., Kaxiras, E. and Jarillo-Herrero, P. (2018). Unconventional superconductivity in magic-angle graphene superlattices, *Nature* **556**, pp. 43–50.

Carr Jr, W. J., Coldwell-Horsfall, R. A. and Fein, A. E. (1961). Anharmonic contribution to the energy of a dilute electron gas-interpolation for the correlation energy, *Phys. Rev.* **124**, p. 747.

Ceperley, D. M. and Alder, B. J. (1980). Ground state of the electron gas by a stochastic method, *Phys. Rev. Lett.* **45**, p. 566.

Chakraborty, D., Berland, K. and Thonhauser, T. (2020). Next-generation nonlocal van der Waals density functional, *J. Chem. Theory Comput.* **16**, pp. 5893–5911.

Chowdhury, M. T., Saito, R. and Dresselhaus, M. S. (2012). Polarization dependence of X-ray absorption spectra in graphene, *Phys. Rev. B* **85**, p. 115410.

Dion, M., Rydberg, H., Schröder, E., Langreth, D. C. and Lundqvist, B. I. (2004). Van der waals density functional for general geometries, *Phys. Rev. Lett.* **92**, p. 246401.

Dirac, P. A. M. (1930). Note on exchange phenomena in the Thomas atom, in *Proc. Cambridge Phil. Soc.*, Vol. 26 (Cambridge University Press), pp. 376–385.

Dresselhaus, M. S., Dresselhaus, G. and Jorio, A. (2008). *Group Theory: Application to the Physics of Condensed Matter* (Springer, Berlin).

Dynes, R. C. (1972). McMillan's equation and the Tc of superconductors, *Solid State Commun.* **10**, pp. 615–618.

Eda, G., Yamaguchi, H., Voiry, D., Fujita, T., Chen, M. and Chhowalla, M. (2011). Photoluminescence from chemically exfoliated MoS_2, *Nano Lett.* **11**, 12, pp. 5111–5116.

Eliashberg, G. M. (1960). Interactions between electrons and lattice vibrations in a superconductor, *Sov. Phys. JETP* **11**, 3, pp. 696–702.

Emery, C., N.and Hérold, Marêché, J. F. and Lagrange, P. (2009). Synthesis and superconducting properties of CaC_6, *Sci. Technol. Adv. Mater* **9**, p. 044102.

Ewald, P. P. (1921). Die berechnung optischer und elektrostatischer gitterpotentiale, *Annalen der Physik* **369**, pp. 253–287.

Fermi, E. (1928). Eine statistische methode zur bestimmung einiger eigenschaften des atoms und ihre anwendung auf die theorie des periodischen systems der elemente, *Z. Physik* **48**, pp. 73–79.

Feynman, R. P. (1939). Forces in molecules, *Phys. Rev.* **56**, p. 340.

Fletcher, R. (1991). A new variational result for quasi-newton formulae, *SIAM J. Optim.* **1**, pp. 18–21.

Fock, V. (1930). Näherungsmethode zur lösung des quantenmechanischen mehrkörperproblems, *Zeitschrift für Physik* **61**, 1-2, pp. 126–148.

Fretigny, C., Saito, R. and Kamimura, H. (1989). Electronic structures of unoccupied bands in graphite, *J. Phys. Soc. Jpn.* **58**, pp. 2098–2108.

Fu, C. L. and Ho, K. M. (1983). First-principles calculation of the equilibrium ground-state properties of transition metals: Applications to Nb and Mo, *Phys. Rev. B* **28**, p. 5480.

Geim, A. K. and Novoselov, K. S. (2007). The rise of graphene, *Nat. Mater.* **6**, p. 183–191.

Gell-Mann, M. and Brueckner, K. A. (1957). Correlation energy of an electron gas at high density, *Phys. Rev.* **106**, p. 364.

Giannozzi, P., Baroni, S., Bonini, N., Calandra, M., Car, R., Cavazzoni, C., Ceresoli, D., Chiarotti, G. L., Cococcioni, M., Dabo, I., Corso, A. D., de Gironcoli, S., Fabris, S., Fratesi, G., Gebauer, R., Gerstmann, U., Gougoussis, C., Kokalj, A., Lazzeri, M., Martin-Samos, L., Marzari, N., Mauri, F., Mazzarello, R., Paolini, S., Pasquarello, A., Paulatto, L., Sbraccia, C., Scandolo, S.,

Sclauzero, G., Seitsonen, A. P., Smogunov, A., Umari, P. and Wentzcovitch, R. M. (2009). QUANTUM ESPRESSO: a modular and open-source software project for quantum simulations of materials, *J. Phys. Condens. Matter* **21**, p. 395502.

Goldfarb, D. (1970). A family of variable-metric methods derived by variational means, *Math. Comp.* **24**, pp. 23–26.

Grimme, S. (2004). Accurate description of van der Waals complexes by density functional theory including empirical corrections, *J. Comput. Chem.* **25**, pp. 1463–1473.

Grimme, S., Antony, J., Ehrlich, S. and Krieg, H. (2010). A consistent and accurate ab initio parametrization of density functional dispersion correction (DFT-D) for the 94 elements H-Pu, *J. Chem. Phys.* **132**, p. 154104.

Hartree, D. R. (1928). The wave mechanics of an atom with a non-coulomb central field. Part I. Theory and methods, in *Math. Proc. Camb. Philos. Soc.*, Vol. 24 (Cambridge University Press), pp. 89–110.

Heine, V. and Abarenkov, I. V. (1964). A new method for the electronic structure of metals, *Phil. Mag.* **9**, pp. 451–465.

Hellmann, H. (1937). *Einführung in die quantenchemie* (Deuticke, Leipzig).

Heyd, J., Scuseria, G. E. and Ernzerhof, M. (2003). Hybrid functionals based on a screened coulomb potential, *J. Chem. Phys.* **118**, pp. 8207–8215.

Hohenberg, P. and Kohn, W. (1964). Inhomogeneous electron gas, *Phys. Rev.* **136**, p. B864.

Huang, Y. L., Chen, Y., Zhang, W., Quek, S. Y., Chen, C. H., Li, L. J., Hsu, W. T., Chang, W. H., Zheng, Y. J., Chen, W. and Wee, A. T. S. (2015). Bandgap tunability at single-layer molybdenum disulphide grain boundaries, *Nat. Commun.* **6**, pp. 1–8.

Jackson, J. D. (1999). *Classical Electrodynamics (3rd ed.)* (Wiley New York).

Kittel, C. (1976). *Introduction to Solid State Physics* (Wiley New York).

Kohn, W. (1959). Image of the fermi surface in the vibration spectrum of a metal, *Phys. Rev. Lett.* **2**, p. 393.

Kohn, W. and Sham, L. J. (1965). Self-consistent equations including exchange and correlation effects, *Phys. Rev.* **140**, p. A1133.

Latzke, D. W., Zhang, W., Suslu, A., Chang, T. R., Lin, H., Jeng, H. T., Tongay, S., Wu, J., Bansil, A. and Lanzara, A. (2015). Electronic structure, spin-orbit coupling, and interlayer interaction in bulk MoS_2 and WS_2, *Phys. Rev. B* **91**, p. 235202.

Lazzeri, M. and Mauri, F. (2003). First-principles calculation of vibrational raman spectra in large systems: Signature of small rings in crystalline SiO_2, *Phys. Rev. Lett.* **90**, p. 036401.

Lazzeri, M. and Mauri, F. (2006). Nonadiabatic Kohn anomaly in a doped graphene monolayer, *Phys. Rev. Lett.* **97**, p. 266407.

Lee, C., Yang, W. and Parr, R. G. (1988). Development of the colle-salvetti correlation-energy formula into a functional of the electron density, *Phys. Rev. B* **37**, p. 785.

Lee, K., Murray, E. D., Kong, L., Lundqvist, B. I. and Langreth, D. C. (2010). Higher-accuracy van der Waals density functional, *Phys. Rev. B* **82**, p. 081101.

Li, H., Zhang, Q., Yap, C. C. R., Tay, B. K., Edwin, T. H. T., Olivier, A. and Baillargeat, D. (2012). From bulk to monolayer MoS_2: evolution of raman scattering, *Adv. Funct. Mater.* **22**, pp. 1385–1390.

Li, Y., Chernikov, A., Zhang, X., Rigosi, A., Hill, H. M., Van Der Zande, A. M., Chenet, D. A., Shih, E. M., Hone, J. and Heinz, T. F. (2014). Measurement of the optical dielectric function of monolayer transition-metal dichalcogenides: MoS_2, $MoSe_2$, WS_2, and WSe_2, *Phys. Rev. B* **90**, p. 205422.

Lieb, E. H. and Oxford, S. (1981). Improved lower bound on the indirect coulomb energy, *Int. J. Quantum Chem.* **19**, pp. 427–439.

Liu, H. L., Yang, T., Chen, J. H., Chen, H. W., Guo, H., Saito, R., Li, M. Y. and Li, L. J. (2020). Temperature-dependent optical constants of monolayer MoS_2, $MoSe_2$, WS_2, and WSe_2: spectroscopic ellipsometry and first-principles calculations. *Sci. Rep.* **10**, p. 15282.

Madelung, O. (1978). *Introduction to Solid-State Theory* (Springer).

Marder, M. P. (2010). *Condensed Matter Physics* (John Wiley & Sons).

Marzari, N., Mostofi, A. A., Yates, J. R., Souza, I. and Vanderbilt, D. (2012a). Maximally localized Wannier functions: Theory and applications, *Rev. Mod. Phys.* **84**, pp. 1419–1475.

Marzari, N., Mostofi, A. A., Yates, J. R., Souza, I. and Vanderbilt, D. (2012b). Maximally localized Wannier functions: Theory and applications, *Rev. Mod. Phys.* **84**, p. 1419.

Marzari, N. and Vanderbilt, D. (1997). Maximally localized generalized Wannier functions for composite energy bands, *Phys. Rev. B* **56**, 20, p. 12847.

Marzari, N., Vanderbilt, D., De Vita, A. and Payne, M. C. (1999). Thermal contraction and disordering of the Al (110) surface, *Phys. Rev. Lett.* **82**, p. 3296.

McMillan, W. L. (1968). Transition temperature of strong-coupled superconductors, *Phys. Rev.* **167**, p. 331.

Mermin, N. D. (1965). Thermal properties of the inhomogeneous electron gas, *Phys. Rev.* **137**, p. A1441.

Methfessel, M. P. A. T. and Paxton, A. T. (1989). High-precision sampling for Brillouin-zone integration in metals, *Phys. Rev. B* **40**, p. 3616.

Monkhorst, H. J. and Pack, J. D. (1976). Special points for brillouin-zone integrations, *Phys. Rev. B* **13**, p. 5188.

Mostofi, A. A., Yates, J. R., Lee, Y. S., Souza, I., Vanderbilt, D. and Marzari, N. (2008). Wannier90: A tool for obtaining maximally-localised Wannier functions, *Comput. Phys. Commun.* **178**, pp. 685–699.

Ohta, T., Bostwick, A., McChesney, J. L., Seyller, T., Horn, K. and Rotenberg, E. (2007). Interlayer interaction and electronic screening in multilayer graphene investigated with angle-resolved photoemission spectroscopy, *Phys. Rev. Lett.* **98**, p. 206802.

Ohta, T., Bostwick, A., Seyller, T., Horn, K. and Rotenberg, E. (2006). Controlling the electronic structure of bilayer graphene, *Science* **313**, pp. 951–954.

Otero-De-La-Roza, A. and Johnson, E. R. (2012). Van der waals interactions in solids using the exchange-hole dipole moment model, *J. Chem. Phys.* **136**, p. 174109.

Parr, R. G. and Yang, W. (1989). *Density-Functional Theory of Atoms and Molecules* (Oxford University Press).

Pauli, W. (1925). Über den zusammenhang des abschlusses der elektronengruppen im atom mit der komplexstruktur der spektren, *Zeitschrift für Physik* **31**, 1, pp. 765–783.

Perdew, J. P., Burke, K. and Ernzerhof, M. (1996a). Generalized gradient approximation made simple, *Phys. Rev. Lett.* **77**, p. 3865.

Perdew, J. P., Chevary, J. A., Vosko, S. H., Jackson, K. A., Pederson, M. R., Singh, D. J. and Fiolhais, C. (1992). Atoms, molecules, solids, and surfaces: Applications of the generalized gradient approximation for exchange and correlation, *Phys. Rev. B* **46**, p. 6671.

Perdew, J. P., Ernzerhof, M. and Burke, K. (1996b). Rationale for mixing exact exchange with density functional approximations, *J. Chem. Phys.* **105**, pp. 9982–9985.

Perdew, J. P. and Levy, M. (1983). Physical content of the exact Kohn-Sham orbital energies: band gaps and derivative discontinuities, *Phy. Rev. Lett.* **51**, p. 1884.

Perdew, J. P. and Wang, Y. (1992). Accurate and simple analytic representation of the electron-gas correlation energy, *Phys. Rev. B* **45**, p. 13244.

Perdew, J. P. and Zunger, A. (1981). Self-interaction correction to density-functional approximations for many-electron systems, *Phys. Rev. B* **23**, p. 5048.

Pickett, W. E. (1989). Pseudopotential methods in condensed matter applications, *Comput. Phys. Rep.* **9**, pp. 115–197.

Pimenta, M. A., Dresselhaus, G., Dresselhaus, M. S., Cançado, L. G., Jorio, A. and Saito, R. (2007). Studying disorder in graphite-based systems by Raman spectroscopy, *Physical Chemistry Chemical Physics* **9**, pp. 1276–1291.

Pizzi, G., Vitale, V., Arita, R., Blügel, S., Freimuth, F., Géranton, G., Gibertini, M., Gresch, D., Johnson, C., Koretsune, T. *et al.*

(2020). Wannier90 as a community code: new features and applications, *J. Phys.: Cond. Matt.* **32**, p. 165902.

Poncé, S., Margine, E. R., Verdi, C. and Giustino, F. (2016). Epw: Electron–phonon coupling, transport and superconducting properties using maximally localized Wannier functions, *Comput. Phys. Commun.* **209**, pp. 116–133.

Rasamani, K. D., Alimohammadi, F. and Sun, Y. (2017). Interlayer-expanded MoS_2, *Mater. Today* **20**, pp. 83–91.

Razado-Colambo, I., Avila, J., Vignaud, D., Godey, S., Wallart, X., Woodruff, D. P. and Asensio, M. C. (2018). Structural determination of bilayer graphene on SiC (0001) using synchrotron radiation photoelectron diffraction, *Sci. Rep.* **8**, pp. 1–10.

Roch, J. G., Froehlicher, G., Leisgang, N., Makk, P., Watanabe, K., Taniguchi, T. and Warburton, R. J. (2019). Spin-polarized electrons in monolayer MoS_2, *Nat. Nanotech.* **14**, pp. 432–436.

Roothaan, C. C. J. (1951). New developments in molecular orbital theory, *Rev. Mod. Phys.* **23**, p. 69.

Sabatini, R., Gorni, T. and De Gironcoli, S. (2013). Nonlocal van der Waals density functional made simple and efficient, *Phys. Rev. B* **87**, p. 041108.

Saito, R., Fujita, M., Dresselhaus, G. and Dresselhaus, M. S. (1992). Electronic structure of chiral graphene tubules, *Appl. Phys. Lett.* **60**, pp. 2204–2206.

Saito, R., Grüneis, A., Cançado, L. G., Pimenta, M. A., Jorio, A., Souza Filho, A. G., Dresselhaus, M. S. and Dresselhaus, G. (2002a). Double resonance Raman spectra in disordered graphite and single wall carbon nanotubes, *Mol. Cryst. Liq. Cryst.* **387**, pp. 287–296.

Saito, R., Grüneis, A., Samsonidze, G. G., Brar, V. W., Dresselhaus, G., Dresselhaus, M. S., Jorio, A., Cançado, L. G., Fantini, C., Pimenta, M. A. and Souza Filho, A. G. (2003). Double resonance Raman spectroscopy of single wall carbon nanotubes, *New J. Phys.* **5**, pp. 157.1–157.15.

Saito, R., Jorio, A., Souza Filho, A. G., Dresselhaus, G., Dresselhaus, M. S. and Pimenta, M. A. (2002b). Probing phonon dispersion relations of graphite by double resonance Raman scattering, *Phys. Rev. Lett.* **88**, p. 027401.

Saito, R., Masashi, M. and Dresselhaus, M. S. (2018). Ballistic and diffusive thermal conductivity of graphene, *Phys. Rev. Appl.* **9**, 2, pp. 024017.

Schrödinger, E. (1926). An undulatory theory of the mechanics of atoms and molecules, *Phys. Rev.* **28**, 6, p. 1049.

Schwarz, K. (1972). Optimization of the statistical exchange parameter α for the free atoms H through Nb, *Phys. Rev. B* **5**, p. 2466.

Setyawan, W. and Curtarolo, S. (2010). High-throughput electronic band structure calculations: Challenges and tools, *Comput. Mater. Sci.* **49**, pp. 299–312.

Sham, L. J. and Schlüter, M. (1983). Density-functional theory of the energy gap, *Phys. Rev. Lett.* **51**, p. 1888.

Sham, L. J. and Schlüter, M. (1985). Density-functional theory of the band gap, *Phys. Rev. B* **32**, p. 3883.

Shanno, D. F. (1970). Conditioning of quasi-newton methods for function minimization, *Math. Comp.* **24**, pp. 647–656.

Slater, J. C. (1929). The theory of complex spectra, *Phys. Rev.* **34**, 10, p. 1293.

Slater, J. C. (1930a). Atomic shielding constants, *Phys. Rev.* **36**, p. 57.

Slater, J. C. (1930b). Note on Hartree's method, *Phys. Rev.* **35**, 2, p. 210.

Slater, J. C. (1951). A simplification of the Hartree-Fock method, *Phys. Rev.* **81**, p. 385.

Slater, J. C. and Johnson, K. H. (1972). Self-consistent-field $x\alpha$ cluster method for polyatomic molecules and solids, *Phys. Rev. B* **5**, p. 844.

Sohier, T., Calandra, M. and Mauri, F. (2017). Density functional perturbation theory for gated two-dimensional heterostructures: Theoretical developments and application to flexural phonons in graphene, *Phys. Rev. B* **96**, p. 075448.

Souza, I., Marzari, N. and Vanderbilt, D. (2001). Maximally localized Wannier functions for entangled energy bands, *Phys. Rev. B* **65**, p. 035109.

Srivastava, G. P. (1990). *The Physics of Phonons* (CRC Press).

Stephens, P. J., Devlin, F. J., Chabalowski, C. F. and Frisch, M. J. (1994). Ab initio calculation of vibrational absorption and circular dichroism spectra using density functional force fields, *J. Phys. Chem.* **98**, pp. 11623–11627.

Sucher, J. (1958). Energy levels of the two-electron atom to order α^3 ry; ionization energy of helium, *Phys. Rev.* **109**, 3, p. 1010.

Tatsumi, Y., Kaneko, T. and Saito, R. (2018). Conservation law of angular momentum in helicity-dependent raman and rayleigh scattering, *Physical Review B* **97**, 19, pp. 195444.

Tatsumi, Y. and Saito, R. (2018). Interplay of valley selection and helicity exchange of light in raman scattering for graphene and mos_2, *Phys. Rev. B* **97**, pp. 115407.

Thijssen, J. (2007). *Computational Physics* (Cambridge University Press).

Thomas, L. H. (1927). The calculation of atomic fields, in *Proc. Cambridge Phil. Soc.*, Vol. 23 (Cambridge University Press), pp. 542–548.

Thonhauser, T., Zuluaga, S., Arter, C. A., Berland, K., Schröder, E. and Hyldgaard, P. (2015). Spin signature of nonlocal correlation binding in metal-organic frameworks, *Phys. Rev. Lett.* **115**, p. 136402.

Tkatchenko, A. and Scheffler, M. (2009). Accurate molecular van der Waals interactions from ground-state electron density and free-atom reference data, *Phys. Rev. Lett.* **102**, p. 073005.

Umrigar, C. J. and Gonze, X. (1994). Accurate exchange-correlation potentials and total-energy components for the helium isoelectronic series, *Phys. Rev. A* **50**, 5, p. 3827.

Vosko, S. H., Wilk, L. and Nusair, M. (1980). Accurate spin-dependent electron liquid correlation energies for local spin density calculations: a critical analysis, *Can. J. Phys.* **58**, pp. 1200–1211.

Wannier, G. H. (1937). The structure of electronic excitation levels in insulating crystals, *Phys. Rev.* **52**, p. 191.

Wu, Q. and Yang, W. (2002). Empirical correction to density functional theory for van der Waals interactions, *J. Chem. Phys.* **116**, pp. 515–524.

Index